Sustainable Aquaculture

Global Perspectives

Editors

B.B. Jana

Carl D. Webster

ATLANTIC®

PUBLISHERS & DISTRIBUTORS

Food Products Press®

New York • London • Oxford

ATLANTIC®

PUBLISHERS & DISTRIBUTORS
7/22, Ansari Road, Darya Ganj
New Delhi – 110 002
Tel: 2327 3880, 2327 5880, 2328 0451
Fax: 91-11-2328 5873
E-mail: info@atlanticbooks.com
Website: www.atlanticbooks.com

Published in India by
Atlantic Publishers and Distributors, New Delhi
in association with
Food Products Press®, an imprint of The Haworth Press, Inc.
10 Alice Street, Binghamton, New York 13904-1580

First published 2003 by Food Products Press®
First Indian Reprint 2005

ISBN 81-269-0490-9

Printed in India by
Nice Printing Press, Delhi

Sustainable Aquaculture: Global Perspectives

CONTENTS

ABOUT THE EDITORS

B. B. Jana, PhD, is Professor of Aquaculture in the Department of Zoology and Coordinator of the International Center of Ecological Engineering at the University of Kalyani, India. He is a founding member of the Asian Fisheries Society, an International Board member of the Ecological Engineering, and a Fellow of the National Academy of Agricultural Sciences. Dr. Jana has authored more than 170 research papers and co-authored several books on aquaculture and ecological engineering.

Carl D. Webster, PhD, is Principal Investigator/Associate Professor at the Aquaculture Research Center, Kentucky State University in Frankfort. He is a widely published author, editor of the *Journal of Applied Aquaculture* (Haworth), and co-editor of *Nutrition and Fish Health* (Haworth) and *Nutrient Requirements and Feeding of Finfish for Aquaculture*. Dr. Webster has served as President and as Secretary/Treasurer of the United States Aquaculture Society, is a program committee member for "Aquaculture America '02," Program Chair for "Aquaculture America '03," and a member of the steering committee for "Aquaculture America '04."

Preface

Providing world food security has become an important issue while planning for future generations because of the alarming increase in the world's population which may exceed 12 billion people within the next three decades. Of these, approximately 500,000,000 people will get less protein from their diet than they need for proper nutrition. Because of a compounding increase in demands for improving the standard of living, pressure on natural resources, particularly on aquatic ecosystems, would also increase in developing countries. Over-fishing, coastal and wetland destruction, habitat and environmental degradation, and conflicts between stakeholders are all problems that have arisen due to the ever-increasing demands people have placed in the environment as they try to make a livelihood for themselves and their families. One of the results has been a drastic decline in the global fish yield from wild fisheries.

Alternatively, there is enormous potential of aquaculture for worldwide household food security and income generation. The biological and technological basis of aqua-farming is well founded. Of the estimated quantities available for human consumption in the coming decade, contribution from aquaculture is estimated to be more than 35 percent. An example of the challenges and opportunities that many countries face in aquaculture is India. Endowed with a bountiful of aquatic resources, rich diversity of cultivable fish species, and high tropical temperatures, India has a long tradition of aquaculture. In terms of relative importance of aquaculture to national aquatic production, India is among the top four countries of the world. Aquaculture has been expanding very fast as an important industry and economic cornerstone by exporting shrimp and other marine products to world markets. Among Asian countries, India ranks second in culture fisheries producing 7% of global aquaculture production. Of the estimated total inland

[Haworth co-indexing entry note]: "Preface." Jana, B. B. Co-published simultaneously in *Journal of Applied Aquaculture* (Food Products Press, an imprint of The Haworth Press, Inc.) Vol. 13, No. 1/2, 2003, pp. xiii-xv; and: *Sustainable Aquaculture: Global Perspectives* (ed: B. B. Jana, and Carl D. Webster) Food Products Press, an imprint of The Haworth Press, Inc., 2003, pp. xi-xiii. Single or multiple copies of this article are available for a fee from The Haworth Document Delivery Service [1-800-HAWORTH, 9:00 a.m. - 5:00 p.m. (EST). E-mail address: getinfo@haworthpressinc.com].

fish production of 2.5 million tonnes in India, nearly 70% is from aquaculture. Further, more than 6 million rural people and fish farmers in India depend on fisheries and aquaculture for their livelihood; a very high number.

Nonetheless, there are problems with aquaculture production in India. Total yield is far from expectation considering the vast natural resources and favorable tropical climate of the region. Under-utilization of the natural resources and lack of adoption of modern scientific methods are the two major causes for the low fish production currently found in India. Research aimed at improving sustainable aquaculture production is being carried out worldwide. Some basic issues of aquaculture have not been addressed adequately in India such as the need to adopt standardized, proven pond fertilization protocols. More research is needed for developing a general pond fertilization protocol for the different agro-climatic regions in India so that the various regions can achieve optimal fish production methods. Further, long-term inbreeding and loss of genetic variability due to genetic drift may have detrimental effect on fish growth. The problem has been frequently experienced in the induced breeding program where inbreeding has been a common practice for many years in India. A sufficiently large and sound breeding population is needed for the selection program of cultured fishes.

The long-term sustainability of the fisheries and aquaculture industries in some countries has been threatened as a consequence of the intensive farming, injudicious use of chemicals and antibiotics, landscape destruction, over-fishing in wild fisheries, habitat degradation, coastal wetland and nursery destruction, environmental degradation, illegal introduction of exotic species, depletion of bio-diversity, and conflicts over land use between agriculture and aquaculture. The multiple advantages of integrating aquaculture into small-holder farming systems have been widely recognized for efficient use of natural resources. The present issue is a compilation of different aspects of aquaculture from expert contributors from different countries. The contributed papers range from freshwater pearl culture in India, fish nutrition, culture methods, breeding programs, pond fertilization regimes, fish diseases in tropical climates, indoor recirculating culture systems, the status of aquaculture in India, and water quality management for shrimp farming.

This publication is dedicated to all the progressive farmers of the world whose experience, insight, and hard work have conclusively boosted the economy of the poor. Special thanks are due to the contributors for their important articles. I sincerely thank Dr. S. Singh, Imprint Coordinator, The Haworth Press, who took interest and initiative for publication of this special volume. I extend my thanks to Dr. Carl Webster for his kind support. I am indebted to the University of Kalyani for providing facilities for compiling this work. Needless to say that it will be my immense pleasure if this volume plays any contributory part for the development of aquaculture in India and the world.

B. B. Jana
Aquaculture and Applied Limnology Research Unit
Department of Zoology
University of Kalyani
Kalyani 741235, West Bengal
India

Indian Aquaculture

K. Gopakumar

SUMMARY. Fish play an important role in the nutrition of people in India. Per capita availability of fish in the country is 9.5 kg with production at 5.8 million tonnes during 1999-2000. This production level makes India the second largest aquaculture producing country in the world, China is the leading aquaculture producer. India cultures many different species including Indian major carps (catla, *Catla catla*; rohu, *Labeo rohita*; naini, *Cyprinus mrigala*; and calbasu, *Labeo calbasu*), marine shrimp, freshwater shrimp, oysters, green and brown mussels, and pearl oysters for domestic and international markets. To meet the increased demand for fisheries products by 2010, India must increase its current level of production to 7-8 million tonnes per year. With marine fishery production/capture steady at 3 million tonnes per year, inland aquaculture production must increase from its current level of 2.5 million tonnes to 5 million tonnes by 2010. India must intensify culture production methods for this increased production values to be achieved. *[Article copies available for a fee from The Haworth Document Delivery Service: 1-800-HAWORTH. E-mail address: <getinfo@haworthpressinc.com> Website: <http://www.HaworthPress.com>*

KEYWORDS. India, aquaculture, production

INTRODUCTION

Fish plays an important role in human nutrition in India, particularly to people of coastal regions. It also contributes to domestic food secu-

---

K. Gopakumar, Deputy Director General (Fisheries), Indian Council of Agricultural Research, Krishi Bhawan, New Delhi-110 001, India.

[Haworth co-indexing entry note]: "Indian Aquaculture." Gopakumar, K. Co-published simultaneously in *Journal of Applied Aquaculture* (Food Products Press, an imprint of The Haworth Press, Inc.) Vol. 13, No. 1/2, 2003, pp. 1-10; and: *Sustainable Aquaculture: Global Perspectives* (ed: B. B. Jana, and Carl D. Webster) Food Products Press, an imprint of The Haworth Press, Inc., 2003, pp. 1-10. Single or multiple copies of this article are available for a fee from The Haworth Document Delivery Service [1-800-HAWORTH, 9:00 a.m. - 5:00 p.m. (EST). E-mail address: getinfo@haworthpressinc.com].

rity of the nation. The per-capita availability of fish in India is now 9.5 kg with 56% of Indians considered fish eaters. It is estimated that by 2010, India's requirement for fish will be around 10 million tonnes. As marine fish capture has been stable at 3 million tonnes during the last decade, all additional requirements will have to be met from aquaculture. India produced 5.8 million tonnes of fish during 1999-2000 and is the world's second leading aquaculture producer. (China is the leading producer.) Of this, only 8% is exported, leaving 92% for domestic consumption. The gross value of fisheries in India was $4.9 billion (US) accounting for 1.5% of the total Gross Domestic Production (GDP) of India.

Aquaculture has made great strides in the 1990s in India with an average growth rate of 6%. Inland production has increased from 1.16 million tonnes in 1985-86 to 2.5 million tonnes 1999-2000. Out of the total inland production, nearly 70% comes from aquaculture. Indian and exotic carp species contribute over 90% of the inland aquaculture production. This sector alone contributes one-third of the total fish production in India. Adoption of improved, modern, scientific farming technologies has increased fish production from a meager level of between 0.3 and 0.5 tonnes/ha/year a decade ago to a current production level of 10-15 tonnes/ha/year. Impressive growth has also been recorded in brackishwater shrimp aquaculture with production of over 0.8 million tonnes of shrimp. In export, India contributes 2.5% of the world fish trade, which is only lower to rice (10.4%), tea (16.4%), animals (11.2%), and animal diets (4.3%).

Aquaculture maintained sustained growth of 6% in the decade although all major agricultural commodities sustained negative growth during 2000. Total fish production also suffered a set-back in 1998-99 due to a fall in production from marine capture fisheries. However, during the period 1999-2000, fish production has showed improved growth in marine capture fisheries and aquaculture. The export of marine products during the period 1999-2000 recorded an increase of 20% in quantity and nearly 24% in value (provisional).

FISHERY RESOURCES

India has vast water resources, both marine and inland. For marine production, about 70% of the potential resources are fully exploited. According to present trends, the commercial capture is steady at 2.8 million tonnes for the past seven years. However, inland production has increased from 0.8 million tonnes ten years ago to 2.8 million tonnes today, equaling the production value of marine capture fisheries. Indian farmers use the practice of composite fish culture using different species of carps of both exotic and indigenous origin. Many farmers particularly in the states of Punjab, Haryana, and Orissa now use integrated fish farming, with other agricultural enterprises.

Fish Species Used for Aquaculture

India possesses a large number of indigenous species for aquaculture. About ninety percent of fish production in the country is carp, both indigenous and exotic species. Air-breathing and non air-breathing catfishes are also grown. Catfish commands higher consumer acceptability and market price as they do not contain embedded bones in the tissues. An average Indian farmer with a productivity of 4 tonnes/ha earns around $1400 (US) per year. Among the freshwater prawns, *Macrobrachium rosenbergii* and *M. malcomsoni* are now extensively cultured in India because of the high export demand. Seabass, *Lates calcarifer*, has emerged as a potential cultivable species. Breeding under hatchery conditions was achieved by the Central Institute of Brachishwater Aquaculture (CIBA), Chennai (ICAR). The institute is now distributing fry for culture. India has signed a contract with the French Government for establishing a commercial hatchery at Chennai for commercial production of seabass fingerlings to farmers. The project will become operational by the end of 2002.

Inland Capture Fisheries

Although fish production from the inland sector is increasing quite significantly, productivity of the rivers, lakes, and reservoirs in India has not shown much increase. There is large-scale poaching of fish in public water bodies. A large portion of fish caught from the rivers and reservoirs are not accounted for and it is extremely difficult to have proper statistical recordings of the landings. The wild stocks in these waters show signs of stock depletion and many species, like the golden mahseer, *Tor putitoria*, are endangered.

Fish Culture in Fresh-Water Ecosystems

Freshwater aquaculture production has made notable strides in the recent years due to development technologies for quality fry production and farming of freshwater fish under commercial conditions. Presently, approximately 2.36 million hectares of ponds, 1.07 million hectares of natural water bodies, and 2.05 million hectares of reservoirs are available in the country for production of quality fry and market-size fish of economically important species. Numerous species of carps and air-breathing fish are available to suit varied ecological conditions of ponds, tanks, derelict water bodies, natural lakes, and reservoirs. In order to utilize production potential of these resources, research efforts have been made for developing methods for culture of the following commercially important species.

Major Carps

Highest priority has been accorded for quality production and culture of carps because India has high consumer acceptance of Indian major carps namely catla, *Catla catla*; rohu, *Labeo rohita*; naini, *C. mrigala*; and calbasu, *L. calbasu.* Carps do not breed in confined water, therefore, efforts have been made for development of induced breeding technique under controlled system. The development of induced breeding by hypophysation, as well as the methods for the hatchery, have changed the scenario of seed production of major carps in the country. Further, advancement has been made in breeding the carps during the non-breeding season, and techniques such as increasing the rate at which carps mature, carp milt cryopreservation, and multiple carp breedings have been developed. The development of cryopreservation of spermatozoa has eliminated the maintenance of broodstock during the breeding season thus reducing the cost of investment of broodstock. Similarly, multiple spawning techniques have enhanced the quality of seed production in the country. These technologies has been widely adopted by farmers, as a result of which the seed production of carps, both endemic and exotic, has increased from a level of 6,321 million fry in 1985-86 to 15,504 million fry during 1997-98. Similarly, for the production of market-size carps such as, catla; rohu; mrigal, *Cirrhinus mrigala*; common carp, *Cyprinus carpio*; silver carp, *Hypophthalmichthys molitrix*; and grass carp, *Ctenopharyngodon idella*; efforts have been made to develop scientific farming systems. Presently, extensive, semi-intensive, intensive, and super-intensive farming systems are available to suit the economic conditions of farmers. The minimum achievable production of carps using different culture systems are: multiple carp culture, 4-6 tonnes/ha/year; intensive pond culture with supplementary feeding and aeration, 10-15 tonnes/ha/year; integrated fish farming with poultry, pigs, ducks or horticulture, 3-5 tonnes/ha/year; pen culture, 3-5 tonnes/ha/year; cage culture, 10-15 kg/m^2/year; and fish culture using running water, 20-50 kg/m^2/year. Presently, composite fish culture technology is widely adopted in smaller water bodies, whereas pen culture, cage culture, and fish culture with running water are used in larger water bodies.

Air-Breathing Fish Culture

The air-breathing fishes command high prices in the domestic market and thrive in adverse ecological conditions. These fish include: singhi, *Heteropneustes fossilis*; magur, *Clarias batrachus*; koi, *Anabus testudineus*; garai, *Channa striatus*; and saura, *C. marulius*. It is possible to breed these fish using pitutary extract, ovaprim, LHRH-a, and pimozide under controlled conditions and produce seed in the hatchery for culture operation. Success has also been achieved in prolonging the breeding period by manipulation of broodstock by hormone treatment with hormone pellets made of LHRH-a and HCG with cholesterol and a binder. These pellets are implanted in to the fish during the preparatory period of the gonadal cycle. Better advancement of maturation has also been achieved by combination of hormonal applications using HCG, LHRH-a, and oral application of pellet implantation with water quality management, balanced diet supply, and photo-thermal applications. For quality seed production of these fish, many hatcheries have been established in the several parts of the country. Similarly, for the production of marketable-size, both mono- and poly-culture of air-breathing fish in ponds, cages, and cisterns have been developed. The air-breathing fish are generally stocked at 5-10 times higher than the carps. With proper management practices, a production rate of 3.8 tonnes/ha/year can be obtained.

Crustacean Culture

The edible crustaceans are considered a delicious food throughout the world and thus, command a good market in both domestic and international markets. In India, there are 18 species of shrimps and 3 species of crabs, which are the commercially important. The jumbo tiger shrimp, *Penaeus monodon*; Indian white shrimp, *P. indicus*; banana shrimp, *P. merguiensis*; green tiger shrimp, *P. semisulcatus*; and yellow shrimp *Metapenaeus brevicornis*, are the major species for culture in the brackish water ecosystems. Amongst all these crustaceans, jumbo tiger shrimp is the most preferred species for farming as it grows to a maximum size of 330 mm while Indian white shrimp is second attaining a length of 175 mm. Of the total cultured shrimp, jumbo tiger shrimp accounts for over 90% of the production in India. Today, cultured shrimp accounts for over 42% in terms of quantity and 70% in terms of value of marine products exported from in India in 2000.

It is estimated that over 1.4 million hectares of brackishwater area are available in India suitable for brackishwater aquaculture (Table 1). Of the total area, 48% percent are concentrated in West Bengal, followed by Andhra Pradesh (18%), Maharashtra (9%), Kerala (8%), Tamil Nadu (7%), and Gujarat (4%). Only 17 percent of the total area are at present shrimp farms. Karnataka and Andhra Pradesh have used 44% of area available and Orissa 36%. However, West Bengal which has the most area available for aquaculture, has so far utilized only around 10% of the available area for shrimp production. More than half of the shrimp production in the country comes from Andhra Pradesh. Production performances vary from state to state. In Andhra Pradesh, production of shrimp is growing at the rate of 20% per year (highest), while in Kerala, it has declined to a rate of 3% per year (lowest). Substantial increase in

TABLE 1. Indian shrimp production in 2000.

State	Potential area (hectacres)	Developed area (hectacres)	Area under culture (hectacres)	Production (tonnes)
West Bengal	405,000	45,525	43,067	18,326
Orissa	31,600	11,332	8,000	6,000
Andhra Pradesh	150,000	60,249	71,000	44,856
Tamil Nadu	56,000	670	1,087	1,820
Pondicherry	800	22	22	27
Kerala	65,000	14,705	14,705	7,660
Karnataka	8,000	3,540	3,564	2,690
Goa	18,500	650	650	590
Maharashtra	80,000	970	426	409
Gujarat	376,000	997	316	256
TOTAL	1,190,900	135,660	141,837	32,634

area under cultivation is due to the new policy of economic liberalization and demand from the shrimp processing industry.

Of the total export earning of $1.2 billion (US) from marine fisheries products, shrimp contributes about 50%. With a production of 82,000 tonnes in 2000, India ranks fourth in world shrimp production, however, yield/ha in India is very low at 750 kg/ha (national average) compared with 2,500 kg/ha in Thailand. The main reason is that majority of coastal fishermen use extensive shrimp farming with low stocking density and small (0.5-1.0 ha) farms. The number of farmers under each category of land, holding in the different maritime states are presented in Table 2. Over 90% of Indian fish farmers are small and marginal farmers having holding less than 2 ha. About 50% of farms used for shrimp farming are under 2 ha in size. Farms with more than 5 ha area comprise 21% in West Bengal; 23% in Kerala; 24% in Karnataka; and 64% in Goa. Only 10 corporate companies from the 500 top-most companies in India are involved in aquaculture. These 10 companies use 1,898 ha of land, of which only 758 ha has been developed into shrimp farms. This constitutes just 0.54% of the total water developed for shrimp culture in the country. Further, there are only 28 companies with

TABLE 2. Number of farmers under each category of land holding in the different maritime states.

	Size of Farm							
	0-2 ha		2-5 ha		5-10 ha		>10 ha	
State	No. of farmers	%	No. of farmers	%	No. of farmers	%	No. of farmers	%
Gujarat	22	15.60	103	73.05	10	7.09	6	4.26
Maharashtra	74	54.00	36	26.00	13	10.00	13	10.00
Karnataka	393	61.69	94	14.76	48	7.54	102	16.01
Goa	41	24.55	19	11.36	97	58.08	10	5.99
Kerala	1,297	50.00	699	27.00	234	9.00	3,620	14.00
Tamil Nadu	352	42.00	349	41.00	107	13.00	36	4.00
Andhra Pradesh	29,738	96.19	2,190	3.02	336	0.46	238	0.33
Orissa	7,580	97.88	100	1.29	51	0.66	13	0.17
West Bengal	4,327	55.00	1,778	29.00	1,077	13.00	646	8.00

farms of 40 ha or more. The area developed by these companies is just 1.25% of the total are developed for shrimp farming in the country.

Shrimp farming in India is now faced with several constraints, particularly in the areas of environment, legal, and disease. Coastal shrimp farming within 500 m from the seafront is banned as well as intensive shrimp farming. Only modified shrimp farming with a production level of 2 to 2.5 tonnes/ha is permitted. Shrimp farming in India is today controlled by the Aquaculture Authority of India, the supreme authority to regulate shrimp farming which has been created by an act in the parliament at the direction of the Supreme Court of India.

Freshwater Prawns

Freshwater prawns, *Macrobrachium rosenbergii* and *M. malcolmsonii,* are abundantly available in the rivers and their demand in the domestic market is extremely high. To provide these food items to the consumer, technology for broodstock management, breeding, hatchery, and grow-out have been developed. Both mono- and poly-culture of freshwater prawns are being done by the farmers. Production of freshwater prawns is in the range of 1-3 tonnnes/ha/year.

Oysters

The seed requirement of the oysters, *Crassostrea madrasensis*, *C. rivularis*, and *C. gryphoides* is met either from natural sources or hatcheries. Generally, oyster seed (or spat) is collected on cultch material from the estuaries and backwaters. The shells of oysters, mussels, and coconuts are arranged on nylon ropes or wire suspended from in the shallow estuaries, bays, and back waters. Generally, oysters are cultured on rafts and racks. In the rack and string method, the spat ladden shell string are enclosed in velon screen bags and suspended from racks in the farming zone. In the rack and tray method, the cultch-free seed oysters are stocked in box-type trays and suspended from racks. The oysters attain marketable size within 7-8 months. Being a filter-feeder, it gives high production per unit area and eliminates the cause for additional expenditure on the supplementary feed. Under rack and ren farming, 80-100 tonnes/ha of oysters are obtained. The annual catch which was estimated at less then 4,000 tonnes was subsequently increased in recent years due to farming practices in the estuarine systems. Currently, production is over 15 tonnes of oyster/year.

Mussel Culture

The seeds of green mussel, *Perna viridis,* and brown mussel, *P. indica,* are collected from the natural beds on nylon ropes. Mussels are cultured on raft or long lines deployed at an unpolluted sites in the open sea and protected bays. Mussels have traditionally been the food of the working class. But it has now become a delicacy for the upper class also. Mussel farming has become very popular in Kerala and Karnataka, in the westcoast of India.

Pearl Production

The demand for the pearls is increasing in the domestic and international markets. In India, pearls are produced either from marine pearl oysters, *Pinctada fucata* and *P. margaritifera*, or the freshwater mussels, *Lamellidens marginalis*, *L. corrianus*, and *Perreysia corrugata.* However, the pearls produced from *Pinctada fucata* are best as compared to other pearl producing bivalves. Today, India imports pearls worth $7-8 million (US). One of the recent developments is that the country started exporting around $20,000 worth pearls in 1996. The natural beds of the pearl oyster, *Pinctada fucata*, are available mostly in the Gulf of Manner, Palk Bay, Gulf of Kutch, Andaman Island, and Nicobar Island.

FUTURE NEEDS

India has the potential need to produce 7-8 million tonnes of fish by 2010 AD. Assuming that the marine production will be steady at 3 million tonnes and the deep sea fishery does not develop, the balance of production has to come from aquaculture. This evidently shows that inland fish production has to be raised from the current level of 2.5 million (1999-2000) to 5 million tonnes. This can only be achieved by bringing more areas under cultivation and also increasing the production rate. Some progressive farmers in two states, Punjab and Haryana, have already shown that they are capable of producing 13-16 tonnes/ha by adopting modern technology. Several other farmers in these two states also have shown average production rate of 6-8 tonnes/ha. India has large area of inland saline water bodies unutilized. The saline of Haryana areas are suitable for shrimp farming and also culture of species like tilapia, *Oreochromis mossambicus*, and pearl spot, *Etroplus suratensis*. The average inland productivity is very low varying from 30-100 kg/ha. Cage culture has not been established in India. In reservoirs, cage culture can be safely started to drastically increase the productivity as well as providing employment and income for rural areas.

Fish consumption is fast expanding in urban areas. Sustained economic growth and increase in per capita income in coming years will enhance fish consumption, which in turn is expected to activate domestic market forces to increase prices. This will be beneficial to farmers. However, shrimp farming will be largely dictated by the international market. At present, there is little export of freshwater farmed fish from India. Aquaculture, particularly brackishwater fish farming, is destined to play a vital role in the economy of India in the new millennium. The present production of 80,000 tonnes/year of shrimp can be doubled in five years. The present production of 1.9 million tonnes of fish by aquaculture can be doubled in five years provided effective strategies and investment are given to this sector. However, aquaculture needs to be supported by the appropriate legal and institutional frameworks in India with active cooperation of stake holders.

Bottom Soil and Water Quality Management in Shrimp Ponds

Claude E. Boyd

SUMMARY. Marine shrimp farming has become an important component of aquaculture. Water and soil conditions have a great influence on the efficiency of shrimp production, and a review of water quality and bottom soil management will be provided. After a shrimp pond has been drained for harvest, organic carbon analyses can indicate if organic matter concentrations are too low (< 0.5%) or too high (> 3 or 4%). Bottoms should be dried for 2 to 3 weeks to enhance oxidation of organic matter and other reduced substances. Soil pH should be measured and agricultural limestone applied to acidic pond bottoms. In soil with excessive organic matter, nitrogen fertilizers may be applied at 200 to 400 kg/ha to enhance microbial activity. Nitrate nitrogen is especially effective as a nitrogen source for treating pond bottoms because it also is a soil oxidant. Empty pond bottoms also may be tilled to 5 to 10 cm depth with a disk harrow to encourage oxidation. Once ponds have been filled with water, chemical fertilizer containing both nitrogen and phosphorus should be used to encourage natural productivity. Applications of 2 to 4 kg/ha of N and P_2O_5 at 2- to 3-day intervals are usually adequate. Within 1 or 2 weeks, natural food should be abundant and postlarvae can be stocked. Fertilization with 1-2 kg N and 0.5-1 kg P_2O_5/ha at intervals of about 1 week should be continued for a few weeks. However, once feeding rates exceed 20 to 30 kg/ha per day, fertilization should be used cautiously because of the danger of causing excessive phytoplankton blooms. Mechanical aeration can be used to increase feeding rates and allow higher

Claude E. Boyd, Department of Fisheries and Allied Aquacultures, Auburn University, AL 36849 USA.

[Haworth co-indexing entry note]: "Bottom Soil and Water Quality Management in Shrimp Ponds." Boyd, Claude E. Co-published simultaneously in *Journal of Applied Aquaculture* (Food Products Press, an imprint of The Haworth Press, Inc.) Vol. 13, No. 1/2, 2003, pp. 11-33; and: *Sustainable Aquaculture: Global Perspectives* (ed: B. B. Jana, and Carl D. Webster) Food Products Press, an imprint of The Haworth Press, Inc., 2003, pp. 11-33. Single or multiple copies of this article are available for a fee from The Haworth Document Delivery Service [1-800-HAWORTH, 9:00 a.m. - 5:00 p.m. (EST). E-mail address: getinfo@haworthpressinc.com].

levels of shrimp production. Water exchange is widely used in shrimp farming, but water exchange often can be reduced to much lower levels than commonly applied. Suggestions on the use of better practices to minimize possible adverse environmental effects of shrimp farming are provided.

KEYWORDS. Shrimp farming, water quality, sediment

INTRODUCTION

Shrimp farming in earthen ponds has become an important commercial endeavor producing nearly 30% of the shrimp on the world market. Shrimp require a high quality pond environment, so much attention is given to maintaining good bottom soil condition and water quality. Most shrimp culture systems are based on the use of manufactured feed to increase production above that possible from natural productivity. It is not uncommon to obtain production of 1 to 2 tons/ha per crop in semi-intensive culture systems and production above 5 tons/ha per crop is often achieved in intensive culture. Excessive phytoplankton blooms may occur in shrimp ponds as a result of feed inputs. Overabundant plankton, low dissolved oxygen concentrations, high ammonia concentrations, and impaired bottom soil quality can stress shrimp and result in low survival and production. Mechanical aeration is used to enhance dissolved oxygen concentrations, improve water and soil quality, and permit greater shrimp production.

The purpose of this report is to discuss shrimp pond soil and water quality to optimize shrimp production in ponds with feeding. Suggestions also will be made on ways to make shrimp culture more environmentally responsible.

POND PREPARATION

When ponds are drained for harvest, nutrients and plankton are discharged, and the flocculent layer of highly organic material and many of the benthic organisms which occur near the soil-water interface are suspended and lost from ponds (Ayub et al. 1994). It is common practice to dry pond bottoms after draining, and many times, bottom soils are baked dry before ponds are refilled. Therefore, when ponds are refilled with water and stocked to start a new crop, plankton and benthos to serve as natural food for postlarvae are scarce. Ponds may be treated by various techniques before stocking in order to improve the abundance of natural food organisms for postlarvae. In this discussion, pond preparation refers to all activities that are done from the time a pond is drained for harvest until the postlarvae are stocked for the next crop.

Organic Carbon

Shrimp farmers are concerned about the accumulation of organic matter in pond soils (Peterson and Daniels 1992). Although this problem is probably less common than often thought, monitoring of bottom soil organic matter concentrations can be useful to management decisions. The best time to obtain a soil sample is soon after a pond has been drained, but before any soil treatments have been applied. Soil samples should be collected from several places in the pond bottom. It usually is adequate to take 10 to 12 random samples of the upper 5-cm layer and combine equal volumes of these samples to provide a composite sample for analysis. The composite sample should be mixed thoroughly, dried in an oven at 60°C or in the sun, and pulverized to pass a 20-mesh screen. The Walkley-Black Method (Boyd and Tucker 1992) can be used to determine organic carbon concentrations. The standard procedure requires a modest amount of laboratory equipment, but the Hach Chemical Company, Loveland, Colorado, sells a portable soil organic carbon kit (Model CEL/700) based on the Walkley Black Method that provides results comparable to the standard Walkley-Black procedure (Queiroz and Boyd 1998a).

Soil organic carbon concentrations in bottom soils of shrimp ponds seldom exceed 1 or 2%, and values in excess of 3 or 4% are probably excessive. Some organic matter in bottom soils is good, because it favors benthic productivity. Ponds with less than 0.5% organic carbon may not have good benthic productivity. Occasionally, shrimp ponds may be constructed on organic soils and organic carbon concentrations may be 20% or more. Such ponds are not good sites for shrimp production and should be avoided. Organic carbon values can be multiplied by a factor of 2 to provide a rough estimate of actual organic matter (Nelson and Sommers 1982). Methods for accelerating the decomposition of soil organic matter are discussed below.

Drying

It is common practice to dry pond bottoms between crops (Clifford 1992). Drying accelerates the decomposition or organic matter accumulated in the bottom during the previous crop by providing a better oxygen supply for bacteria. Better aeration of the soil oxidizes reduced inorganic and organic compounds in the soil to improve soil condition. Drying also kills disease organisms and their carriers that may exist in the soil. A drying period of 2 to 3 weeks usually is adequate. In wet weather, it may be impossible to adequately dry soil, but as a rule, pond bottoms should be dried well once per year (Boyd 1995a).

Liming and Disinfection

Application of agricultural limestone should be made to pond bottoms that are acidic (pH below 7). Samples for soil pH may be collected and processed as described above for organic matter samples. To measure pH, mix 10 or 20 g of dry, pulverized sample with 10 or 20 ml of distilled water, stir intermittently for 20 min and measure the pH with a glass electrode. The lime requirement of soil may be determined by the pH change in a buffered solution (Pillai and Boyd 1985), but most shrimp farmers will not be able to make lime requirement tests. The application rate for agricultural limestone may be estimated according to the following soil pH scale:

Soil pH	Agricultural limestone (kg/ha)
Above 7.0	0
6.9-6.5	500
6.4-6.0	1,000
5.9-5.5	2,000
5.4-5.0	3,000

Of course, where soil pH is below 5 because of acid-sulfate soils, 5,000 to 10,000 kg/ha or even more agricultural limestone may be needed to increase soil pH to an acceptable level.

Agricultural limestone should be spread uniformly over the pond bottom. It should be applied within 3 or 4 days after ponds are drained and before the bottom soil becomes extremely dry. The limestone must dissolve in soil pore water to increase soil pH and stimulate microbial activity.

In ponds where disease was a serious problem in the previous crops, some farmers treat the pond bottom with a chemical agent to kill disease organisms and reduce the possibility of disease in the next crop. An economical way of disinfecting a pond is to apply burnt lime (CaO) or hydrated lime [$Ca(OH)_2$] to increase the soil pH above 10 and kill disease organisms (Snow and Jones 1959). A treatment rate of 1,000 kg/ha of burnt lime or 1,500 kg/ha of hydrated lime usually is sufficient for disinfecting pond bottoms (Boyd 1995a). The lime must be spread uniformly over pond bottoms while they are still wet so that the lime will dissolve and penetrate the soil mass to raise pH and kill pathogens and their carriers. Chlorination has been used to disinfect pond soils. However, organic matter quickly reduces chlorine residuals to non-toxic chloride (White 1992). High concentrations of calcium hypochlorite may be needed for disinfection (Potts and Boyd 1998), and chlorination is much more expensive than lime treatment.

In recent years, there have been considerable problems with the white spot syndrome virus (WSSV) disease and other viral diseases in shrimp farming. Viruses can enter ponds as virus particles in the water or in carrier organisms such as small shrimp and microcrustaceans. It has become popular to try to kill virus particles and carriers by treating recently-filled ponds with strong chemical agents such as formalin, chlorine, insecticides, and lime. These products must be used at relatively high concentrations to effect disinfection. In small, intensive production ponds, farmers can afford disinfection. However, the cost will be a much greater factor in large, semi-intensive shrimp ponds. Also, it will be difficult to apply disinfectants effectively to large ponds, and it is doubtful that a complete kill of vectors will result. Insecticides such as Dipterex (an organophosphorus compound) and Sevin (a carbamate compound) are probably the most cost-effective means of pond disinfection. Dipterex and Sevin both seem acceptable for killing vectors of WSSV in ponds because they will degrade quickly and probably will not cause environmental or food safety problems. Nevertheless, environmentalists will complain about the use of even highly-degradable insecticides, and these complaints could lead to problems in the market. All potential pond disinfectants are potentially hazardous to humans and all could cause environmental problems if used incorrectly (Boyd and Massaut 1999).

Oxidants

Nitrogen fertilizers can be applied to pond bottoms to enhance the decomposition of soil organic matter (Peterson and Daniels 1992). Such treatment is possibly beneficial in ponds where carbon:nitrogen ratios in soils are greater than 15 or 20. Nitrate compounds such as sodium nitrate are particularly good for soil treatments. They dissolve in soil moisture and penetrate into anaerobic zones in the soil where the nitrate serves as a source of oxygen for bacteria (Avnimelech and Zohar 1986; Boyd 1995b) as illustrated in the following generalized equation:

$$4NO_3 + 5CH_2O \text{ (organic matter)} \rightarrow 4H^+ + 2N_2 + 5CO_2 + 7H_2O$$

Oxygen from nitrate is used by bacteria to mineralize organic matter to carbon dioxide and water. The nitrogen gas produced in denitrification is lost to the atmosphere. Nitrates have a basic reaction in ponds because carbon dioxide and water combine to form bicarbonate. This reaction reduces acidity, and use of nitrate in ponds with acidic bottom soils can lessen the amount of liming materials needed.

For most ponds, an application rate of 25 to 50 kg/ha of nitrogen to pond bottoms in fertilizer is sufficient, but if organic carbon concentrations are high (above 3 or 4%), twice as much nitrogen fertilizer should be broadcast uniformly over pond bottoms while the soil is still moist. Nitrogen fertilizers can be applied with agricultural limestone in acidic ponds, but they should not be applied simultaneously with hydrated or burnt lime.

Calcium peroxide is sometimes used in Asia to oxidize pond bottoms (Chamberlain 1988). This compound releases oxygen spontaneously even in presence of dissolved oxygen because of its high oxygenation potential. Nitrate only releases oxygen when the dissolved oxygen concentration and redox potential are low. Therefore, nitrate is more effective as a soil oxidant than peroxide, because it does not release its oxygen until the oxygen is needed to prevent low redox potential.

Tilling

Tilling of pond bottoms during the drying period can enhance aeration of the soil. Heavy-textured soils (clays and clay loams) will benefit more from tilling than will light-textured soils (sands, sandy loams, and loams). Tilling should be done with a disk harrow and limited to a depth of 5 to 10 cm (Boyd 1995a). A rotor tiller also is suitable for tilling pond bottoms, but it is destructive of soil structure. Turning plows (breaking plows) require much more energy to use than disk harrows. They are useful when there is an excessive concentration of organic matter on the soil surface, because this plow turns the soil over to bury the surface layer under soil of lower organic matter concentration from deeper layers. Tilling should be done while bottom soils are still moist, but bottoms should be dry enough to support the weight of the tractor tires and prevent formation of ruts in the pond bottom.

Sediment Removal

It usually is not necessary to remove sediment from ponds. However, if the shallow interior canals for enhancing complete draining of a pond fill in, or if particularly large amounts of sediment accumulate in deeper parts of ponds causing them to lose volume, sediment removal may be necessary. Sediment disposal should be done in a way to prevent the sediment from washing into ponds or canals after heavy rains and to avoid adverse ecological impacts outside of ponds. Site specific methods of sediment disposal should be developed for each farm (Donovan 1998).

Fertilization

Once the pond bottom has been dried and necessary soil treatments applied, the pond can be refilled. At this time it usually is necessary to apply nutrients to encourage plankton and benthos. The two key nutrients are nitrogen and phosphorus. The common source of phosphorus is orthophosphate, but nitrogen may be supplied as urea, ammonia nitrogen, or nitrate. Urea quickly hydrolyzes to ammonia. Ammonia is undesirable in ponds for three reasons: (1) it can be toxic to shrimp at relatively low concentrations; (2) it is converted to nitrate by nitrifying organisms that produce hydrogen ion and lower the pH in the process; (3) nitrification requires a large amount of dissolved oxygen. Thus, sodium, potassium and calcium nitrates have an advantage as nitrogen fertilizers because nitrate is non toxic, it does not form acidity, and it does not have an oxygen demand (Boyd and Tucker 1998). Moreover, nitrate is a source of oxygen to bacteria, and when it is denitrified in ponds, it raises the pH slightly.

The best rates of nitrogen and phosphorus application for establishing a plankton bloom will vary with the availability of these two nutrients in pond soil and source water. A good application rate for general purposes is 2 to 4 kg/ha both of N and P_2O_5. Farmers can purchase mixed fertilizer that already contains both nitrogen and phosphorus in the proper ratios rather than using fertilizer source compounds to mix fertilizers on the farm. Fertilizer applications should be made at 2- or 3-day intervals until a good plankton bloom is established. Granular fertilizer should be premixed in pond water for a few minutes and the resulting slurry splashed over the pond surface. Within 1 to 2 weeks, the pond should have a good plankton bloom and benthos will begin to grow. At this time, postlarvae should be stocked.

Some shrimp farmers like to apply manure to ponds to enhance plankton blooms. In my opinion, manure should not be routinely used in aquaculture ponds. Manure can cause low dissolved oxygen concentrations and deterioration of bottom soil condition. It also may contain high concentrations of heavy metals, and it may contain antibiotics that can contaminate shrimp. Of course, additions of organic matter can enhance the availability of benthos in ponds with low concentrations of organic matter in soil, and it is effective in encouraging rapid blooms of zooplankton (Geiger 1983). If one wants to use organic fertilizers, it is better to use plant meals rather than manures.

MAINTENANCE OF PRODUCTIVITY

Most shrimp ponds are supplied manufactured feed from stocking until harvest. The amount of feed applied depends upon the shrimp biomass, and the feeding rate increases as the grow-out period progresses. However, in semi-intensive shrimp production, feeding rates seldom exceed 20 kg/ha until the later part of the grow-out period, and fertilizers can be beneficial in stimulating natural productivity to supplement manufactured feed. Good levels of natural productivity in ponds also help maintain water quality, especially by providing dissolved oxygen through photosynthesis and removal of ammonia.

In intensive ponds, it is often not necessary to fertilize after the first 6 or 8 weeks. In fact, fertilization of ponds with feeding rates above 20 to 30 kg/ha per day may encourage excessive phytoplankton blooms.

Fertilization

One objective of water quality management should be to try to maintain a moderate but stable phytoplankton bloom (Boyd and Tucker 1998). The best way to accomplish this is with a fertilization program of

1 to 2 kg N and 0.5 to 1 kg P_2O_5/ha per week. Some farmers prefer a high proportion of diatoms in the phytoplankton communities of shrimp ponds. There is evidence that a high nitrogen:phosphorus ratio encourages diatoms, and it is common practice to use only nitrogen fertilizer or a fertilizer with a wide nitrogen:phosphorus ratio of 15 or 20 to encourage diatoms. There is evidence that nitrate is more efficient in promoting diatoms than other nitrogen fertilizers (Daniels and Boyd 1993). There also is evidence that applications of silicate and chelated iron can stimulate diatoms in shrimp ponds. Although this is probably true in some situations, the limiting concentrations of silicate and iron to diatoms in pond water are not known, and it is difficult to make recommendations on where these treatments are needed and data on effective application rates of silicate and iron are not available. Nevertheless, shrimp farmers may want to try applications of silicate and iron in order to see if they increase diatom abundance.

Pond fertilizers should be applied according to the Secchi disk visibility in order to conserve nutrients, reduce costs, and prevent excessive phytoplankton. In my opinion, the best Secchi disk visibility range is 25 to 40 cm. Fertilizer application rates can be adjusted for Secchi disk visibility as illustrated below:

Secchi disk (cm)	Fertilizer (kg/ha)
20	0
25	2.5
30	5.0
35	7.5
40	10.0 (full application rate)

Liming

Unless the total alkalinity of pond water falls below 60 mg/liter as equivalent calcium carbonate, agricultural limestone should not be applied to pond waters during the production period (Boyd and Tucker 1998). Burnt or hydrated lime should never be applied to pond water during grow-out because of the danger of high pH. In areas where pond waters have low alkalinity, total alkalinity should be monitored monthly. Water analysis kits are available for measuring total alkalinity. If total alkalinity drops below 60 mg/liter, agricultural limestone should be broadcast over pond surfaces at 500 kg/ha.

Water Exchange

Water exchange has traditionally been used in shrimp ponds at rates of 10% to 15% of pond volume per day. It is difficult to justify the use of routine water exchange, because if pond water is of adequate quality, renewal of a portion of the pond water daily is of no benefit. Also, water exchange flushes out nutrients and plankton, so it reduces natural productivity in ponds. It is counterproductive to apply fertilizers to ponds for the purpose of enhancing phytoplankton productivity and then flush the nutrients from the ponds by exchanging water. Water exchange is necessary in ponds where salinity may rise to unacceptable concentrations during the dry season. Water exchange at 2 to 5% of pond volume per day can minimize the increase in salinity during the dry season. In intensive ponds, water exchange may sometimes be necessary to flush out excessive ammonia.

It is significant to note that water exchange is seldom used in other kinds of pond aquaculture. For example, in channel catfish farming in the United States, farmers once used water exchange just like many shrimp farmers do today. However, research and practical experience demonstrated that water exchange was not necessary. Today, catfish farmers do not exchange any water, and they harvest fish with seines to conserve water in ponds. The only time water is discharged from ponds is after heavy rains or when ponds must be drained to remove large fish that have escaped capture for several years or to repair wave damage to inside slopes of embankments. Channel catfish ponds are drained every 5 to 10 years (Boyd and Tucker 1998).

There obviously are great benefits to reducing water use in shrimp ponds (Boyd 1999). Water conservation can reduce the amount of pump capacity needed on a farm and the amount of energy used for pumping water thereby lowering costs. Retention of water in ponds lowers the quantity of fertilizer nutrients necessary to maintain natural productivity, and this also lowers the cost of shrimp production. A longer water retention time in ponds allows for greater assimilation of nutrients and organic matter within ponds, and this leads to enhanced effluent quality. Lower water velocities will reduce erosion of earthwork and lessen concentrations of suspended solids in effluent. Conservative use of water in shrimp ponds reduces production costs and protects against pollution of coastal waters by reducing the volume and enhancing the quality of effluents.

Other Treatments

In Asia, a large number of chemical, physical, and biological treatments have been used for the purpose of enhancing water quality in ponds, and this practice is increasingly frequent in the Americas. These products include formalin, chlorine, benzylchromium chloride, provodone iodine, zeolite, peroxides, bacterial inocula, enzyme preparation, etc. There also has been some use of bacterial inocula, enzyme preparations, and grapefruit seed extracts in shrimp culture. To date, there is little evidence that these substances can significantly enhance soil and water quality in ponds or improve natural productivity. However, there is evidence that bacterial inocula and grapefruit seed extracts can improve survival of culture species (Queiroz and Boyd 1998b). Much additional research is needed to elucidate the modes of action of these products and to determine how and when they can be used for the most benefit to shrimp production (Boyd and Gross 1998).

MECHANICAL AERATION

Aerators perform two basic functions: (1) they oxygenate water and (2) they induce water circulation (Boyd 1998). Both functions are important in pond aquaculture. Pond water normally contains plenty of dissolved oxygen during daylight hours, but in ponds with high feeding rates and abundant phytoplankton blooms, dissolved oxygen concentrations may decline to dangerously low levels at night (Boyd and Tucker 1998). Aeration can be used during the night to stabilize dissolved oxygen concentrations within safe ranges. A moderate degree of water circulation in ponds is beneficial; it prevents stratification of the water to provide more uniform water quality throughout the pond volume (Fast et al. 1983). Also, organic matter from uneaten feed, feces, and dead plankton settles to pond bottoms and decomposes. This can lead to anaerobic conditions in the superficial layers of bottom sediment with the release of toxic microbial metabolites into the overlaying pond water (Boyd 1995a). Induced circulation of pond water increases water flow and delivery of dissolved oxygen to the surface of sediment to lower the possibility of anaerobic conditions at the sediment-water interface.

Because of better water and soil quality in response to aeration, shrimp eat better, have greater resistance to diseases, suffer less mortality, and achieve better feed conversion efficiency in aerated ponds than in ponds without aeration. Aeration has become a standard practice to improve yields and profitability in shrimp culture and other kinds of aquaculture (Boyd 1998).

Types of Aerators

Aerators may be separated into two broad categories: (1) splasher aerators and (2) bubbler aerators. Splasher aerators include three major types as follows: vertical-pump aerators, pump-sprayer aerators, and paddlewheel aerators. Vertical pump aerators employ a propeller to jet water vertically into the air to effect aeration. Such aerators have a limited zone of influence because they tend to recirculate the water around the aerator. To be effective, they must be used in small ponds or multiple aerators must be positioned so their zones of influence overlap. Pump-sprayer aerators spray water from holes or nozzles into the air at high velocity to effect aeration. They cause horizontal flow and produce strong water currents. However, head loss is great because high velocity water is discharged through relatively small openings. Paddlewheel aerators rely on slow speed paddles to splash large volumes of water into the air for aeration. They are generally more efficient than other types of splasher aerators (Boyd 1998).

The two main types of bubbler aerators are diffused-air systems and propeller-aspirator-pump aerators. Diffused-air aerators employ an air blower and tubing to deliver air to diffusers mounted at the pond bottom. Small air bubbles released by the diffusers rise through the water and oxygen in the air within bubbles diffuses into the water. Diffused-air aeration systems must have diffusers at many places in the pond to effect uniform aeration and mixing. They are limited by water depth unless air is released very slowly from diffusers (Boyd and Moore 1993). Propeller-aspirator-pump aerators rely on a high-speed propeller to propel water rapidly away from the end of a diffuser mounted on the end of a housing that is open to the atmosphere. The Venturi principle causes air to be drawn from the shaft and pass through the diffuser to enter as fine air bubbles into the highly-turbulent water in front of the propeller. These aerators produce strong horizontal mixing and they can be highly efficient (Boyd and Ahmad 1987).

Propeller-aspirator-pump and paddlewheel aerators are by far the most widely used in shrimp ponds. This results from the following factors: simplicity, relatively low cost and maintenance, strong horizontal mixing, adequate oxygen transfer, and aggressive marketing. Diffused-air aeration systems are widely used in hatchery operations because one air blower can service diffusers in many tanks.

Aerator Tests

Aerators for use in aquaculture usually are built according to modifications of wastewater aerator designs. The modifications in design usually are made in order to reduce costs. Aerators for wastewater treatment have been subjected to much testing to improve designs and enhance oxygen-transfer efficiency. Design modifications imposed to make aerators less expensive for aquaculture often result in lower oxygen-transfer efficiency. The art of aerator testing has not been widely applied to aquaculture aerators. Still, manufacturers have been able to sell many types of aerators to fish and shrimp farmers. Farmers have little knowledge of the principles of aeration or the aerator design features that influence oxygen transfer efficiency. Thus, pond aeration equipment sales usually are decided by the cost per unit and the ability of the salesperson. For this reason, improvements in most pond aeration equipment have been slow.

Catfish farming in the United States provides an excellent example of how aerator design can be improved by cooperation between researchers and manufacturers. In the early 1980s, an aerator design and testing program was initiated at Auburn University. All major types of aerators used in the industry were tested for oxygen transfer efficiency (Boyd and Ahmad 1987). The data included the standard oxygen transfer rate (SOTR) in kilograms of oxygen per hour for individual aerators, the standard oxygen efficiency (SAE) in kilograms of oxygen per kilowatt hour of each aeration unit, and the cost of transferring oxygen in dollars per kilogram for each aerator. Aerator tests are conducted in aeration basins of known volume. Water is deoxygenated with sodium sulfite and cobalt chloride and the aerator is used to reoxygenate the water. During reoxygenation, dissolved oxygen concentrations are monitored in the tank at frequent and known intervals. These data can then be treated mathematically to determine the oxygen transfer coefficient of the aerator. Data on power input, tank volume, water temperature, and other factors can then be used to compute SOTR, SAE, and cost of transferring oxygen (Boyle 1992). Values for SOTR and SAE are presented for standard conditions of 20°C, 0 mg/L dissolved oxygen, and clean, freshwater. In ponds, standard conditions do not exist, but equations are available for estimating actual oxygen transfer from SOTR and SAE under actual operating conditions. Nevertheless, SOTR and SAE data can be extremely useful because they permit one to compare the abilities of different aerators to transfer oxygen to water.

It was quickly learned that most aerators commonly used in aquaculture transferred less than 1.2 kg O_2/kW·hr. Studies were then conducted to modify aerator design to improve efficiency. It was possible to increase the SAE of some types of aerators to more than 2.5 kg O_2/kW·hr (Boyd 1998). These designs were adopted by manufacturers of aerators for catfish farming, and a remarkable improvement in catfish pond aerators was achieved. The new aerators are slightly more expensive than the older designs, but they were more economical to operate and they had a much longer service life.

Paddlewheel aerators used in shrimp farming in most countries have been subjected to little objective testing. It is obvious that many of the paddlewheel aerators currently used in shrimp farming could be greatly improved. The paddle designs are badly flawed, and aerator shaft speed often is not optimal (Boyd 1998). This does not mean that aerators currently used are not beneficial. Rather, it suggests that these aerators are lower in efficiency than some other types of aerators.

The propeller-aspirator-pump aerators are of two types: (1) an original model made in the United States (Aire-O_2, Aeration Industries, Chaska, Minnesota) and (2) copies of this model. Studies conducted by Ruttanagosrigit et al. (1991) revealed that the original model made in the United States had markedly greater SAE values than a copy of it made in Asia. The loss in efficiency in the copies apparently resulted from modifications made to reduce the cost of the copies below that of the aerator made in the United States. The results of the study by Ruttanagosrigit et al. (1991) also demonstrated that propeller-aspirator-pump aerators increased in oxygen-transfer efficiency in brackishwater as compared to freshwater.

Aerator testing can provide data for comparing the efficiencies of different aerators. It also may be used to determine how changes in aerator design influence efficiency. Aerator testing is expensive, but it can provide enormous benefits. Procedures for oxygen-transfer rate testing are provided by Boyle (1992) and Boyd (1998). The methods for determining water circulation are provided by Boyd and Martinson (1984) and Howerton and Boyd (1992).

Aeration-Production Relationships

The oxygen dynamics of a pond are complex. Phytoplankton and other aquatic plants produce oxygen and all plants and animals living in ponds use oxygen. Water temperature and dissolved oxygen concentration change continuously, and it is virtually impossible to compute the actual oxygen transfer rate of an aerator in a pond. It is not possible to know exactly how much of the oxygen provided by aeration is used in respiration by shrimp and how much is consumed by other pond organisms. It also is difficult to estimate precisely how much additional shrimp production can be achieved by a given amount of aeration. However, experience suggests that each horsepower of aeration will allow for 400 to 500 kg/ha of shrimp production above that possible in un-aerated ponds (Boyd and Tucker 1998). It is not unusual for intensive shrimp ponds to be aerated at 10 to 20 kg/ha.

Water Circulation

Aeration causes water movement in ponds and mixing of the water column and produces horizontal water flow across pond bottoms (Boyd and Tucker 1998). The ability of an aerator to direct water away from it has another benefit. The oxygen-transfer efficiency of an aerator decreases as the dissolved oxygen concentration in the water increases. Ideally, water entering the aerator should have the lowest dissolved oxygen concentration in the pond in order to favor high oxygen transfer. All aerators tend to move water away from them, but an aerator that does not mix water well tends to recirculate the water in the zone immediately around it. This results in an increase in dissolved oxygen concentration in water around the aerator and a decrease in oxygen transfer efficiency. The vertical-pump aerator is a good example of an aerator that tends to recirculate water within a relative small zone of influence. Paddlewheel and propeller-aspirator-pump aerators force water away from them and do not create small zones of influence with resulting loss of aeration efficiency.

There is considerable interest among shrimp farmers regarding equipment that directs most of its energy to circulating pond water instead of aerating it (Thiriez 1999). The low-energy water circulation devices could be used to circulate pond water during the daytime when dissolved oxygen concentrations are high. This would blend oxygen-supersaturated surface waters with deep layers of low dissolved oxygen content to provide a high dissolved oxygen concentration from top to bottom in the evening. Aerators could be turned on once dissolved oxygen concentrations begin to decline during the night (Busch 1980; Busch and Flood 1980; Fast et al. 1983; Howerton et al. 1993). Although water circulators can prevent stratification and improve water quality in bottom layers of pond water, there is no conclusive evidence that they reduce the need for mechanical aeration or increase production (Tucker and Steeby 1995).

The water circulation capacity of aerators and other mechanical devices also can be tested. Boyd and Martinson (1984) used a dye test to determine the length of time for a highly-visible dye to be spread over a pond surface by aerator-induced mixing. They also used a salt-mixing test in which a brine solution was poured into the water in front of an aerator and the time to mix the salt to a uniform concentration in all parts of the pond was measured. The dye test is limited to surface mixing only, and the salt test is only applicable to freshwater. Howerton and Boyd (1992) showed the rate of weight loss to dissolution by small gypsum (calcium sulfate) blocks placed in ponds was a reliable index of water velocity. The procedure, called the gypsum clod-card method, can be used to test mixing patterns in both freshwater and brackishwater.

Sediment Resuspension

Because aerators induce water circulation and cause greater velocity of water across pond bottoms, they resuspend sediment. Solids suspended by aerator-induced currents can increase the turbidity of pond water. Also, solids can settle in areas of ponds where water currents are less to create sediment mounds (Boyd 1992). These sediment mounds consist primarily of mineral soil particles, but organic solids suspended by aeration also settle with the soil particles (Boyd et al. 1994). As a result, the sediment mounds in ponds have a high oxygen demand. In highly-intensive shrimp farming, the superficial layer of sediment mounds may become anaerobic even though aeration maintains sufficient concentrations of dissolved oxygen in pond waters (Boyd 1995a).

ENVIRONMENTAL ISSUES

Although shrimp farming has been quite successful and the technology has continually improved, it has caused some environment damage in certain places, and in recent years it has been the subject of much criticism by environmental-concern groups (Boyd 1999). Environmentalists have taken examples of adverse environmental effects from specific, isolated locations and used these examples to support sweeping generalizations that shrimp farming is causing large amounts of ecological damage in many nations. A fairly balanced view of the environmental issues related to shrimp farming was recently published in the Scientific American (Boyd and Clay 1998). The authors showed that the environmental damage that has been caused in some regions is not inherent to shrimp farming but the result of poor planning and bad management. They recognized that many shrimp farmers are aware of the potential adverse environmental impacts of their operations, and the industry is adopting more "environmentally-friendly" production procedures. However, much remains to be done regarding improvements in site selection, design and construction of farms, production methods, and effluent management.

Environmental Management

Environmental management can be effected in a variety of ways. The most sophisticated approach is to require environmental impact assessment (EIA) to identify potential negative environmental impacts and to prepare an approved mitigation plan for preventing the negative effects. Regulations may exclude projects from certain habitats, e.g., wetlands, mangroves, or other ecologically-sensitive areas. Permits may be required for certain items, e.g., wells, roads, effluents, etc. Water pollution control may be effected through discharge permits with specific limits on certain water quality variables. Monitoring often is required to show compliance with permits or to demonstrate that the mitigation plan actually provides the desired level of environmental protection.

In some instances, best management practices are formulated for projects to help meet regulatory requirements or to avoid ecological damage by projects where strict regulatory requirements and monitoring are not mandated. Best management practices (BMPs) are practices determined to be the most effective practical means of reducing pollution or other ecological impacts to levels compatible with water quality or other resource management goals (Wood 1995; Hairston et al. 1995). Best management practices may be mandated by governmental agencies as part of regulatory efforts, or they may be prepared by industry groups and voluntarily accepted by individuals or companies within the industry. In the case of shrimp farming, there has been relatively little research on environmental management and few governments have initiated strict regulatory programs. Thus, there is not much experience with BMPs or other types of regulation. It seems certain that in the future, governments of most nations will impose environmental regulations on shrimp farming. Thus, it is in the best interest of the shrimp-farming industry to develop BMPs for voluntary adoption. Adoption of BMPs is proactive both from the standpoint of deflecting criticism from environmentalists and by providing a way to have input into the development of future governmental regulations. Admittedly, there is not enough information upon which to form BMPs for all aspects of shrimp farming, but we certainly know enough to adopt better management practices that can eventually evolve into true BMPs. Shrimp farmers in some nations, e.g., Australia, Belize, and Thailand, have developed shrimp-farming codes of conduct that contain BMPs, and several countries currently are preparing such documents. The Global Aquaculture Alliance, a shrimp farming industry advocacy group, also has prepared a code of conduct with general suggestions about BMPs to guide national groups interested in formulating more specific BMPs (Boyd 1999). This organization is now planning to implement an environment stewardship program that can lead to "certified ecolabelling of shrimp."

Best Management Practices

The major environmental complaints about shrimp farming are as follows: destruction of mangroves and other wetlands; eutrophication and sedimentation in coastal waters; salinization of freshwater supplies; spread of disease to native shrimp; introduction of non-native species; use of toxic and bioaccumulative chemicals; adverse effects on biodiversity. Therefore, BMPs for site selection, design and construction, pond operations, and effluent management should be formulated with the goals of reducing these possible adverse impacts to an acceptable level. To be realistic, it is probably not possible to develop a shrimp farm or most any other aquacultural or agricultural project without causing some changes in the environment. However, through the use of BMPs in shrimp farming, impacts should be reduced to a degree that local ecosystems can continue to function normally and without significant changes in community composition and biodiversity. This may not be acceptable to the more extreme environmentalists, but it would satisfy the goals of coastal resource management and possibly prevent governments from imposing unnecessarily strict regulations on shrimp farmers.

Space constraints prevent listing all the specific BMPs the author considers appropriate for shrimp farming. Thus, only some general BMPs can be provided to the reader with an overview of the general approach to environmental management in shrimp farming through BMPs.

One of the major reasons for negative environmental impacts is location of shrimp farms on inferior sites. Some examples of inferior sites are: mangrove forests and other wetlands; sites with inferior soils, e.g., sandy, high-sulfur-content, or organic soils; coastal areas where development including shrimp farms have already polluted the water; areas where coastal waters are stagnant. A good site evaluation procedure and EIA can identify undesirable site features and predict adverse impacts. In some situations, site limitations can be corrected through special design and construction techniques, but in other cases, it may be impossible or too expensive to correct limitations and the site should be rejected (Hajek and Boyd 1994). For example, the bottoms of ponds in sandy areas may be covered with a layer of soil of sufficient particle-size composition to retard seepage. Although ponds should not be constructed in mangroves, they can be located behind the mangrove, and mangrove removal necessary for intake and discharge canals can be mitigated by replanting in other areas.

Shrimp farm design and construction also are critical factors in environmental protection. Standard design and construction methods should be used to provide canals of adequate size and avoid excessive water velocities and scouring, to slope and compact levees and other earthwork properly, to allow controllable inlet and exit structures in ponds, to prevent impoundment of freshwater by roads and other earthwork, to seal pond bottoms properly, to limit construction activities to the immediate areas, and to prevent turbid runoff during construction.

Many of the negative effects of shrimp farm operations on the environment can be largely prevented or mitigated by the following practices:

- Maintain grass cover on exposed earthwork to reduce erosion and control turbidity in runoff.
- The industry should strive to develop systems for producing broodstock and increase the use of hatchery-reared postlarvae. This system would be more reliable and protect natural shrimp populations from excessive capture.
- Pump intakes should be screened to prevent impingement of larger aquatic animals.

- It is not necessary to dilute brackishwater in ponds with freshwater from wells. This practice is seldom used today, but it should be stopped entirely.
- To prevent salinization, brackishwater from ponds should not be discharged in freshwater or onto agricultural land.
- Shrimp health management should focus on disease prevention instead of disease treatment. Disease free larvae should be used and disease reduced through better handling, moderate stocking densities, good nutrition, and optimal environmental conditions.
- Stocking rates should be moderate so that feed inputs do not exceed the assimilative capacity of ponds.
- High quality feeds should be applied in quantities that can be consumed by the shrimp and avoid overfeeding because it is wasteful and causes water quality deterioration.
- Use minimum rates of water exchange to permit maximum assimilation of organic matter and nutrients within ponds by natural processes.
- Ponds should be drained in a manner to prevent excessive water velocity and suspension of soil particles from canal walls and bottoms.
- Settling basins can greatly improve effluent quality and they are relatively inexpensive to build and operate.
- Effluent should be discharged in such a matter that concentrations of nutrients, organic matter, and suspended solids increase only in the mixing zone. Discharge into stagnant water should be avoided in order to minimize the size of the mixing zone.
- Toxic or bioaccumulative chemicals to include therapeutic agents should only be used for a specific, diagnosed problem and according to recommended procedures.
- Sediment should be disposed of in a manner to prevent turbid runoff, leaching of salt into freshwater, and ecological nuisances.

The BMPs for pond operations listed above make sense both from production and environmental viewpoints. Many shrimp farmers already have adopted many of these practices, and many more will do so in the future. The major constraints on adoption of new practices are lack of awareness, insufficient technical knowledge, and expense. Shrimp farming associations should strive to make farmers aware of better procedures and to develop technical guidelines for the use of BMPs. Shrimp are sensitive to environmental conditions, and shrimp farms de-

pend upon a supply of high quality water from coastal waters. Therefore, improvements in coastal waters through implementation of BMPs should greatly benefit conditions for shrimp production in ponds. Although the initial implementation of BMPs may be expensive and require changes in management philosophy and training of workers, the long-term benefits of BMPs will no doubt make shrimp farming more profitable, environmentally-friendly, and sustainable (Boyd and Haws 1999).

REFERENCES

Avnimelech, Y., and G. Zohar. 1986. The effect of local anaerobic conditions on growth retardation in aquaculture systems. Aquaculture 58:167-174.

Ayub, M., and C. E. Boyd. 1994. Comparison of different methods for measuring organic carbon concentrations in pond bottom soil. Journal of the World Aquaculture Society 25:322-325.

Boyd, C. E. 1992. Shrimp pond bottom soil and sediment management. Pages 166-181 *in* J. Wyban, ed. Proceedings Special Session on Shrimp Farming. World Aquaculture Society, Baton Rouge, Louisiana.

Boyd, C. E. 1995a. Bottom Soils, Sediment, and Pond Aquaculture. Chapman and Hall, New York, New York.

Boyd, C. E. 1995b. Potential of sodium nitrate to improve environmental conditions in aquaculture ponds. World Aquaculture 26(2):38-40.

Boyd, C. E. 1998. Pond water aeration systems. Aquacultural Engineering 18:9-40.

Boyd, C. E. 1999. Codes of Practice for Responsible Shrimp Farming. Global Aquaculture Alliance, St. Louis, Missouri.

Boyd, C. E., and T. Ahmad. 1987. Evaluation of Aerators for Channel Catfish Farming. Bulletin 584, Alabama Agricultural Experiment Station, Auburn University, Alabama.

Boyd, C. E., and J. W. Clay. 1998. Shrimp aquaculture and the environment. Scientific American, June 1998, 278(6):42-49.

Boyd, C. E., and A. Gross. 1998. Use of probiotics for improving soil and water quality in aquaculture ponds. Pages 101-106 *in* T. W. Flegel, ed. Advances in Shrimp Biotechnology. The National Center for Genetic Engineering and Biotechnology, Bangkok, Thailand.

Boyd, C. E., and M. C. Haws. 1999. Good management practices (GMPs) to reduce environmental impacts and improve efficiency of shrimp aquaculture in Latin America. Page 93 *in* B. W. Green, H. C. Clifford, M. McNamara, and G. M. Moctezuma (eds.). Simposio 5 Centroamericano de Acuacultura, San Pedro Sula, Honduras.

Boyd, C. E., and D. J. Martinson. 1984. Evaluation of propeller-aspirator-pump aerators. Aquaculture 36:283-292.

Boyd, C. E., and L. Massaut. 1999. Risks associated with use of chemicals in pond aquaculture. Aquacultural Engineering 20:113-132.

Boyd, C. E., and J. M. Moore. 1993. Factors affecting the performance of diffused-air aeration systems for aquaculture. Journal of Applied Aquaculture 2:1-12.

Boyd, C. E., and C. S. Tucker. 1992. Water Quality and Pond Soil Analyses for Aquaculture. Alabama Agricultural Experiment Station, Auburn University, Alabama.

Boyd, C. E., and C. S. Tucker. 1998. Aquaculture Water Quality Management. Kluwer Academic Publishers, Boston, Massachusetts.

Boyd, C. E., P. Munsiri, and B. F. Hajek. 1994. Composition of sediment from intensive shrimp ponds in Thailand. World Aquaculture 25:53-55.

Boyle, W. C. (editor). 1992. Measurement of Oxygen Transfer in Clean Water. ASCE Standard, ANSI/ACSE 2-91, American Society of Civil Engineers, New York, New York.

Busch, C. D. 1980. Water circulation for pond aeration and energy conservation. Proceedings of the World Mariculture Society 11:93-101.

Busch, C. D., and C. A. Flood, Jr. 1980. Water movement for water quality in catfish production. Transactions of the American Society of Agricultural Engineers 23:1040.

Chamberlain, G. 1988. Rethinking Shrimp Pond Management. Coastal Aquaculture, Vol 2. Texas A&M University/Texas Agricultural Extension Service, College Station, Texas.

Clifford, H. C., III. 1992. Marine shrimp pond management: a review. Pages 110-156 *in* J. Wyban, ed. Proceedings of the Special Session on Shrimp Farming. World Aquaculture Society, Baton Rouge, Louisiana.

Daniels, H. V., and C. E. Boyd. 1993. Nitrogen, phosphorus, and silica fertilization of brackishwater ponds. Journal of Aquaculture in the Tropics 8:103-110.

Donovan, D. J. 1998. Environmental Code of Practice for Australian Prawn Farmers. Pacific Aquaculture and Environment, The Gap, Queensland, Australia.

Fast, A. W., D. K. Barclay, and G. Akiyama. 1983. Artificial Circulation of Hawaiian Prawn Ponds. University of Hawaii, UNIH-SEAGRANT-CR-84-01, Honolulu, Hawaii.

Geiger, J. C. 1983. A review of pond zooplankton production and fertilization for the culture of larval and fingerling striped bass. Aquaculture 35:353-369.

Hairston, J. E., S. Kown, J. Meetze, E. L. Norton, P. L. Dakes, V. Payne, and K. M. Rogers. 1995. Protecting Water Quality on Alabama Farms. Alabama Soil and Water Conservation 1966 Committee, Montgomery, Alabama.

Hajek, B. F., and C. E. Boyd. 1994. Rating soil and water information for aquaculture. Aquacultural Engineering 13:115-128.

Howerton, R. D., and C. E. Boyd. 1992. Measurement of water circulation in ponds with gypsum blocks. Aquacultural Engineering 11:141-155.

Howerton, R. D., C. E. Boyd, and B. J. Watten. 1993. Design and performance of a horizontal, axial-flow water circulator. Journal of Applied Aquaculture 3:163-183.

Nelson, D. W., and L. E. Sommers. 1982. Total carbon, organic carbon, and organic matter. Pages 539-579 *in* A. L. Page, R. H. Miller, and D. R. Keeney, eds. Methods of Soil Analysis: Part 2, Chemical and Microbiological Properties. American Society of Agronomy and Soil Science Society of America, Madison, Wisconsin.

Peterson, J., and H. Daniels. 1992. Shrimp industry perspectives on soil and sediment management. Pages 182-186 *in* J. Wyban (ed.). Proceedings of the Special Session on Shrimp Farming. World Aquaculture Society, Baton Rouge, Louisiana.

Pillai, V. K., and C. E. Boyd. 1985. A simple method for calculating liming rates for fish ponds. Aquaculture 46:157-162.

Potts, A. C., and C. E. Boyd. 1998. Chlorination of channel catfish ponds. Journal of the World Aquaculture Society 29:432-440.

Queiroz, J. F., and C. E. Boyd. 1998a. Evaluation of a kit for estimating organic matter concentrations in bottom soils of aquaculture ponds. Journal of the World Aquaculture Society 29:230-233.

Queiroz, J. F., and C. E. Boyd. 1998b. Effects of a bacterial inoculum in channel catfish ponds. Journal of the World Aquaculture Society 29:67-73.

Ruttanagosrigit, W., Y. Musig, C. E. Boyd, and L. Sukhareon. 1991. Effect of salinity on oxygen transfer by propeller-aspirator-pump and paddle wheel aerators used in shrimp farming. Aquacultural Engineering 10:121-131.

Snow, J. R., and R. O. Jones. 1959. Some effects of lime applications to warmwater hatchery ponds. Proceedings of the Annual Conference Southeastern Association Game and Fish Commission 13:95-101.

Thiriez, G. 1999. Mechanical pond water circulation: a new approach to shrimp farming. Global Aquaculture Advocate 2(4/5):74-75.

Tucker, C. S., and J. A. Steeby. 1995. Daytime mechanical water circulation of channel catfish ponds. Aquacultural Engineering 14:15-28.

White, G. F. 1992. The Handbook of Chlorination. Van Nostrand Reinhold, New York, New York.

Wood, C. 1995. Environmental Impact Assessment: A Comparative Review. Longman Scientific and Technical, Essex, United Kingdom.

Pond Fertilization Regimen: State-of-the-Art

S. K. Das
B. B. Jana

SUMMARY. Pond fertilization has assumed an important role to supplement nutrient deficiency and augment biological productivity through autotrophic and heterotrophic pathways. This is especially important in the extensive and semi-intensive culture systems by promoting the functioning of natural ecosystems in a benign environment. The composition of inorganic and organic fertilizers forms the basis for selection of dose and quality of fertilizer application. While inorganic fertilizers produce perceptible results within a short period, organic manure is extremely cheap and is of considerable significance in developing countries. Nitrogen demand in fish ponds can be compensated through nitrogen fixation, as well as from accumulated humus from bottom sediments, especially from old fish ponds. The frequency of fertilizer application should be economical, though it is accepted that the lower the frequency, the better the productivity. In aquaculture ponds, the optimum N:P ratio was suggested between 4:1 to 8:1, whereas the optimum C:N ratio for composting was between 20 and 40. The exchange properties and equilibrium phosphorus concentration between soil and water influence water quality, nutrient status, and primary productivity of the pond ecosystem.

S. K. Das, Department of Aquaculture, West Bengal University of Animal and Fishery Sciences, Mohanpur-741 252, West Bengal, India.

B. B. Jana, Aquaculture and Applied Limnology Research Unit, Department of Zoology, University of Kalyani, Kalyani-741 235, West Bengal, India.

Address correspondence to: B. B. Jana, Aquaculture and Applied Limnology Research Unit, Department of Zoology, University of Kalyani, Kalyani-741 235, West Bengal, India.

[Haworth co-indexing entry note]: "Pond Fertilization Regimen: State-of-the-Art." Das, S. K., and B. B. Jana. Co-published simultaneously in *Journal of Applied Aquaculture* (Food Products Press, an imprint of The Haworth Press, Inc.) Vol. 13, No. 1/2, 2003, pp. 35-66; and: *Sustainable Aquaculture: Global Perspectives* (ed: B. B. Jana, and Carl D. Webster) Food Products Press, an imprint of The Haworth Press, Inc., 2003, pp. 35-66. Single or multiple copies of this article are available for a fee from The Haworth Document Delivery Service [1-800-HAWORTH, 9:00 a.m. - 5:00 p.m. (EST). E-mail address: getinfo@haworthpressinc.com].

These act as buffers to stabilize environmental conditions in ponds. Pond soils may exert negative influence on aquaculture production if one or more of their properties are outside the optimum range for aquaculture. The present study reviews state-of-the art pond fertilization in relation to the role of pond soils; different inorganic fertilizers such as phosphorus (P), nitrogen (N), potassium (K); fertilizer dose and frequency; P:N ratio; organic manure; aquatic food web; optimal manuring; decomposition of organic manures; mineralization; production efficiency; and limitations of organic manures. More studies on pond fertilization in the context of nutrient dynamics and fertilizer-microbial interactions under different agroclimatic regions are necessary for an effective, appropriate, and economic fertilization program. The environmental consequences of overfertilization resulting in pollution and subsequent hazards to public health should be taken into consideration.

KEYWORDS. Pond fertilization, organic manures, inorganic fertilizers, aquatic food web

INTRODUCTION

Modern aquaculture of larval fish and planktivorous fish demands an adequate amount of fertilizers for increasing biological productivity through a corresponding increase in available food resources of the grazing and detrital food chain of cultured fishes. Since biological productivity is often limited by the nutrients in least supply, pond fertilization has assumed an important role to supplement nutrient deficiency and augment biological productivity through autotrophic and heterotrophic pathways (Schroeder 1974, 1980; Moav et al. 1977; Martyshev 1983; Green et al. 1989; Debeljak et al. 1990; Jana and Sahu 1994; Jhingran 1995; Das and Jana 1996; Jana and Chakrabarty 1997a). In essence, pond fertilization is based on Liebig's law of the minimum; under limited nutrient supply, the rate of nutrient uptake by phytoplankton is concentration dependent, and the total phytoplankton is directly proportional to the initial concentration of limiting nutrient. However, such a simple functional relationship hardly exists in a pond ecosystem during pond fertilization, which can be extremely complex due to the dynamics of intrinsic and extrinsic factors.

While inorganic fertilizers with specific chemical formulations are instantly soluble in water and produce perceptible results within a short period, they are expensive and may have adverse effects on structure and microbial composition of soil. As a result, organic manuring along with inorganic fertilizers has emerged as an effective tool for nutrient management in aquaculture ponds. Organic manure is composite in nature, inexpensive, and very significant in India, due to the fact that it can make use of various wastes derived from production processes of livestock and agricultural resources and that it is cost effective (Edwards 1983; Pretto 1996). Manures are used as the principal nutrient input in aquaculture in many Asian countries including India and China (Buck et al. 1979; Edwards 1980, 1983; Wolhfarth and Schroeder 1979; Schroeder 1980), Europe (Schroeder 1974, 1975; Moav et al. 1977; Rappaport and Sarig 1978). However, manures are of limited use in the United States (Buck et al. 1978, 1979).

In essence, the productivity of cultured animals in aquaculture ponds largely depends on the conversion efficiency of the autochthonous and allochthonous organic matter into biomass through various production processes in the food chain and food web. In an allochthonous organic carbon-fed polyculture fish pond, much of the production is derived from the heterotrophic bacterial detrital food chains. It is stated that in these ecosystems, bacteria play a central role; the magnitude of daily bacterial production is comparable with the daily primary production; the available organic carbon pool is almost doubled every day (Olah 1986). This raises doubts about calculating fish production efficiencies based on primary production values. In some countries of Southeast Asia, livestock, crops, human waste, and fish are integrated either in two- or three-tier systems, closely integrated and productive. The subject of organic fertilization in fish culture has been reviewed by Wolfarth and Schroeder (1979), Edwards (1980), Wolfarth and Hulata (1987), Pekar and Olah (1990), and Das and Jana (1996a).

Though there has been a major shift in the current trend of aquaculture from fertilizer-based natural-functioning ecosystems to energy-intensive formulated diet-dependent intensive farming, the former is still a better approach because of environmental hazards and loss of biodiversity caused by the latter. India is endowed with vast natural, agricultural, and human resources but is economically constrained and needs to emphasize natural functioning production systems through rational and scientific utilization of more than 5.7 million ha of inland and 2.2 million ha of brackishwater resources. As a consequence, pond fertilization, which has become less important in a diet-based aquaculture system, assumes greater importance in extensive and semi-intensive culture methods.

The present study attempts to review state-of-the art pond fertilization under Indian conditions in relation to different extrinsic factors of soil criteria, types of fertilizers, doses and frequency of fertilizers, management parameters, culture techniques, etc.

THE ROLE OF POND SOILS

Due to continual mud-water nutrient exchange properties, pond bottom soil plays a key role in the nutrient dynamics of fish pond ecosystems. The exchange properties and equilibrium phosphorus concentration between soil and water influence water quality, nutrient status, and primary productivity of the pond ecosystem, and they act as buffers to stabilize environmental conditions in ponds (Bostrom et al. 1982; Andersson et al. 1988).

Pond soils of India, being in the tropical zone, are poor in organic matter, due to continual oxidation. There are eight categories of land soils: alluvial, black, red, laterite, forest, desert, saline alkaline, and peat. In general, alluvial soils, though deficient in nitrogen (N), contain adequate quantities of alkalis and phosphoric acid; black soils are highly argillaceous with a high proportion of calcium and magnesium, carbonates, and iron but are low in phosphorous (P), N, and organic matter. Red and lateritic soils are generally low in N, P, lime, iron oxide, and humus but rich in the mineral kaolinite. The soil particles of pond mud contains organic matter, much of which exists as humus. The humus derived from acid and peaty soils is a mixture of colloidal acids, whereas that derived from fertile soils with neutral or weakly alkaline reactions contains calcium salt of the colloidal complex of weak acids (Jhingran 1995). The saline soils are alkaline, and it is the carbonate that renders the soil sterile. Alkaline and peaty soils have an accumulation of large quantities of water soluble salts (Brady 1995).

Research has shown that lime application in fish ponds can bring a major shift in biogeochemical cycling bacterial populations and activities attributed to high pH stress, as well as a major shift in the CO_2-HCO_3-CO_3 equilibrium system (Ganguly et al. 1999).

A survey of soil and water quality in relation to fish production from a large number of ponds in eastern and central India showed a wide variability in pond productivity: highly productive, average productive, and unproductive (Table 1). It is generally agreed that productivity of the pond is considerably higher in ponds located in the states of West Bengal, Andhra Pradesh, and Gujarat, which have a high percentage of alluvial soils.

TABLE 1. Relationship between soil condition in India and productivity of fish (data adapted from Jhingran 1995).

Parameter (range)	No. of ponds			
	Total	Productive	Average	Unproductive
pH				
5.5 (Highly acidic)	2	-	-	2
5.5-6.5 (Moderately acidic)	22	2	9	11
6.5-7.5 (Near neutral)	25	13	4	8
7.5-8.5 (Moderately alkaline)	25	7	11	7
8.5 (Highly alkaline)	6	-	3	3
Available N (mg/100 g)				
25	39	4	21	14
25-50	27	10	6	11
50-75	11	6	-	5
75	3	1	1	1
Available P_2O_5 (mg/100 g)				
3	28	-	-	28
3-6	27	-	26	1
6-12	18	14	3	1
12	7	7	-	-
Organic carbon (%)				
5	20	1	8	11
0.5-1.5	37	6	19	12
1.5-2.5	17	14	-	3
2.5	6	1	2	3
C:N ratio				
5	4	-	-	4
5-10	18	13	6	9
10-15	27	11	12	4
15	31	7	11	13
Exchangeable calcium (mg/100 g)				
100	17	4	4	9
100-200	18	7	6	5
200-300	18	7	4	7
300	27	3	15	9

FERTILIZER TYPES

Fertilizers used in fish ponds fall into two categories: inorganic and organic. The composition of inorganic and organic fertilizers forms the basis for selection of dose and quality of fertilizer to be used in aquaculture ponds deficient in specific nutrients (Das and Jana 1996).

Inorganic Fertilizers

Inorganic nutrients are often broadcast in different combinations of N, P, and/or potassium (K) for agricultural crop production. However, the concept of N-P-K fertilizers is not equally applicable for aquaculture ponds, as potash has not been proved to be a limiting element in most of the fish ponds in India (Das and Jana 1996). Pond productivity, on the other hand, is influenced by the availability of different forms of carbon such as dissolved and particulate organic carbon derived from autochthonous and allochthonous sources. The trophic dynamic structure of freshwater ecosystems is mostly dependent for its energy upon the detrital dynamic structure of carbon cycling (Wetzel 1983) and therefore influences ecosystem productivity.

P Fertilization

Many studies (Metzger and Boyd 1980; Albinati et al. 1983; Bombeo-Tuburan et al. 1989; Das 1992; Jana and Das 1992; Jana and Sahu 1994; Das and Jana 1996b; Das et al. 1999) have attested the importance of phosphate fertilizer as the most critical single factor in maintaining pond fertility. The carbon (C):N:P ratio required by most phytoplankton species is approximately 106:16:1, indicating the potential of even trace levels of P to influence the primary productivity in the presence of sufficient concentrations of the two other elements (Stickney 1994).

One of the advantages of P fertilization is that the concentrations of orthophosphate in water increase almost immediately after ponds are fertilized with inorganic P fertilizers and decline sharply to pretreatment levels (Boyd 1982), which is attributed to the complexity of factors such as water retention time, dilution, assimilation by macrophytes, planktonic algae, attached algae, bacteria and fungi, and adsorption in the bottom sediments (Boyd 1986). Studies with radioactive-P demonstrated that phytoplankton can absorb orthophosphate very quickly, with a large percentage of the total uptake occurring within a few minutes to 24 hours and exhibiting decline logarithmically over time (Boyd and Musig 1981). According to Kimmel and Lind (1970), 90% of the orthophosphate added to undisturbed mud-water systems was absorbed by mud within 4 days. It is estimated that 0.4 g of dry mud could absorb 0.05 mg of orthophosphate in less than 30 minutes (Fitzgerald 1970).

P dynamics in muds of fertilized ponds are also influenced by the phosphate content of bottom soil, oxygen tension, redox potential, pH, relative iron concentrations (Olsen 1958a, 1958b), and orthophosphate equilibrium (Eren et al. 1977). It is demonstrated that when the overlying water is deficient in P, the nutrient is released from the bottom mud until the equilibrium concentration is attained (Hepher 1963). Eren et al. (1977) estimated that pond water attains equilibrium with bottom soil at soluble inorganic P concentration of about 100 mg/L but is likely to vary under different conditions. Phosphate ions in bottom soil form insoluble compounds with aluminum and iron under acidic conditions and with calcium under alkaline condition. With increase in soil pH from 4.0 to 8.9, P fixation by calcium steadily increases and that by iron steadily decreases. Thus, the total amount of P fixed in the soil is maximum in alkaline soil and minimum in neutral soil. A considerable fraction of phosphate ions remain absorbed on colloidal complexes. Since both these forms are rendered soluble under acidic and reducing conditions, the soluble fraction of phosphate in water is higher in concentration in acid soil than in neutral or alkaline soil.

Phosphorus Dose

Selection of optimum dose of fertilizer is one of the important aspects of pond fertilization. The proper dose may vary from one pond to another, depending upon the locality of soil criteria, pond fertility, initial nutrient status, and other characteristics of the pond. Researchers have shown that the concentration of P and N in water should not exceed 0.5 mg/L and 2.0 mg/L, respectively, in a fish culture pond (Hepher 1963). Using various combinations of rock phosphate charged with organic manures and single superphosphate, Jana and Das (1992) demonstrated that net fish yield tended to rise with increasing concentrations of orthophosphate in water up to 0.34 mg/L but declined at concentrations of 0.52 mg/L. Two-fold increase in fish production compared to low dose, suggesting applicatior of low dose rock phosphate, is suitable low-cost P fertilizer option for India (Das et al. 1999). Muller (1990) found 177% increase in carp, *Cyprinus carpio*, production attributable to 25-30 kg P_2O_5/ha (11-13.2 kg P/ha) as compared with control ponds.

The optimum rate of fertilization suggested by Hickling (1962) was 44.8 kg P_2O_5/ha for maximum fish production. According to Boyd (1982), standard fertilizer dose is 60 kg/ha superphosphate (11 kg/ha of P_2O_5) and 60 kg/ha of ammonium sulphate (13 kg/ha of N). In European fish ponds, the suggested rate of fertilization was 15-20 kg P_2O_5/ha and 30-40 kg ammonium sulphate at biweekly intervals or 15-20 kg at weekly intervals along with P fertilizer (Stickney 1994). Dobbins and Boyd (1976) and Lichtkopler and Boyd (1977) demonstrated that 9 kg/ha/application of P_2O_5 as triple superphosphate was suitable fertilization for heavily-fished ponds in wooded watersheds of the southern United States, whereas a monthly dose of 20 kg/ha of superphosphate (Bishara 1978) was said to be the best and least expensive method of fertilization for increasing the growth of mullets, *Mugil cephalus*.

The recommended dose of NPK fertilizer was 50 kg/ha of 16-20-0 fertilizer where the pond soil is rich in K for US fish ponds (Stickney 1994). On the other hand, in Southeast Asian brackishwater ponds, 50-100 kg of NPK (18:46:0) or 100-150 kg of NPK (16:20:0) is used, depending upon the soil condition, for development of algal complexes, such as 'lab-lab' (Pillay 1995). Though the general recommended dose of fertilizer in India is 40-50 kg/ha/month of single superphosphate, such dose selection has little basis in scientific investigation. Recent studies have indicated variability of fertilizer dosage for achieving the desired level of productivity (Bhakta and Jana, in press) under different initial P status of the soil. Further investigations are needed under different geoclimatic conditions for soil productivity mapping as well as for ascertaining the basis of dosage selection.

Frequency of Fertilization

It is recommended that the frequency of fertilizer application should be economical, though, in general, it is accepted that the lower the frequency, the better the productivity. Nevertheless, the frequency of pond fertilization should be cost effective. According to Ling (1986), fertilizer should be applied at an interval of 3-4 days. However, Ball (1949) found no advantage of weekly or biweekly fertilization for plankton production over every 3-week applications in fish culture ponds. According to some investigators (Hepher 1963; Bombeo-Tuburan et al. 1989; Debeljak et al. 1990) biweekly fertilizer application is more economical than weekly application (Yashouv and Halevy 1972). MacLean and Ang (1994) observed that greatest marketable yield of freshwater prawn, *Macrobrachium rosenbergii,* was achieved with fortnightly application of manure. Using weekly, fortnightly, and monthly fertilization regimens and high (200 kg P_2O_5) and low (100 kg P_2O_5) doses of rock phosphate application, Jana and Sahu (1994) observed distinctly higher growth rates (23 to 76%) of mrigal, *Cirrhinus mrigala,* in weekly application, as opposed to fortnightly and monthly application systems.

It is suggested that application of N:P:K at the concentration of 1:5.5:25 was suitable for plankton production. Application of NPK fertilizer (8:8:2) at the rate of 112 kg/ha in 8-14 installments has been suggested by Swingle and Smith (1947) for a culture period of one year. Less productive ponds with light or moderate fishing require 8-12 periodic applications of fertilizer to maintain plankton blooms and should receive 4.5 kg P_2O_5/ha/application, whereas in the case of infertile ponds with heavy fishing, the application rate should be 9 kg P_2O_5/ha (Boyd 1982). Sudden temporary increase in primary production (Dickman and Efford 1972) accompanied by increase of pH up to 10 (Ahonen and Niemitalo 1990) was observed in fish ponds due to a single highdose fertilizer application.

Stickney (1994) suggested that fertilizer application should be repeated at 10-14 day intervals until a Secchi disc transparency of approximately 30 cm is obtained. For mass production of zooplankton, *Daphnia carinata,* in the ponds, Jana and Chakrabarty (1997b) used manure mixtures such as cattle manure, poultry droppings, and mustard oil cake (1:1:1) at the rate of 12.5% for 2 days or 25% for 4 days, for a total dosage of 600 g over 16 days, and found significantly higher abundance and biomass of *Daphnia* than those achieved at high-dosage treatment every 16 days, suggesting low dosage at frequent intervals was more favourable than high dosage.

In India, the optimal requirement of fertilizers under different types of soil conditions, as well as their interactions with various factors of the pond ecosystem, has not been investigated. The fertilizer frequency is again variable under different situations. Studies on fertilizer-microbial interactions are necessary to determine the optimal requirement and judicious application of fertilizers under different agroclimatic conditions.

Nitrogen

Nitrogen cycle in aquaculture ponds mainly consists of biological transformations of N input through inorganic and organic fertilizers, and formulated diets. Application of excess N in pond results in deterioration of water quality with accumulation of ammonia and nitrite toxic to biota. Interactions between sediment and water are of considerable importance in the N cycle of fish ponds. Sediment acts as a source of ammonia and sink for nitrite and nitrate (Hargreaves 1998). Different forms of N fertilizers are applied, depending on the soil conditions of aquaculture ponds. Loss of ammonical form of N in soils with neutral reaction is less than that in highly alkaline soils. Recovery of added N from ammonium sulphate is highest in acidic and neutral soils but lowest from potassium nitrate. The reverse is true in alkaline soils. Thus, soil pH plays an important role for selecting ammonical forms of N in ponds with acid and neutral soils and of nitrate form of N in ponds with alkaline soil. Though there is considerable loss of N from ammonical form in alkaline soil, the use of ammonium sulphate in low doses is generally recommended in India because sulphate aids in reducing soil alkalinity.

Evaluation of the use of N fertilizers in fish ponds has shown contradictory results. Several investigators (Rabanal 1960; Hickling 1962) inferred that N fertilizer was a poorer producer than phosphate. This is perhaps due to the fact that the N demand in fish ponds can be compensated through N fixation, as well as from accumulated humus from bottom sediments, especially from old fish ponds.

Potassium

Most studies on the role of K on fish production have indicated either negative impact or insignificant increase in fish production (Mortimer and Hickling 1954; Hickling 1962; Pan et al. 1994). This was due to the fact that most of the fish ponds exhibit higher amounts of organic matter than upland soils (Singh and Ram 1971; Saha 1979), resulting in higher amounts of K in fish pond soils than is actually required. Boyd (1984) stated that the level of 1.3 mg/L water-soluble K may be considered the critical limit for application of K fertilizer in U.S. fish ponds. Mandal and Chattopadhyay (1992) observed that under tropical conditions where many fish ponds become dry during summer, the available K status of the bottom soil increases during the subsequent wet season. Besides, alkaline range of pH of both water and bottom soil of the most productive ponds helps to maintain high levels of available K in pond soils (Pattrick and Mikkelsen 1971). Liming of fish ponds also influences availability of K in water phase, through cation exchange mechanism (Nolan and Pritchett 1960). Because of the above-stated facts, the concept of NPK fertilizer used for agricultural crop production may not be equally applicable to aquaculture ponds. However, information on the role of K on fish production is very scarce for fish ponds under the different geochemical soil conditions of India. Application of K fertilizer seems to be useful in newly constructed ponds or in renovated ones and also in ponds located in lateritic soil zones (Chattopadhyay 1995) where it is deficient in most cases.

P:N Ratio

The ratio of N and P seems to be important in limiting the efficacy of nitrogenous fertilizers. Evaluations of N fertilizer in fish ponds have not shown uniform results. Hickling (1962) infers that N fertilizer is a poorer producer than phosphate. Similar results were also derived from some other studies (Rabanal 1960; Swingle 1954). This is due to the fact that the N demand of fish ponds is perhaps compensated by the nitrogen fixation of blue green algae and accumulation of organic matters in the bottom sediments of the ponds, more so in old fish ponds.

In pond aquaculture, the optimum N:P ratio as suggested by Winberg and Liakhnovich (1965) was of 4:1 to 8:1. Swingle and Smith (1939) found N:P ratio best at 4:1. But in Israel, the N:P ratio was found to be much wider (Hepher 1963). In the US the most widely used NPK fertilizer has a ratio of 8:4:2 (Martyshev 1983); N:P ratio is 2:1 in China (Pan et al. 1994). Ghosh et al. (1979) reported the N:P ratio of 1.4-1.5:1 as optimal for plankton production. However, extensive studies need to be done in different agroclimatic regions for such generalization. Wolny (1967) opines that the determination of optimum N:P ratio is a difficult proposition in pond culture, since both the bicarbonate level in the water and the acidity level in the bottom sediments are subject to constant variation during the vegetative season, which, in turn, may affect the variation in N:P ratio. Though in general it is accepted that a narrow N:P ratio gives better yield, no detailed information about the optimal N:P ratio in fish ponds under Indian conditions is available. Because of large variability of soil criteria and pond fertility, the optimal N:P ratio would differ from one pond to another. Research need to be carried out on these aspects, under different soil conditions.

Organic Manure

Organic manure as a class are composite and contain almost all the major elements (Table 2) required in pond ecosystems (Martyshev

TABLE 2. Composition of fresh manure from various animal species (adapted from Pillay 1995).

Composition	Horse	Cattle	Sheep	Pig
Water	71.3	77.3	64.6	72.4
Organic matter	25.4	20.3	31.8	25.0
Total nitrogen (N)	0.58	0.45	0.83	0.45
Proteinic nitrogen	0.35	0.28	–	–
Ammoniacal nitrogen	0.19	0.14	–	–
Phosphorus (P_2O_5)	0.28	0.23	0.23	0.19
Potassium (K_2O)	0.63	0.50	0.67	0.60
Calcium (CaO)	0.21	0.40	0.33	0.18
Magnesium (MgO)	0.14	0.11	0.18	0.09
Sulphuric acid (SO_3^{-2})	0.07	0.06	0.15	0.08
Iron and aluminium sesquioxides (R_2O_3)	0.11	0.05	0.24	0.07

1983; Jhingran 1995). Organic manure upon decomposition can enrich soil structure and increase fertility of the aquatic system by releasing nutrients (Jhingran 1995) and often f\acilitate the use of chemical fertilizers under appropriate conditions (Pillay 1995). Because of some definitive advantages over inorganic fertilizers, organic manure are preferable and are extensively used in grow-out ponds. Integrated crop-livestock-fish farming systems are an age old practice in Asia (Kapur and Lal 1986; Yingxue et al. 1986). They are also practiced in Israel, Europe, North and South America (Hill et al. 1997; Pekar and Olah 1998), and Africa (Egna and Boyd 1997).

Manure is applied to fish farming ponds by growing livestock over the water so that their excreta is continually added to the ponds, hence avoiding environmental pollution through direct accumulation of manure (Cruz and Shehadeh 1980; Hopkins and Cruz 1980; Barash et al. 1982). For example, polyculture fish farming integrated with duck (Ullah 1989; Pekar and Olah 1998), cattle (Schroeder 1977; Schroeder et al. 1990), buffalo (Edwards et al. 1996), pig (Burns and Stickney 1980; Hopkins and Cruz 1980, Sharma and Olah 1986), and chickens (Hopkins and Cruz 1980; Knud-Hansen et al. 1993) has proved useful in terms of fish yield and long-term sustainability.

Alternatively, manure from different livestock are transported to ponds without direct integration of the livestock and ponds. Organic manure are classified according to source and carbohydrate content

(Mortimer and Hickling 1954; Pillay 1995). In the final analysis, manure act in the fish pond by releasing nutrients upon decomposition. It seems, therefore, apparent that the basis for manure classification is indirectly indicative of the N:P:C ratio of added manure (Table 3). The composition of organic manure varies widely, depending upon the nature, source, nutrition, and physiological state of farm animals. A vast array of extrinsic factors largely determines the extent of decomposition and regeneration as well. Thus, there are some inherent limitations in organic manure in formulating a comprehensive pond fertilization schedule with a view to providing the specific demand of a pond system at a particular time and place.

The economic efficiency of pig-cum-fish farming in India is poor and low in profit. Buffalo manure is also a relatively poor nutritional input for ponds in Thailand (Edwards 1983). On the other hand, cow manure is often used in fish culture ponds in India. Among various livestocks integrated with fish farming, raising ducks on the surface of fish ponds has been found to be most economically efficient. A symbiotic relationship develops in the case of geese or duck. Hence, fish-cum-duck farming in India shows highest economic profit (Sharma and Olah 1986).

Qualitatively different fertilizers were found to exert profound influences on the microbial biogeochemical activities of water and sediments in fish ponds and thereby strongly affect the grazing and detrital food chain of pond ecosystems (Chatterjee 1997). In an extensive study, it was clearly demonstrated that pond fertilization with mixed fertilizers

TABLE 3. C:N and N:P ratio of different manure.

Manure	C:N ratio	N:P ratio
Night soil	6-10	2
Urine	0.8	11
Cow manure	18	3
Buffalo manure	-	2
Poultry manure	15	2
Goat and sheep	-	2
Horse manure	25	-
Pig manure	-	2
Duck manure	-	2

was the best of four treatments; cattle manure, poultry droppings, inorganic fertilizers and their mixed forms. Post-lime organic fertilization had some advantages over inorganic fertilization, as the former caused significantly higher levels of nutrients, as well as favorable N:P ratio and water quality conducive to fish farming (Ganguly et al. 1999).

Organic Manuring and Aquatic Food Web

Generally, a series of complicated processes between the input of manure and output of fish in the pond ecosystem (Xianzhen et al. 1986; Singh and Sharma 1999) is involved via production of food organisms, material cycle, and energy flow. Apart from being the nutrient source, organic manures may also enhance fish production through detrital formation (Wolhlfarth and Schroeder 1979; Coleman and Edwards 1987) and impact the microbial-detrital food chain of the aquatic ecosystem (Rappaport and Sarig 1978; Schoreder et al. 1978; Pillay 1995). Thus their role in autotrophic and heterotrophic food chains has been emphasized in several recent studies (Schroeder and Hepher 1979; Pekar and Olah 1998). Lin et al. (1997) stressed polyculture of microphagous and macrophagous fishes in manure-fertilized ponds for optimal utilization of both heterotrophic and autotrophic production pathways. Schroeder et al. (1978) observed that production rates of large-sized pelagic and benthic organisms to be consumed by fish were insufficient to account for the measured fish yield in organically-manured fish ponds. As microorganisms are too small to be individually selected and consumed by fish, it may be that fish consume manure where microbial growth is present (Schroeder et al. 1978; Nandeesha et al. 1984). In another study, Schroeder et al. (1990) emphasized that more than 90% of the fish yield was based on food webs derived directly or indirectly from algal nutrients. In recent years, radioisotopic tracer methods have been employed in the measurement of different production pathways in manured fish ponds (Pekar and Olah 1998). Delta C index was applied in mass spectrographic analysis of the ratio of two naturally occurring stable carbon isotopes (^{12}C and ^{13}C) in organic and inorganic matter to trace the flow of carbon in organically manured systems (Xianzhen et al. 1986; Yingxue et al. 1986). The results showed that the flow of carbon from manure sources to fish depends largely upon their feeding habits. In common carp 50-70% of the carbon originated from manure food webs, and 30-50% originated with microalgae, whereas, in crucian carp, *Carassius carassius*, only 22% carbon was contributed by manure, and

60-80% carbon originated from microalgal production (Xianzhen et al. 1986).

Optimal Manuring Rate

Optimal manuring rate is the highest amount of organic matter that can be utilized in a pond ecosystem without showing any adverse effect on fish growth and water quality. Using regression analysis, Edwards et al. (1996) indicated that a more efficient use would be a daily input of a maximum of 1 kg dry matter (DM) per 200 m^2 pond. The buffalo manure conversion efficiency would almost double from 1.0 to 1.9% with a recommended reduced buffalo manure loading rate from 5 to 2 kg DM with further increase in fish yield. This suggests a limited value of buffalo manure as a major pond fertilizer, due to competition from the vertically integrated agro-industry.

The results of different experiments and the conversion values per unit of input (Table 4) indicated an optimal manuring rate of about 5 g/m^2/day carbon in ponds treated with higher doses. This is equivalent to 100 kg DM of manure/ha/day, which may be accepted as a standard manuring rate in Europe (Olah 1986). In Thailand, manuring rate exceeding 20 g DM/m^2/day caused water quality deterioration in terms of erratic diurnal dissolved oxygen fluctuations and high concentrations of ammonia and nitrate (Villacorta 1989). In the management of carp nurs-

TABLE 4. Optimal manuring rate for different fish pond management systems. Multiply the g/m^2/day carbon values by 100 to get kg of fish/ha/day and by 20 to get kg of dry matter/ha/day.

Manure	Culture type	Density (fish/ha)	Fish production (g/m^2/day)	Manuring rate (g/m^2/day)	References
Poultry	Tilapia	8,000	0.162	5.0	Burns and Stickney (1980)
Pig	Polyculture	10,000	0.220	1.65	Buck et al. (1978)
	Polyculture	8,500	0.184	2.0	Sharma and Olah (1986)
	Biculture	3,500	0.180	2.0	Sharma and Olah (1986)
Cow	Polyculture	8,000	0.215	3.4	Noriega-Curtis (1979)
	Polyculture	5,000	0.106	1.2	Schroeder (1974)
	Polyculture	18,000	0.326	5.0	Moav et al. (1977)
	Polyculture	16,000	0.300	5.0	Schroeder and Hepher (1979)
Chicken	Polyculture	17,500	0.310	5.0	Olah (1986)
Mixed	Polyculture	23,000	0.210	1.7	Tang (1970)

ery ponds of India, spaced manuring with fresh cattle manure at an initial dose of 10,000 kg/ha 15 days before stocking and 5,000 kg/ha 7 days post-stocking is useful for sustained production of zooplankton during the culture period of 2-3 weeks (Jhingran 1995). Considering 75% moisture content of the fresh cattle manure, the applied dose is equivalent to about 106 kg DM of manure/ha/day, which is similar to the calculated dose suggested by Olah (1986). Following the same criterion, the corresponding dose of cattle manure for grow-out ponds is 140-150 kg DM/ha/day, along with N:P:K fertilizers (11:5:1) at the rate of 2.7-4.2 kg/ha/day (Jhingran 1995).

Fertilizing with cow manure, the maximum loading level in carp culture has been determined to be 10,000 kg/ha/year (Kapur and Lal 1986). Dinesh et al. (1986) recommended that 2,000 kg/ha poultry manure, along with 100 kg/ha urea, is the safe and economic rate for carp culture ponds. Even from these data, no generalized rate can be recommended, in view of the great variability of pond productivity, initial soil fertility, and high metabolic rate of tropical fish ponds due to continual oxidation and lack of distinct chemical stratification.

A number of experiments have evaluated the fertilizer values of different manure such as pig (Tang 1970; Stickney 1979; Sharma and Olah 1986), cow (Schroeder 1974), chicken (Tang 1970; Wohlfarth et al. 1980; Green 1990), and green manure (Morrissens et al. 1996; Pretto 1996). Fish yield from different manured ponds was found to range from 4 to 40 kg/ha/day (Table 5). Further, fish yield was density-dependent up to certain limits but remained stationary, even after increase in stocking density to above 10,000 fish/ha (Pekar and Olah 1990). The manured, non-fed fish ponds have approximately the same maximum fish yield of 25 to 35 kg/ha/day averaged over the growing season in China, Israel, and Philippines (Moav et al. 1977; Shan et al. 1985). There was a two-fold increase in fish yield for ponds receiving pig or chicken manure, as compared to those receiving inorganic fertilizers (Hopkins and Cruz 1982). While comparing different manure, Varghese and Shankar (1981) found optimal growth of catla, *Catla catla*, and rohu, *Labeo rohita*, with poultry manure, and mrigal, *Cirrhinus mrigala* with cattle, pig, and sheep manure.

Decomposition of Manure and Release of Nutrients

The decomposition of organic manure in fish ponds is carried out by different groups of bacteria, fungi, and actinomycetes (Boyd 1995; Gour et al. 1995). In the decomposition process bacteria emerge as the

TABLE 5. Fish yields from manured ponds (Data adapted from Pekar and Olah 1990).

Source of nutrients	Fish stocked	Stocking rate (ha^{-1})	Daily yield (kg/ha)	References
Pig manure	Chinese and common carp	10,700	17-22	Buck et al. (1979)
	Chinese and common carp	18,000	40	Shan et al. (1985)
	Chinese carp and tilapia	15,500	36	Behrends et al. (1983)
Duck manure	Common carp, silver carp, and tilapia	10,000-20,000	36	Barash et al. (1982)
Pig, duck, chicken manure	Tilapia and common carp	10,000-20,000	19-20 10-15	Cruz and Shadelah (1980) Hopkins (1982) Hopkins and Cruz (1980, 1982)
Cattle manure	Common and chinese carp, and tilapia	9,000-18,000	32	Moav et al. (1977)
	Tilapia	10,000	16	Collis and Smitherman (1978)
Chicken manure	Common and chinese carp, tilapia	8,000-16,000	29-35	Wohlfarth et al. (1980)
	Common carp	2,100	7	Bok and Jongblood (1984)
	Tilapia	10,000	4.1-17.7	Teichert-Coddington and Green (1990)
Chicken manure + pellet	Tilapia	20,000	28.8-31.6	Green (1990)
Nightsoil, rice bran	Chinese carp, common carp, and mullet	23,500	24	Tang (1970)

first link between the living world and the abiotic factors (Maun 1972; Pekar and Olah 1990). Boyd (1995) observed that actinomycetes and fungi are less efficient than bacteria in the decomposition of organic matter. Aerobic decomposition of organic matter results in the production of ammonia and carbon dioxide in the soil water interface, whereas production of methane, hydrogen sulfide, nitrogen gases, soluble organic compounds, etc., takes place in the deeper anaerobic layer where oxygen as the terminal electron acceptor is scarce (Sorokin and Kadota 1972; Boyd 1995; Polprasert 1996).

It was demonstrated that percentage decomposition of different organic residues in one year were sugar 99%, hemicellulose 90%, cellulose 75%, lignin 50%, waxes 25%, and phenols 10% (Egna and Boyd 1997). The decomposition of cellulose in acid soils proceeds more slowly than in natural and alkaline soils. Similarly, lignin decomposes slowly than cellulose (Sahai 1990). Moreover, organic matter associated with microparticles (< 0.25 nm) decomposes slowly, but that with macroparticles decomposes rapidly (Elliot 1986). Therefore, organic matter associated with sand decomposes more readily than that associated with silt and clay (Gregorich et al. 1989).

Decomposition of Manures and Dissolved Oxygen Profile

Organic loading to a pond influences bacterial activity in waste stabilization utilizing oxygen (Polprasert 1996). An increase in the values of biological oxygen demand and corresponding decrease in dissolved oxygen following organic manuring has been demonstrated in several studies (Romaire et al. 1978; Chattopadhyay and Mandal 1980; Debeljak and Fasaic 1985; Boyd 1990; Jhingran 1995; Pillay 1995). Mean early morning dissolved oxygen was inversely correlated with loading rate of chicken litter (Batterson et al. 1989; Green et al. 1989; Teichert-Coddington et al. 1990). Bhattari (1985) proposed an empirical oxygen model at dawn (DOd) for manure fed ponds, which might be used as a tool for easily determining the occurrence of critical dissolved oxygen so that contingency measures like aeration or temporary suspension of manuring can be adopted. The proposed model is as follows:

$$\text{DOd} = 10.745 \exp\{-(0.017\,t + 0.002\,\text{Lc})\}$$

where, t = time of septage loading in days;
Lc = cumulative organic loading to fish ponds up to time t, kg COD (per 200 m^3 pond volume)

Mineralization of Organic Manure

The rate of nutrient released from animal manure over time normally determines the fertilization schedule to be adopted in a given pond (Egna and Boyd 1997). The results of percentage of total-N and total-P released from different manure (Table 6) ranged from 43-90% and 35-73%, respectively. The availability of all the inorganic nutrients (Table 7) from poultry droppings was reported to be considerably higher than that from manure from pig, goat, and cow during a 20-day experiment at 35°C (Kapur and Lal 1986). A number of factors have

TABLE 6. Dissolved inorganic fractions from different manures after 2-20 days of application. (DIN = Dissolved inorganic nitrogen; TN = Total nitrogen; TP = Total phosphorus).

	N		P		Period (day)	Reference
	DIN	TN	DIP	TP		
Chicken manure	~66% (> 90% as ammonia)	–	–	–	2	Knud-Hansen et al. (1991)
	–	43%	–	–	6	Knud-Hansen et al. (1991)
Chicken manure	–	51-57%	–	–	3	Egna and Boyd (1997)
	–	–	–	68-73%	20	Egna and Boyd (1997)
Duck manure	~100%	–	~100%	–	4	Ullah (1989)
Buffalo manure	–	90% (kj)	–	–	3	Shevgoor et al. (1994)
	–			35%		Shevgoor et al. (1994)

TABLE 7. Nutrients released on 20th day at 35°C from different organic manure.

	Nutrient released as percentage of total		
Types of manure	Nitrate	Phosphate	Potassium
Poultry	37.3±4.67	55.8±3.06	47.9±5.83
Pig	34.1±1.34	46.3±4.99	54.4±5.24
Goat-sheep	33.2±3.37	40.0±2.86	41.7±4.85
Cow	27.4±2.69	36.0±2.15	36.8±3.72

been demonstrated to be responsible for mineralization of nutrients from organic manure. Ghosh and Mohanty (1981) demonstrated that aeration induced mineralization of manurial nitrogen of cow dung by 28% during a 90-day experiment (Table 8). Salinity was found to have an adverse effect on the mineralization process of organic manure (Ghosh 1975; Chattopadhya and Mandal 1980, 1982).

C:N:P Ratio and Mineralization

The C:N ratio of different organic wastes varies widely (Table 3). The C:N ratio of manure (Almazan and Boyd 1978) and also the C:N:P ratio of pond soil (Saha 1995) can greatly influence the rate of decomposition and regeneration from organic manure through regulation of

TABLE 8. Mineralization of organic nitrogen from cow dung with and without aeration.

Treatment			No. of days after treatment 0	7	15	30	45	60	90
Cow manure without aeration	Water	NH_4-N	2.60	17.09	2.52	2.19	2.88	2.70	2.49
		NO_3-N	1.70	5.08	0.18	1.11	2.27	0.87	1.29
	Soil	NH_4-N	2.40	16.45	13.33	7.05	4.20	3.93	5.38
		NO_3-N	0.85	1.10	3.56	2.67	1.07	1.34	0.34
	Total - N (mg/100 g soil)		7.55	39.72	19.59	13.02	10.42	8.84	9.50
Cow manure with aeration	Water	NH_4-N	3.00	24.98	4.98	3.64	3.12	3.22	2.86
		NO_3-N	2.25	7.29	0.80	1.20	2.39	2.05	1.46
	Soil	NH_4-N	2.20	20.92	14.38	9.73	4.32	6.05	5.63
		NO_3-N	0.07	1.69	3.56	3.13	1.07	1.34	0.84
	Total - N (mg/100 g soil)		8.15	54.88	23.72	17.70	10.90	12.66	10.79

microbial activity. This is because microbes remove nitrate or ammonia-N from the soil solution and use it to supplement their N requirement when N-deficient residue is added to the soil (Boyd 1995). Therefore, unless the environment contains abundant inorganic nitrogen, decomposition of manure with a wide C:N ratio will be slowed down and incomplete. Goldman et al. (1987) suggested that bacterial growth efficiency decreases with increasing C:N and C:P ratio in the substrate. As a result, the transformation of nitrogen from organic to inorganic form with C:N ratio of 40% C and 0.5% N decomposes slowly, whereas, materials with a smaller C:N ratio of 40% C and 4% N decay quickly (Boyd 1982, 1995).

The results of mineralization of organic N of fresh cow manure in soil with varying C:N ratios (5-40) under waterlogged condition revealed the declining rate was rapid, moderate, and slow in the ranges of < 10, 10-20, and > 20, respectively (Mohanty et al. 1994). However, Polprasert (1996) considered that C:N ratio between 20 and 40 as optimum for composting, which appears to be much greater than that of the earlier observation of Hora and Pillay (1962), who concluded that C:N ratio of 10:1 is necessary for rapid decay of vegetable matter.

While examining the influence of different C:N:P (12:2:1 to 151:12:1) applied at the rate of 0.043 g/week on growth of certain biogeochemical cycling bacterial populations and nutrient status of the system, Jana et al. (2001) observed that the values of primary productivity were the direct function of the values of mineralization indices for C, N, and P. From the results of the experiment, it was concluded that mixed fertilizer (C:N:P ratio of 88.6:7.5:1) comprising cattle manure (95%), poultry droppings (2.5%), urea (2%) and single superphosphate (0.5%) applied at the rate of 23,000 kg/ha/year was considered a suitable cost effective fertilization option for aquaculture practices in a tropical climate.

Production Efficiency

Researchers have shown that common carp production in fertilized ponds in temperate regions was 2 to 10 times higher than that in unfertilized ponds (Mortimer and Hickling 1954). In tropical Africa, production of tilapia in fertilized ponds (5,135-9000 kg/ha) was 2 to 4 times higher than that in unfertilized ponds (2,240 kg/ha). Fish yield potential from manured ponds were reported from 7 to 36 kg/ha/day (Buck et al. 1979; Wolhfarth et al. 1980; Barash et al. 1982; Wolhfarth and Hulata 1987; Schroeder et al. 1990; Egna and Boyd 1997). The yield potential of fish widely varies (Table 5) according to types of wastes, variable loading rate, geo-climatic conditions, and species of culture (Table 9). Most of these extremely high fish yields were reported from Israel and ranged from 20 kg/ha/day to 30 kg/ha/day with chicken manure and cat-

TABLE 9. Summary of fish production (kg/ha/year) under different organic manuring schedule in different countries.

Country	Type of manures	Production	References
Israel	Cattle manure	10,950	Schroeder (1977)
	Chicken manure	10,768	Schroeder et al. (1990)
	Cattle manure + Suppl. feeding	11,225	Schroeder and Hepher (1979)
	Duck manure	14,600	Wohlfarth and Schroeder (1979)
Indonesia	Livestock waste	7,500	Djajadiredja and Jangkaru (1978)
Hungary	Duck manure	1,100-2,800	Pekar and Olah (1994)
Philippines	Biogas slurry	8,000	Maramba (1978)
Taiwan	Night soil	6,893-7,786	Tang (1970)
Thailand	Biogas slurry	3,700	Edwards et al. (1988)
	Composted nightsoil	2,800-5,600	Polprassert et al. (1982)
		5,700	Villacorta (1989)
India	Duck dropping	4,300	Sharma et al. (1979)
Bangladesh		2,800	Miah et al. (1996)
North America (Illinois)	Pig manure	6,400-8,200	Buck et al. (1979)

tle manure, respectively, to 40 kg/ha/day with duck manure. Again, these high yields reported were obtained from polyculture of tilapia with Chinese carps, where common carp and silver carp contributed largely towards total production, whereas production of tilapia monoculture under a manured system was much less and ranged from 8.6 to 19.2 kg/ha/day (Collis and Smitherman 1978; Hopkins and Cruz 1982; Diana et al. 1991; Knud-Hansen and Lin 1993; Green et al. 1994). A parabolic relationship has been observed between the amount of chicken and pig manure input, and fish yield in Honduras (Green et al. 1994) and Philippines (Hopkins 1982), respectively.

Limitations of Organic Manure

Though organic manuring has many advantages in increasing fish production, organic enrichment of water due to excessive manuring results in deterioration of water quality rendering it favorable for the growth, sustenance, and multiplication of pathogenic bacteria. Therefore, microbiological quality of fish pond water should be considered as the criterion for assessment of organically manured pond water. Buras et al. (1985) experimentally found critical concentration of standard plate count bacteria to be 5×10^4/mL, beyond which microbes appeared in the fish meat.

The occurrence of fecal coliform bacteria in fish has been referenced as an index of pollution level of water because coliforms are not the normal flora of fish (Rao et al. 1968; Cohen and Shuval 1973; Evison and James 1973). Within the organic loadings up to 150 kg COD/ha/day, the total coliform and fecal coliform bacteria, as well as *Escherechia coli* bacteriophages, were found to be absent from fish organs such as blood, bile, and meat but were encountered in high densities (up to 10^9 cells/mL) in fish intestines (Polprasert et al. 1982; Edwards et al. 1984). Similar observations have also been reported by Cloete et al. (1984). Hejkal et al. (1983) found that even when levels of bacteria exceeded 10^5/100 g in the fish guts, very little penetrated into the fish muscle; a maximum of 25 fecal streptococci/25 g was found in the fish meat. Buras et al. (1985) defined a "threshold concentration" as the minimum of bacteria that, when inoculated into the fish, causes its appearance in the fish meat (Table 10).

Public health concerns related to helminths requiring fish or other aquatic organisms as intermediate hosts in organic manured ponds have been well documented (Larsson 1994; Santos 1994; Polprasert 1996; Egna and Boyd 1997). It seems apparent that organic manure, if used, should be applied with caution in view of the above facts.

TABLE 10. Threshold concentration of microorganisms inoculated to fish (data adapted from Buras et al. 1985).

Microorganisms	Threshold concentrations (no. fish)	
	Blue tilapia	Common carp
Bacteria		
E. coli	2.5×10^6	1.5×10^6
Clostridium freundii	9.3×10^3	–
Streptococcus faecalis	1.9×10^4	4.0×10^4
Streptococcus monotevideo	1.8×10^4	3.7×10^4
Bacteriophages		
T_2 virus	4.0×10^3	4.6×10^3
T_4 virus	2.0×10^4	–

REFERENCES

Ahonen, M., and V. Niemitalo. 1990. Manipulation of whitefish fingerling production in natural food ponds in Northern Finland. Pages 199-206 *in* R. Berka and V. Hilge, eds. Proceedings of the FAO-EIFAC Symposium on Production Enhancement in Still Water Pond Culture. Research Institute of Fish Culture and Hydrobiology, Vodnany, Czechoslovakia.

Albinati, R.C.B., J.A.F. Veloso, G.C.A. Filho, and F.L. Albinati. 1983. A gain in weight of *Sarotherodon niloticus* and the carp (*Cyprinus carpio*) in a polyculture system in tanks treated with chicken manure and tripple superphosphate. Archive Brasilian Medical Veterinary Zootechnology 35 (5):699-708.

Almazan, G., and C.E. Boyd. 1978. Effects of nitrogen levels on rates of oxygen consumption during decay of aquatic plants. Aquatic Botany 5:119-126.

Andersson, G., W. Graneli, and J. Stenson. 1988. The influence of animals on phosphorus cycling in lake ecosystems. Hydrobiologia 170:267-284.

Ball, R.C. 1949. Experimental use of fertilizer in the production of fish food organisms and fish. Technical Bulletin No. 223, Michigan State College Agricultural Experimental Station, East Lansing, Michigan.

Barash, H., I. Plavnik, and R. Moav. 1982. Integration of duck and fish farming: Experimental results. Aquaculture 27:129-140.

Batterson, T.R., C.D. McNabb, C.F. Knud-Hansen, H.M. Eidman, and K. Sumatadinata. 1989. Indonesia: Cycle III of the Global Experiment. Page 135 *in* H.S. Egna, ed. Pond Dynamics/Aquaculture CRSP Data Reports.

Behrends, L.L., J.B. Kingsley, J.J. Maddox, and E.L. Waddel. 1983. Fish production and community metabolism in an organically fertilized fish ponds. Journal of the World Mariculture Society 14:510-522.

Bhakta, J.N. and B.B. Jana (in press). Influence of sediment phosphorus on utilization efficiency of phosphate fertilizer: A mesocosm study. Aquaculture Research.

Bhattarai, K.K. 1985. Septage Recycling in Waste Stabilization Ponds, Doctoral dissertation No. EV-85-1, Asian Institute of Technology, Bangkok, Thailand.

Bishara, N.F. 1978. Fertilizing fish ponds. II. Growth of *Mugil cephalus* in Egypt by pond fertilization and feeding. Aquaculture 13:361-367.

Bok, A.H., and H. Jongblood. 1984. Growth and production of sharp tooth catfish *Clarius gariepinus* (Pisces, Clariidae) in organically fertilized ponds in the Cape province, South Africa. Aquaculture 36:141-155.

Bombeo-Tuburan, I., R.F. Agbayani, and P.F. Subusa. 1989. Evaluation of organic and inorganic fertilizers in brackish water milk fish ponds. Aquaculture 76:227-235.

Bostrom, B., M. Jansson, and C. Forsberg. 1982. Phosphorus release from lake sediments. Archives für Hydrobiology Ergebnisse der Limnologie 18:5-59.

Boyd, C.E. 1982. Water Quality Management for Pond Fish Culture. Elsevier Scientific Publishing Company, New York, New York.

Boyd, C.E. 1984. Water Quality in Warm Water Fish Ponds. Auburn University, Alabama.

Boyd, C.E. 1986. Water quality and fertilization. Pages 282-295 *in* R. Billard and J. Marcel, eds. Aquaculture of Cyprinids, INRA, Paris, France.

Boyd, C.E. 1990. Water Quality in Ponds for Aquaculture. Alabama Agricultural Experiment Station, Auburn University, Alabama.

Boyd, C.E. 1995. Bottom Soils, Sediment, and Pond Aquaculture. Chapman and Hall, New York, New York.

Boyd, C.E., and Y. Musig. 1981. Orthophosphate uptake by phytoplankton and sediment. Aquaculture 22:165-173.

Brady, N.C. 1995. The Nature and Properties of Soil. Prentice Hall of India Private Limited, New Delhi, India

Buck, H., R.J. Baur, and C.R. Rose. 1978. Polyculture of Chinese carps in ponds with swine wastes. Pages 90-106 *in* R. Smitherman, W. Shelton, and J. Grover, eds. Proceedings of the Symposium on Culture of Exotic Fishes. American Fisheries Society, Atlanta, Georgia.

Buck, H., R.J. Baur, and C.R. Rose. 1979. Experiments in recycling swine manure in fish ponds. Pages 489-492 *in* T.V.R. Pillay and W.A. Dill, eds. Advances in Aquaculture. Fishing News Books Ltd., Farham, England.

Buras, N., L. Duck, and S. Niv. 1985. Reactions of fish to microorganisms in wastewater. Applied Environmental Microbiology 50:989-995.

Burns, R.P., and R.R. Stickney. 1980. Growth of *Tilapia aurea* in ponds receiving poultry wastes. Aquaculture 20:117-121.

Chatterjee, J. 1979. Responses of Some Biogeochemical Cycling Bacteria and Their Activities to Qualitatively Different Fertilizers and Nitrate Population in Pond System. Doctotal dissertation, University of Kalyani, Kalyani, India.

Chattopadhyay, G.N. 1995. Use of potassium fertilizers in aquaculture. Pages 38-41 *in* G.N. Chattopadhyay, ed. Nutrient Management in Aquaculture. Visva Bharati, Sriniketan, India.

Chattopadhyay, G.N., and L.N. Mandal. .1980. Effect of different levels of water salinity on the decomposition of organic manures in a brackishwater fish pond soil. Hydrobiologia 72:287-292.

Chattopadhyay, G.N., and L.N. Mandal 1982. Concept of fertilizing brackishwater fish ponds. Fertiliser News 27:15-19.

Cloete, T.E., D.F. Toerien, and A.J.H. Pieterse. 1984. The bacteriological quality of water and fish of a pond system for the treatment of cattle feed effluent. Agricultural Wastes 9:1-15.

Cohen, J., and H.I. Shuval. 1973. Coliforms, fecal coliforms and fecal streptococci as indicatiors of water pollution. Water Air Soil Pollution 2:85-95.

Coleman, J.A., and P. Edwards. 1987. Feeding pathways and environmental constraints in waste-fed aquaculture: Balance and optimization. *In* D.J.W. Moriarty and R.S.V. Pullin, eds. Detritus and Microbial Ecology in Aquaculture, ICLARM, Manila, Philippines.

Collis, W.J., and R.O. Smitherman. 1978. Production of *Tilapia aurea* with cattle manure or a commercial diet. Pages 43-54 *in* R.O. Smitherman, W.L. Shelton, and J.H. Grover, eds. Symposium on Culture of Exotic Fishes. Fish Culture Section, American Fisheries Society, Auburn, Alabama.

Cruz, E.M., and Z.H. Shehadeh. 1980. Preliminary results of integrated pig-fish and duck-fish production tests. Pages 225-238 *in* R.S.V. Pullin and Z.H. Shehadeh, eds. Integrated Agriculture-Aquaculture Farming Systems. ICLARM Conference Proceeding 4, International Center for Living Aquatic Resources Management, Manila, Philippines.

Das, S.K. 1992. Evaluation of the Fertilizer Value of Phosphate Rock in Carp Culture: Water Quality and Biological Productivity. Doctoral dissertation, University of Kalyani, Kalyani, India.

Das, S.K., and B.B. Jana. 1996a. Pond fertilization through inorganic sources: An overview. Indian Journal of Fisheries 43 (2):137-155.

Das, S.K., and B.B. Jana. 1996b. Does rock phosphate fertilization adequately sustain fish production in the next year following its application during the first? Journal of the Inland Fisheries Society of India 28 (1): 7-75.

Das, S.K., D. Chakrabarty, and B.B. Jana. 1999. Growth responses of carps to phosphate rock fertilizer in simulated fish ponds and *in situ*. Proceedings of the Zoological Society, Calcutta 52 (1):16-28.

Debeljak, L., and K. Fasaic. 1985. Hydrochemical regime in common carp fingerling fish ponds under the conditions of organic and mineral fertilization. Ekologia 20:37-46.

Debeljak, L., M. Turk, K. Fasaic, and J. Popovic. 1990. Mineral fertilizers and fish production in carp ponds. Pages 187-193 *in* R. Berka and V. Hilge, eds. Proceedings of the FAO-EIFAC Symposium on Production Enhancement in Still Water Pond Culture. Research Institute of Fish Culture and Hydrobiology, Vodnany, Czechoslovakia.

Diana, J.S., C.K. Lin, and P.J. Schneeberger. 1991. Relationship among nutrient inputs, water nutrient concentrations, primary production and yield of *Oreochromis niloticus* in ponds. Aquaculture 92:323.

Dickman, M., and I.E. Efford. 1972. Some effects of artificial on enclosed plankton populations in Marion Lake, British Columbia. Journal of the Fisheries Research Board of Canada 29:1595-1604.

Dinesh, K.R., T.J. Varghese, and M.C. Nandeesha. 1986. Effects of a combination of poultry manure and varying doses of urea on the growth and survival of cultured carps. Pages 565-568 *in* J.L. MacLean, L.B. Dizon, and L.V. Hosillos, eds. Proceedings of the First Asian Fisheries Forum, Manila, Philippines.

Djajadiredja, R., and Z. Jangkaru. 1978. Small Scale Fish/Crop/Livestock/Home Industry Integration, A Preliminary Study in West Java, Indonesia. Inland Fisheries Research Institute, Bogor, Indonesia.

Dobbins, D.A., and C.E. Boyd. 1976. Phosphorus and potassium fertilization in sunfish ponds. Transactions of the American Fisheries Society 105:536-540.

Edwards, P. 1980. A review of recycling organic wastes into fish with emphasis on the tropics. Aquaculture 21:261-279.

Edwards, P. 1983. The future potential of integrated farming systems in Asia. Pages 273-281 *in* I. Takashi, ed. Proceedings of the Fifth World Conference on Animal Production, Vol.1, Japanese Society of Zootechnical Science, Tokyo, Japan.

Edwards, P., C. Pacharaprakiti, K. Kaewpaitoon, V.S. Rajput, P. Ruamthaveesub, S. Suthirawut, M. Yomjinda, and C.H. Chao. 1984. Reuse of cesspool slurry and cellulose agricultural residues for fish culture. AIT Research Report No. 166. Asian Institute of Technology, Bangkok, Thailand.

Edwards, P., C. Polprasert, V.S. Rajput, and Pacharaprakiti, C. 1988. Integrated biogas technology in the tropics-2. Use of slurry for fish culture. Waste Management Research 6:51-61.

Edwards, P., H. Demaine, N.I. Taylor, and D. Turongruang. 1996. Sustainable aquaculture for small scale farmers: Need for a balanced model. Outlook on Agriculture 25 (1):19-26.

Egna, S., and C.E. Boyd. 1997. Dynamics of Pond Aquaculture. CRC Press, Boca Raton, Florida.

Elliot, E.T. 1986. Aggregate structure and carbon, nitrogen and phosphorus in native and cultivated soils. Journal of the Soil Science Society of America 50:627-633.

Eren, Y., T. Tsur, and Y. Avnimelech. 1977. Phosphorus fertilization of fish ponds in the Upper Galilee. Bamidgeh 29:87-93.

Fitzgerald, G. P. 1970. Aerobic lake muds for the removal of phosphorus from lake waters. Limnology and Oceanography. 15:550-555.

Evision, L.M., and A. James. 1973. A comparison of the distribution of intestinal bacteria in British and African water resources. Applied Bacteriology 36:109-118.

Ganguly, S., J. Chatterjee, and B.B. Jana. 1999. Biogeochemical cycling bacterial activity in response to lime and fertilizer applications in pond systems. Aquaculture International 7:413-432.

Gaur, A.C., K.S. Dargan, and K.S. Neelakantan. 1995. Organic Manure. Publication and Information Division, Indian Council of Agricultural Research, New Delhi, India.

Ghosh, S.R. 1975. A study on the relative efficiency of organic manures and the effect of salinity on its mineralization in brackishwater fish farm soil. Aquaculture 5:359-366.

Ghosh, S.R., and A.N. Mohanty. 1981. Observations on the effect of aeration on mineralization of organic in fish pond soil. Bamidgeh 33(2):50-56.

Ghosh, S.R., N.G.S. Rao, and A.N. Mohanty. 1979. Studies on the relative efficiency of organic manures and inorganic fertilizer in plankton production and its relation with water quality. Pages 107-108 *in* Proceedings of the Symposium on Inland Aquaculture, Central Inland Fisheries Research Institute, Barrackpore, India.

Goldman, J.C., D.A. Caron, and M.R. Dennett. 1987. Regulation of gross growth efficiency and ammonium regeneration in bacteria by substrate C:N ratio. Limnology and Oceanography 32:1239-1252.

Green, B.W. 1990. Substitution of organic manure for pelleted feed in tilapia production. Pages 165-171 *in* R. Berka and V. Hilge, eds. Proceedings of the FAO-EIFAC Symposium on Production Enhancement in Still Water Pond culture. Research Institute of Fish Culture and Hydrobiology, Vodnany, Czechoslovakia.

Green, B.W., R.P. Phelps, and H.R. Alvarenga. 1989. The effect of manures and chemical fertilizers on the production of *Oreochromis niloticus* in earthen ponds. Aquaculture 76:37.

Green, B.W., R.P. Phelps, and H.R. Alvarenga. 1990. Hondurus: Cycle II of the CRSP Global Experiment, Pond Dynamics CRSP Data Reports, Oregon State University, Corvallis, Oregon.

Green, B.W., D.R. Teichert-Coddington, and R.P. Phelps. 1994. Development of semi-intensive aquaculture technologies in Hondurus: Summary of freshwater aquacultural research conducted from 1983 to 1992. Research and Development Ser. No. 39. Auburn University, Alabama.

Gregorich, E.G., R.G. Kachanoski, and R.P. Vorney. 1989. Carbon mineralization in soil size fractions after various amounts of aggregate disruption. Journal of Soil Science 40:649-659.

Hargreaves, J.A. 1998. Nitrogen biogeochemistry of aquaculture ponds. Aquaculture 166:181-212.

Hejkal, T., C.P. Gerba, S. Henderson, and M. Freeze. 1983. Bacteriological, virological and chemical evaluation of a wastewater aquaculture system. Water Research. 17:1749-1755.

Hepher, B. 1963. Ten years of research in fish pond fertilization in Israel. II. Fertilizer dose and frequency of fertilization. Bamidgeh 15:87-92.

Hickling, C.F. 1962. Fish Culture, Faber and Faber, London, England.

Hill, S.J., J.D. Sedlacek, P.A. Watson, J.H. Tidwell, K.D. Davis, and W.L. Knight. 1997. Effects of diet and organic fertilization on water quality and benthic macroinvertebrate populations in ponds used to culture freshwater prawn, *Macrobrachium rosenbergii*. Journal of Applied Aquaculture 7(3):19-32.

Hopkins, K.D. 1982. Outstanding yields and profits from livestock-tilapia integrated farming. International Center for Living Aquatic Resources Management Newsletter 5 (3):13.

Hopkins, K.D., and E.M. Cruz. 1980. High yields but still questions: Three years of animal-fish farming. International Center for Living Aquatic Resources Management Newsletter 3(4):12-13.

Hopkins, K.D., and E.M. Cruz. 1982. The ICLARM-CLSU integrated animal-fish farming project: Final report. International Center for Living Aquatic Resources Management, Manila and the Freshwater Aquaculture Center, Central Luzon State University, Nueva Ecija, Philippines.

Hora, S.L., and T.V.R. Pillay. 1962. Handbook on fish culture in the Indo-Pacific region. FAO Fish Biology Technical Paper 14:204.

Jana, B.B., and S.K. Das. 1992. The fertilizer value of phosphate rock in carp culture. Bamidgeh 44(1):13-23.

Jana, B.B., and S.N. Sahu. 1994. Effects of frequency of rockphosphate application in carp culture. Aquaculture 122:313-321.

Jana, B.B., and D. Chakrabarty. 1997a. Relative status and contribution of sediment phosphorus and nitrogen in carp culture system fertilized with various combinations of rockphosphate. Aquaculture Research 28:853-859.

Jana, B.B., and L. Chakrabarty. 1997b. Effect of manuring rate on *in situ* production of zooplankton *Daphnia carinata*. Aquaculture 156:85-99.

Jana, B.B., P. Chakrabarti, J.K. Biswas, and S. Ganguly. 2001. Biogeochemical cycling bacteria as indices of pond fertilization: Importance of CNP ratios of input fertilizers. Journal of Applied Microbiology 90(5):733-740.

Jhingran, V.G. 1995. Fish and Fisheries of India. Hindustan Publishing Corporation (India), Delhi, India.

Kapur, K., and K.K. Lal. 1986. The chemical quality of waste treated waters and its relation with patterns of zooplankton population. Pages 129-132 *in* J.L. Maclean, L.B. Dizon, and L.V. Hosillos, eds. The First Asian Fisheries Forum. Asian Fisheries Society, Manila, Philippines.

Kimmel, B.L., and P.T. Lind. 1970. Factors influencing orthophosphate concentration decline in the water of laboratory mud water systems. Texas Journal of Science 21:339-445.

Knud-Hansen, C.F., and C.K. Lin. 1993. Strategies for stocking Nile Tilapia (*Oreochromis niloticus*) in fertilized ponds. Page 275 *in* H. S. Egna, J. Bowman, B. Goetze, and N. Weidner, eds. Pond Dynamics/Aquaculture Collaborative Research Support Pogram, Oregon State University, Corvallis, Oregon.

Knud-Hansen, C.F., T.R. Batterson, C.D. McNabb, I.S. Harahat, K. Sumantadinata, and H.M. Eidman. 1991. Nutrient input, primary productivity and fish yield in fertilized freshwater ponds in Indonesia. Aquaculture 94:49-58.

Larsson, B. 1994. The overviews on environment and aquaculture in the tropics and subtropics. ALCOM Field Document No. 27, Food and Agriculture Organization of the United Nations, Rome, Italy.

Lichtkopler, F.R., and C.E. Boyd. 1977. Phosphorus fertilization of sunfish ponds. Transactions of the American Fisheries Society 106:634-636.

Lin, C.K., D.R. Teichert-Coddington, B.W. Green, and K.L. Veverica. 1997. Fertilization regimes. Pages 73-103 *in* H.S. Egna and C.E. Boyd, eds. Dynamics of Pond Aquaculture, CRC Press, Boca Raton, Florida.

Ling, C. 1986. Preliminary study on culturing fish with chemical fertilizers. Journal of Aquaculture in the Tropics 1(1):43-48.

MacLean, M.H., and K.J. Ang. 1994. An enclosure design for feeding and fertilization trials with the freshwater prawn, *Macrobrachium rosenbergii* (de Man). Aquaculture 120:71-80.

Mandal, L.N., and G.N. Chattopadhyay. 1992. Nutrient management in fish ponds. Page 144 *in* H.L.S. Tandon, ed. Non-Traditional Sectors for Fertilizer Use. FDCO, New Delhi, India.

Maramba, F.D. 1978. Biogas and Waste Recycling. The Philippines Experience. Maya Farms Division, Liberty Flour Mills Inc., Manila, Philippines.

Martyshev, F.G. 1983. Pond Fisheries. A.A. Balkema, Rotterdam, The Netherlands.

Maun, K.K. 1972. Microphytoplankton and detritus food chains in coastal waters. Memorial 1st. Italian Idrobiologia. 29:359-383.

Metzger, R.J., and C.E. Boyd. 1980. Liquid ammonium polyphosphate as a fish pond fertilizer. Transactions of the American Fisheries Society 109:563-570.

Miah, M. S., M.S. Alam, M. Neazuddin, M.V. Gupta, and M.S. Shah. 1996. Studies on production of carps under polyculture system in farmer's ponds. Indian Journal of Fisheries 43(4):375-379.

Moav, R., G.W. Wohlfarth, G.L. Schroeder, G. Hulata, and H. Barash. 1977. Intensive polyculture of fish in freshwater ponds. Aquaculture 10:25-43.

Mohanty, A.N., D.K. Chatterjee, P.K. Saha, and K.C. Pani. 1994. Effect of varying C/N ratios on the mineralization of organic nitrogen in fish pond soils. Journal of Aquaculture in the Tropics 9(1):9-14.

Morissens, P., M. Oswald, F. Sanchez, and S. Hem. 1996. Designing new fish farming models adopted to rural Cote d Ivoire. Proceedings of the Third International Symposium on Tilapia in Aquaculture 41:118-128.

Mortimer, C.H., and C.F. Hickling. 1954. Fertilizers in Fish Ponds. Fisheries Publication, London, England.

Muller, W. 1990. Aspects of pond fertilization in high-intensive carp culture. Pages 207-211 *in* R. Berka and V. Hilge, eds. Proceedings of the FAO-EIFAC Symposium on Production Enhancement in Still Water Pond Culture. Research Institute of Fish Culture and Hydrobiology, Vodnany, Czechoslovakia.

Nandeesha, M.C., P. Keshavanath, and K.S. Udupa. 1984. Evaluation of organoleptic qualities of fish grown in ponds treated with different organic manures. Fishery Technology (India) 21:94-97.

Nolan, C.N., and W.L. Pritchett. 1960. Certain factors affecting the leaching of potassium from sandy soils. Proceedings of Soil and Crop Science Society 20:130-145.

Noriega-Curtis, P. 1979. Primary productivity and related fish yield in intensely manured fish ponds. Aquaculture 17:335-344.

Olah, J. 1986. Carp production in manured ponds. Pages 293-303 *in* R. Billard and J. Marcel, eds. Aquaculture of Cyprinids, INRA, Paris, France.

Olsen, S. 1958a. Phosphate absorption and isotopic index in lake muds. Experiments with P^{32}; preliminary report. Verhandlungen Internationale Vereinigung Limnologie. 13:915-922.

Olsen, S. 1958b. Fostat balancen mellem bund og vand i Fureso. Forsberg med radioactivt fosfor. Folia Limnologia Scandinavica 10:39-96.

Pan, Q., F. Qixue, and Z. Bangke. 1994. The growth of silver carp and bighead in polyculture ponds fertilized mainly with ammonium chloride. Acta Hydrobiologia Sinica (China) 18(2):116-127.

Pattrick, W.H., and D.S. Mikkelsen. 1971. Plant nutrient behaviour in flooded soil. Pages 187-215 *in* R. A. Olsen, ed. Fertilizer Technology and Use, Soil Science Society of America, Madison, Wisconsin.

Pekar, F., and J. Olah. 1990. Organic fertilization. Pages 116-122 *in* R. Berka and V. Hilge, eds. Proceedings of the FAO-EIFAC Symposium on Production Enhancement in Still Water Pond Culture. Research Institute of Fish Culture and Hydrobiology, Vodnany, Czechoslovakia.

Pekar, F., and J. Olah. 1998. Fish pond manuring studies in Hungary. Proceedings of a Workshop on Integrated Fish Farming, Wusei, Jiangsu Province, China. CRC Press Inc., Boca Raton, Florida.

Pillay, T.V.R. 1995. Aquaculture-Principles and Practices. Fishing News Books, Cambridge, England.

Polprasert, C. 1996. Organic Waste Recycling. John Wiley and Sons, Chichester, England.

Polprasert, C., P. Edwards, C. Pacharaprakiti, V.S. Rajput, and S. Suthirawat. 1982. Recycling rural and urban night soil in Thailand. AIT Research Report No. 143. Asian Institute of Technology, Bangkok, Thailand.

Pretto, R. 1996. Objectives and indicators for aquaculture Development. Report, Expert Consultation on Small-scale Rural Aquaculture. FAO Fisheries Report, No. 548, Rome, Italy.

Rabanal, H.R. 1960. The Effect of No Fertilization and Non-Nitrogenous Fertilization Upon the Chemistry of Water, the Plankton, Bottom Organism and Fish Production in Ponds That Have Received Continued Complete (NPK) Fertilizers During the Preceding 15 Year Period. PhD Dissertation, Auburn University, Alabama.

Rao, D.U., N.H. Parhad, C.S. Rao, and K.S. Rao. 1968. Coliform as indicators of fecal contamination. Environmental Health 10:21-24.

Rapapport, U., and S. Sarig. 1978. The results of manuring on intensive growth fish farming at the Ginosar Station ponds in 1977. Bamidgeh 30(3):27-36.

Romaire, R.P., C.E. Boyd, and W.J. Collis. 1978. Predicting nighttime dissolved oxygen decline in ponds used for *Tilapia* culture. Transactions of the American Fisheries Society 107:804-808.

Saha, G.N. 1979. Techniques of pond fertilization and use of fertilizers in freshwater aquaculture for increased fish production. Fertilizer News 24(2):3-6.

Saha, P.K. 1995. Nutrient management of freshwater rearing and stocking ponds. Pages 66-68 *in* G.N. Chattopadhyay, ed. Nutrient Management in Aquaculture, Institute of Agriculture, Sriniketan, India.

Sahai, V. 1990. Fundamentals of Soils. Kalyani Publishers, New Delhi, India.

Santos, C.A.L. 1994. Prevention and control of food borne trematode infections in cultured fish. Food and Agriculture Organization Aquaculture Newsletter 8:11-15.

Schroeder, G.L. 1974. Use of fluid cowshed manure in fish ponds. Bamidgeh 26(1): 84-96.

Schroeder, G.L. 1975. Cow manure in fish culture. FAO Aquaculture Bulletin, FAO, Rome, Italy.

Schroeder, G.L. 1977. Agricultural wastes in fish farming-a commercial application of the culture of single celled organisms for protein production. Water Research 11:419-420.

Schroeder, G.L. 1980. Fish farming in manure-loaded ponds. Pages 73-86 *in* R.S.V. Pullin and Z.H. Shehadeh, eds. ICLARM Conference Proceedings 4, International Centre for Living Aquatic Resource Management, Manila and Southeast Asian Regional Centre for Graduate Study and Research in Agriculture, College Los Banos, Philippines.

Schroeder, G.L., and B. Hepher. 1979. Use of agricultural and urban waste in fish culture. Pages 487-489 *in* T.V.R. Pillay and W.A. Hill, eds. Advances in Aquaculture. Fishing News Books Ltd. Farham, England.

Schroeder, G.L., J.E. Halver, and K. Tiews. 1978. Microorganisms as the primary diet in fish farming. Pages 379-386 *in* G.L. Schroeder, J.E. Halver, and K. Tiews, eds. Finfish Nutrition and Fishfeed Technology. Heenemann verlagsgesellschaft, Berlin, Germany.

Schroeder, G.L., G. Wohlfareth, A. Alkon, H. Halevy, and H. Krueger. 1990. The dominance of algal-based food webs in fish ponds receiving chemical fertilizers plus organic manures. Aquaculture 86:219-229.

Shan, J., and S. Wu. 1994. Study on the Effects of Fish Production in Polyculture Ponds with Silver Carp and Bighead Carp as the Main Species. Science Press, Beijing, China.

Shan, J., L. Chang, X. Gua, Y. Fang, Y. Zhu, X. Chan, F. Zhou, and G.L. Schroeder. 1985. Observations on feeding habits of fish in ponds receiving green and animal manures in Wuxi, People's Republic of China. Aquaculture 46:111-117.

Sharma, B.K., and J. Olah. 1986. Integrated fish-pig farming in Hungary. Aquaculture 54:135-139.

Sharma, B.K., D. Kumar, M.K. Das, and D.P. Chakrabarty. 1979. Observations on swine dung recycling in composite fish culture. Abstract of the Symposium on Inland Aquaculture. Central Inland Fisheries Research Institute, Barrackpore, India.

Shevgoor, L., C.F. Knud-Hansen, and P. Edwards. 1994. An assessment of the role of buffalo manure for pond culture of tilapia. III. Limiting factors. Aquaculture, 126:107.

Singh, S., and H. Ram. 1971. A comparative study of pond and adjoining cultivated soils. Indian Journal of Chemistry 4(1):13-20.

Singh, V.K., and A.P. Sharma. 1999. Comparative effect of three organic manures viz. cowdung, pigdung and poultry excreta on the growth of *Labeo rohita* (Ham.). Journal of the Inland Fisheries Society of India 31(1):1-5.

Sorokin, Y.I., and H. Kadota. 1972. Techniques for the assessment of microbial production and decomposition in freshwaters. International Biological Programme. Blackwell Scientific Publications, London, England.

Stickney, R.R. 1979. Principles of Warmwater Aquaculture. John Wiley & Sons, Inc., New York, New York.

Stickney, R.R. 1994. Principles of Aquaculture. John Wiley & Sons, Inc., New York.

Swingle, H.S. 1954. Experiments on commercial fish production in ponds. Proceedings of the Conference on Southeast Association of Game and Commissioners. 8:69-74.

Swingle, H.S., and E.V. Smith. 1939. Fertilizer for increasing the natural food for fish in ponds. Transactions of the American Fisheries Society 68:126-135.

Swingle, H.S., and E.V. Smith. 1947. Management of farm fish ponds. Bulletin 264. Alabama Polytechnic Institute Agricultural Experiment Station. Auburn, Alabama.

Tang, Y.A. 1970. Evaluation of balance between fishes and available fish foods in multispecies fish culture ponds in Taiwan. Transactions of the American Fisheries Society 99:708-718.

Teichert-Coddington, D.R., L. Behrends, and R. Smitherman 1990. Effects of manuring regime and stocking rate on primary production and yield of tilapia using liquid swine manure. Aquaculture 88:61-69.

Teichert-Coddington, D.R., and B.W. Green. 1990. Influence of primary productivity, season and site of tilapia production in organically fertilized ponds in two central American countries. Pages 137-145 *in* R. Berka and V. Hilge, eds. Proceedings of the FAO-EIFAC Symposium on Production Enhancement in Still Water Pond Culture. Research Institute of Fish Culture and Hydrobiology, Vodnany, Czechoslovakia.

Ullah, A. Md. 1989. Nutrient release characteristics of duck manure for Nile tilapia production. AIT thesis AE-89-43, Asian Institute of Technology, Bangkok, Thailand.

Varghese, T.J., and K.M. Shankar. 1981. A review of fertilization experiments conducted at the College of Fisheries, Mangalore. Pages 67-76 *in* D.J. Macintosh, ed. Proceedings of the Seminar on Some Aspects of Inland Aquaculture in Karnataka. Mangalore, Karnataka, India.

Villacorta, L.G. 1989. Comparison of Constant and Variable Loading Rates of Organic Manure in Fish Culture. Master's. Thesis, Asian Institute of Technology, Bangkok, Thailand.

Wetzel, R.G. 1983. Limnology. Saunders College Publishing, New York, New York.

Winberg, G.G., and W.P. Liakhnovich. 1965. Udobrenie Prudov. Moskova, USSR.

Wolhfarth, G.W., and G.L. Schroeder. 1979. Use of manure in fish farming–A review. Agricultural Wastes 1:279.

Wolhfarth, G.W., and G. Hulata. 1987. Use of manures in aquaculture. Page 353 *in* D.J.W. Moriarty and R.S.V. Pullin, eds. Detritus and Microbial Ecology in Aquaculture. International Center for Living Aquatic Resources Management, Manila, Philippines.

Wolhfarth, G.W., G. Hulata, and R. Moav. 1980. Use of manure in aquaculture–Some experimental results. Symposium on Aquaculture in Wastewater. Pretoria, South Africa.

Wolny, P. 1967. Fertilization of warm water fish ponds in Europe. Food and Agricultural Organization Fisheries Report 44:64-81.

Xianzhen, G., F. Tingxue, W. Jikum, F. Xiuzheng, and L. Zhiyum. 1986. A preliminary study on sources of fish growth in manured ponds using delta C analysis. Pages 125-128 *in* J.L. Maclean, L.B. Dizon and L.V. Hosillos, eds. The First Asian Fisheries Forum. Asian Fisheries Society, Manila, Philippines.

Yashouv, A., and A. Halevy. 1972. Experimental studies of polyculture in 1971. Bamidgeh 24 (2):31-39.

Yingxue, F., G. Xianzhen, W. Jikum, F. Xiuzheng, and L. Zhiyum. 1986. Effects of different animal manures on fish farming. Pages 117-120 *in* J.L. Maclean, L.B. Dizon, and L.V. Hosillos, eds. The First Asian Fisheries Forum. Asian Fisheries Society, Manila, Philippines.

The Otelfingen Aquaculture Project: Recycling of Nutrients from Waste Water in a Temperate Climate

Juerg Staudenmann
Ranka Junge-Berberovic

SUMMARY. A wastewater-fed, partly indoor aquaculture plant (36 basins, 360 m^2 and 420 m^3 in total) was designed in Otelfingen/Zurich, Switzerland. It was charged with the effluent from a methanization plant processing organic household waste and started operation in spring 1998. The aim of the successive arrangement of the different modules and steps was to efficiently recycle water-borne nutrients in the form of aquatic biomass products, such as floating (ornamental) macrophytes, fish, zoo- and phytoplankton, suitable for selling on the Swiss market. Besides treating the effluent (total organic carbon [TOC], total nitrogen, nitrate [NO_3-N], ammonium [NH_4-N], and total phosphorus concentrations being 670 g/m^3, 255 g/m^3, 150 g/m^3, 95 g/m^3, and 52 g/m^3, respectively) according to Swiss law requirements, the research focused on the search for suitable aquatic organisms and their testing at different environmental conditions. During the 16-week experimental period, a total of 2,150 kg fresh weight (FW) of biomass (97% as floating macrophytes) was harvested. This way, 176 g/week nitrogen and 47 g/week phospho-

---

Juerg Staudenmann and Ranka Junge-Berberovic, Department of Horticulture and Environment, University of Applied Sciences, P.O. Box 335, CH-8820 Waedenswil, Switzerland.

Address correspondence to: Ranka Junge-Berberovic, Department of Horticulture and Environment, University of Applied Sciences, P.O. Box 335, CH-8820 Waedenswil, Switzerland.

[Haworth co-indexing entry note]: "The Otelfingen Aquaculture Project: Recycling of Nutrients from Waste Water in a Temperate Climate." Staudenmann, Juerg, and Ranka Junge-Berberovic. Co-published simultaneously in *Journal of Applied Aquaculture* (Food Products Press, an imprint of The Haworth Press, Inc.) Vol. 13, No. 1/2, 2003, pp. 67-101; and: *Sustainable Aquaculture: Global Perspectives* (ed: B. B. Jana, and Carl D. Webster) Food Products Press, an imprint of The Haworth Press, Inc., 2003, pp. 67-101. Single or multiple copies of this article are available for a fee from The Haworth Document Delivery Service [1-800-HAWORTH, 9:00 a.m. - 5:00 p.m. (EST). E-mail address: getinfo@haworthpressinc.com].

rus were eliminated by assimilation, corresponding to 25-35% of the system's inflow. Due to relatively high evapotranspiration rates (on average 35.4 mm/week) and for water reconditioning purpose in the fish stocking basins, fresh water was added. Nevertheless, the system's final effluent was very low (21% of total inflow plus rainfall) and was carrying only about 2% and 0.5% of the input loads of nitrogen and phosphorus, respectively. Hence, the elimination rate was significantly above the average performance of a conventional system normally applied in Middle Europe, although the concentration values of most parameters in the outflow were comparable. Macrophyte production (and thus nutrient assimilation) was close to theoretical maxima in basins with high nutrient levels. Both plankton and fish growth were, at their best, only moderately satisfying. The semi-continual planktonic microalgae culture, and therefore also zooplankton culture, could be improved if the light absorbing humic substances were removed in a pre-treatment. Under given conditions (i.e., temperate climate) fish would rather play an accompanying role in the ecological production process. A wastewater-fed aquaculture facility resembles an integrated production plant rather than a wastewater disposal site. In addition to that, it has potential to prove advantageous over the highly developed conventional wastewater treatment plants established in Middle Europe. Further research in this field is essential and also recommended, considering political programs like Agenda 21, which contemplate the need for sustainable strategies for handling resources.

KEYWORDS. Wastewater-fed aquaculture, nutrient recycling, aquatic macrophytes, *Eichhornia* spp., *Pistia* spp., microalgae

INTRODUCTION

A wastewater-fed aquaculture system is a constructed aquatic ecosystem consisting of one or several water bodies with an integrated food web, which is charged with wastewater. The central aim of the system is the assimilation of dissolved nutrients into biomass. Simultaneously organic compounds are either consumed or mineralized, and as a consequence the waste water is purified.

Several political programs, such as Agenda 21 (Anonymous 1992), contemplate the need for sustainable strategies for handling resources.

Their declared target is to close resource pathways of food chains and minimize energy losses. In this context, the combination of wastewater treatment and biomass production is a very interesting feature of wastewater-fed aquaculture as compared to conventional treatment systems, where effluents are treated only with regard to degradation and elimination processes (Junge-Berberovic and Staudenmann 1998). In Switzerland, and in the rest of Europe, conventional technology is normally applied, though, producing sludge, which has to be deposited, instead of valuable biomass. Compared to the extensive research of conventional treatment methods research into productive wastewater treatment methods is still rare (VSA 1995; Wissing 1995; Lange and Otterpohl 1997).

Some authors have investigated the nutrient elimination potential of water plants and constructed wetlands (DeBusk et al. 1983; Reddy and Smith 1987; Gumbricht 1993; Landolt 1996; Greenway and Woolley 1999). But the important consecutive step, namely use and marketing of obtained biomass, was rarely addressed. Few experimental wastewater-fed aquaculture facilities exist in temperate regions, like Stensund (Guterstam 1996) or "living machines" (Todd and Josephson 1996). Moreover, most of them have never left the experimental stage.

The main focus of this research project was to develop further the existing ideas on wastewater-fed aquaculture (Guterstam 1991; Staudenmann et al. 1996; Todd and Josephson 1996; Jana 1998) and to adapt them to Swiss conditions, which are representative to a high degree of most European countries. The challenge in adaptation was twofold: it concerned the transfer both from tropical to temperate climate (adaptation in a biological and technical sense) and from non-industrialized to western economy (adaptation of product palette as well as development of new ideas on use and marketing of products).

If the new approach is to compete with common technical solutions, the implementation must take into account a plethora of ecological and economical requirements (Staudenmann and Junge-Berberovic 2000). Among others: More efficient use of resources and less impact on the environment than conventional plants; better quality of the effluent, which must meet the regulations concerning either the discharge in aquifers or for use as irrigation water; equal or lower total costs for water purification; and higher quality of products leaving the plant.

The Otelfingen Pilot Project

In many parts of Switzerland, organic household and gardening waste is collected separately for biological degradation and recovery of the inherent energy content. In Otelfingen/ZH, 10 km outside of Zurich, a methanization plant designed for the processing of some 10,000 tons/year was built in 1996/97. The Kompogas® system implemented is an innovative approach, gaining growing attention in the past decade (Schmid 1992; Edelmann et al. 1993). It consists of a one-step anaerobic, thermophilic reactor followed by an aerobic polishing unit. Besides biogas and solid residues, a liquid fraction with high organic and inorganic loads is released from the anaerobic reactor.

The high nutrient content of this processed water raised the twofold question about its treatment and the elimination of the compounds on one hand and about its potential as a liquid fertilizer for biomass production on the other hand. Together with the University of Applied Sciences of Waedenswil, a research project began in 1997. The aims of the project were:

a. Develop an innovative strategy to use processed water *in situ* as a resource for biomass production, by emulation of natural self-purification and assimilation processes.
b. Search for organisms to be grown that either have a market value or can be utilized as a secondary resource in the system itself.
c. Aspire high nutrient recycling rates.
d. Strive to achieve resource and energy efficiency.
e. Treat waste water to reach levels required by Swiss law (Anonymous 1998).

The main research goals resulting from the above aims included:

a. Search for potentially valuable aquatic macrophytes with high growth rates and a preference for or a tolerance of high nutrient, salt, and suspended solids levels.
b. Test production methods of several fish species based on internally produced feed (e.g. mono- vs. polyculture, feed composition, stocking density, etc.).
c. Raise micro-algae polyculture with cells of suitable size for filter feeders, such as *Daphnia*.
d. Investigate the growth performance of target organisms at different environmental conditions.

Several technical and process engineering challenges had to be faced. Solutions had to be found for zooplankton harvesting, water and wastewater distribution, aeration, marketing requirements for ornamental plants, etc. The appropriate technological tools had to be integrated with the "living part" of the system. This whole process can be best summarized as a "semi-tech ecologically engineered approach" (Staudenmann and Junge-Berberovic 2000).

SITE DESCRIPTION AND EXPERIMENTAL DESIGN

A multi-step aquaculture system was designed and started operation in May, 1998. The pilot plant, unique of its kind in central Europe, consists of a reinforced concrete basin of 65 m × 7 m, divided by brick walls into 36 single basins arranged in three rows (Figure 1). Each basin measured 2 m × 5 m, with a depth of either 0.5 m (24 basins at the east end) or 1.5 m (12 basins at the west end), resulting in a total water surface and volume of 360 m^2 and 420 m^3, respectively. A greenhouse covered 12 low-level basins on the very east. The basins were interconnected to each neighboring one by one or two underwater connection pipes (120 mm in diameter), depending on the basins' depth. These ducts could individually be sealed with lids. This way, the basins could be connected in various combinations with a high degree of flexibility. From June to October 1998 they were arranged in seven major groups or steps, each containing between one and four modules (Figure 1; Table 1).

Aeration with common air (compressor performance 8 kW) was installed in most of the basins (Table 1). Except for the basins where fish and zooplankton were grown, the compressor served for mixing, rather then for oxygen supply. Into some basins in steps A (algae) and P (polishing), exhaust carbon dioxide from the biogas driven power generators was infused as an additional carbon-source for photosynthesis and to test the potential of recycling of CO_2.

The effluent from the biowaste fermentation plant was introduced into the first basin of step T (Pre-treatment) at the east end of the system, where decomposition processes (including denitrification) occurred along with nutrient assimilation by floating plants. The pretreated water was introduced into step M (Macrophytes) and mixed with water from step S (Storage). Water hyacinth, *Eichhornia crassipes*; water lettuce, *Pistia stratiotes*; duckweed, *Lemna* spp.; and other floating macrophytes have been stocked in these basins for nutrient uptake and biomass production (Todt 1998; Saxer 1999). The effluent from step M was pumped daily into the batch basins of module A1. Step A was intended to produce micro-algae as food for filter feeders (zooplankton and fish). In first half of the season there were four indoor and four out-

FIGURE 1. Lay-out of the Otelfingen Aquaculture Pilot Plant with detailed view of the basins' hydraulic interconnection by individually lidded submerged water ducts.

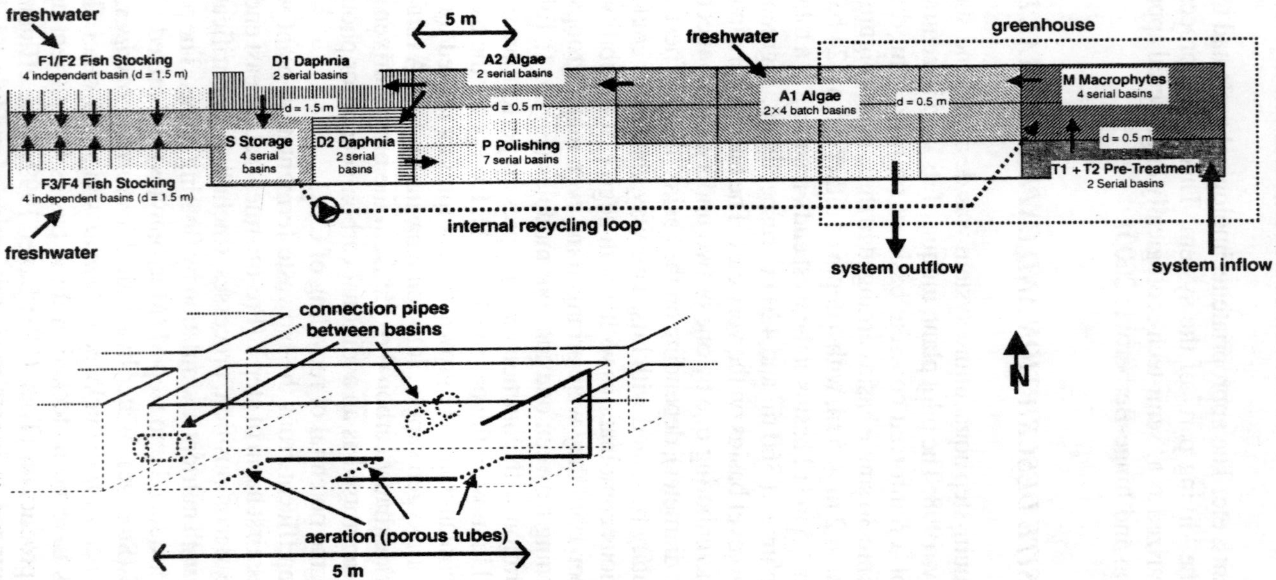

door basins in module A1. The indoor algal culture was discontinued after August 1, 1998. Module A2 was intended for mixing and further development of algal cultures. Algae were not harvested but floated with the water stream from module A2 into step D (*Daphnia*). Step D consisted of two parallel modules containing two deep basins each. Sixty-five percent of the algae-rich water from step A was pumped into D1, the rest into D2, where *Daphnia* consumed the incoming micro-algae. The effluents from modules D1 and D2 were discharged into steps S and P, respectively (Figure 1).

Fish were stocked at initial densities between 10-30 individuals/m^2 (i.e., 0.5-14 kg/m^2) in the basins of step F (Fish) (Graber 1998). Species stocked were common carp, *Cyprinus carpio*; tench, *Tinca tinca*; silver carp, *Hypophthalmichthys molitrix*; and Nile tilapia, *Oreochromis niloticus*. To obtain high-quality edible fish, this module was excluded from the main wastewater course. Duckweed (mainly *Lemna*) and zooplankton (mainly *Daphnia*) were fed manually to the fish. The fresh water necessary to compensate for evapotranspiration of the entire system was also introduced into this module.

The effluent from step F containing organic detritus was discharged into step S, where it was partly degraded and assimilated. Effluent from step S, mixed with the outflow from module D2, was recycled as dilution water into step M at the beginning of the system. This internal loop allowed for degradation and mineralization of organics from these steps (mainly F), making nutrients available for assimilation by macrophytes. Step P was stocked with native macrophytes for the purpose of nutrient residues assimilation. The effluent from step P, i.e., the system outflow, was discharged into an artificial pond (80 m^3) stocked with fish and native helophytes. From there the water either evaporated or precipitated into the ground. Several basins of steps S and P contained also freshwater shrimp, *Astacus astacus,* as detritus feeder, but the results are discussed elsewhere (Mueller 1999).

Monitoring and Analysis

The experimental period was from June, 18 to October, 1 1998 (16 weeks). The monitoring activity was organized in four monthly intervals. All major waste- and freshwater movements were monitored by mechanical water counters. The flows assessed were system inflow (step T and module A1) and system outflow (step P), as well as every transfer from step M to the batch basins in module A1, and from step S back to step M (recycling loop). The distribution factor between mod-

TABLE 1. Technical data and distinct features of the different modules in the Otelfingen Aquaculture Pilot Plant.

Step	Module(s)					Basins				Main Species and Function
		Volume (m^3)	Surface (m^2)	inflow from	outflow to	number	flow type	indoor	aerated	
T (Pre-Treatment) 2 modules in series	T1	5	10	in[1]	T2	1	plug flow	1	0	*Pistia stratiotes*, heterotrophic microorganisms Anaerobic degradation of organic compounds, denitrification
	T2	5	10	T1	M T1 (pump[3])	1	turbulent	1	1	*Pistia stratiotes*, heterotrophic microorganisms Nitrification, aerobic degradation of organic compounds, oxygenation
M (macrophytes) 1 module	M	20	40	T2	A1	4 in series	3 mixed[5] last plug flow	4	3	*Eichhornia crassipes*, *Pistia stratiotes*, *Lemna* spp. Nutrient uptake, primary production
A (single-cell algae) 2 modules in series	A1	40	80	M	A2	8 batch	turbulent	4	8	Microalgae polyculture (mostly *Scenedesmus* spp., *Ankistrodesmus* spp., other green algae). Nutrient uptake and growth of algae
	A2	10	20	A1	D1 (60%) D2 (40%)	2 in series	turbulent	0	2	Microalgae polyculture Nutrient uptake and growth of algae
D (*Daphnia*) 2 parallel modules	D1	30	20	A2 (60%)	S	2 in series	mixed[5]	0	2	*D. magna*, partly covered by *Lemna* spp. Consumption of algae and suspended matter
	D2	30	20	A2 (40%)	P	2 in series	mixed[5]	0	2	*D. magna*, partly covered by *Lemna* spp. Consumption of algae and suspended matter
F (fish stocking) 4 parallel modules	F1	15	10	fresh water	S	3 parallel	aerated	0	3	Three mono-cultures with tench, common carp, and Nile tilapia. Consumption of *Lemna* and *Daphnia*

	F2	15	10	fresh water	S	3 parallel	aerated	0	3	Three mono-cultures with silver carp Consumption of micro algae
	F3	15	10	fresh water	S	1	aerated	0	3	One polyculture (tench/common carp/silver carp/Nile tilapia) Consumption of *Lemna* and *Daphnia*
	F4	15	10	fresh water	S	3 parallel	aerated	0	1	Three different polycultures (common carp/tench/Nile tilapia; common carp/silver carp/tench; common carp/tench) Consumption of *Lemna* and *Daphnia*
S (storage) 1 module	S	60	40	F1-F4 D1	M (pump[4])	4 in series	mixed[5]	0	4	Plankton-polyculture, *Astacus astacus* Mineralization of organic residues from fish basins
P (polishing) 1 module	P	35	70	D2	out[2]	7 in series	6 mixed[5] 1 laminar	1	6	Temperate climate aquatic macrophytes, *Daphnia*, Nile tilapia in basins 2, 3, and 7 Polishing step/internal recycling loop between basins 4 and 1

1 system inflow (wastewater from fermentation process).
2 system effluent.
3 internal pump for back-flow to module T1.
4 recycling pump to step M.
5 smooth mixing by air-bubble injection.

ules D1 or D2 for the effluent from module A2 was set mechanically and the precision controlled every few weeks. Precipitation was determined with a common rain collector. Evapotranspiration was estimated indirectly from water balance data in two steps (in the greenhouse and outdoors), taking into consideration a three-week period around each data point. A weekly average evaporation value in relation to the whole system ("mix") was calculated.

Temperature was measured on-line *in situ* for air (outdoor and indoor) and for water (outdoor steps F and P, indoor steps T and M) and recorded by an analogue line plotter. Daily minima and maxima were identified, and the three-day moving average value was calculated. pH value, conductivity (EC), dissolved oxygen (DO) were measured in every module at least weekly *in situ* with a multi-electrode meter (Multiline P4, WTW GmbH, Germany[1]). Water samples were collected weekly and stored for a maximum of 24 hours at 8 °C for later processing. Total organic carbon (TOC) and total nitrogen content (TN) were determined by combustion and infrared detection (CHNS-Analyzer Vario EL, Elementar GmbH, Germany; HighTOC, Gerber Instruments AG, Germany[1]). From the same samples, chemical oxygen demand (COD), ammonium-nitrogen (NH_4-N), nitrate-nitrogen (NO_3-N), nitrite-nitrogen (NO_2-N), and total-phosphate-phosphorous (PO_4-P) content were determined photometrically (Cadas 30, Dr. Lange AG, Switzerland[1]).

Sporadically, biomass composition of macrophytes, *Daphnia*, and fish was determined. Analyses included dry weight, ash content, carbon, nitrogen, and phosphorus content. All biomass harvest and transfers between the basins were quantified by fresh weight. Standing crops of macrophytes and zooplankton, as well as fish fresh weight, were determined at least monthly. Fishes were anaesthetized before weighing and measuring (Graber 1998).

Algae species composition and abundance in the basins were determined at least weekly in step A and at least monthly also in steps D, P, F and S. Phytoplankton was collected by sampling in the middle of a basin into a 15 mL vial and was conserved with three drops of Lugol-Solution (Schwoerbel 1986). The samples were cooled and enumerated within four weeks. As the goal of the project was to cultivate dense algal cultures, suitable for the *Daphnia* nutrition, only the dominant algae species of the main modules were determined and enumerated. The numbers of cells and colonies were counted under a microscope in an improved Neubauer counting chamber with a volume of 0.1 μL. At least

1. Use of trade or manufacturer's name does not imply endorsement.

four squares were counted. The lowest cell count that could be thus assessed was 2,500 cells/mL.

Single cell sizes of dominant algae species were measured and the volume of cells and colonies calculated.

Water Balance and Environmental Conditions

The monthly average water flows between the different basins were calculated, based on the monitored water exchange rates between the steps, the wastewater inflow, and treated water effluent, considering the average atmospheric water exchange rate (Table 2). Figure 2 shows weekly precipitation (measured), estimated amount of water loss via atmosphere for an average of indoor and outdoor basins (evaporation mix), and calculated net atmospheric water exchange (AWE) for the whole system. On average, weekly rainfall and evapotranspiration rates amounted to 14.6 mm and 35.4 mm, respectively.

Daily minima and maxima of outdoor and indoor temperature (air and water), together with the three-day moving average are presented in Figure 3. The monthly difference between indoor and outdoor temperature on average came up to 5.7-6.7°C for air and 9.1-9.8°C in the water, reaching peak values of 8.4°C and 12.3°C, respectively.

Water Analysis

Typical physico-chemical parameters in the wastewater inflow and the aquaculture effluent are given in Table 3. Electric conductivity (EC) decreased from 4,250 ± 750 μS/cm in the effluent of the pre-treatment step T to values around 1.650 ± 250 μS/cm in the outlet of M, corresponding more or less to the dilution effect by recycled water from S (800-1,100 μS/cm, continually increasing during the measuring period). The expected reduction of EC as a result of assimilation processes of plants could not be observed in this step, contrary to the values in the system outflow (step P). Compared to the EC in the wastewater inflow, the values in the effluent were significantly lower than expected by the effect of dilution with freshwater input and evapotranspiration rates (expected: 3,500 μS/cm, measured: 700 μS/cm) (Table 3). A similar picture was observed concerning COD. The values in outflow of M (275 ± 25 g/m^3) compared to T (1,00 ± 200 g/m^3) and the recycled water from S (115 ± 35 g/m^3) could be explained to approximately 90% by the

TABLE 2. Weekly water flows between the treatment steps (pathways) with theoretical retention time corrected for the atmospheric water exchange (AWE) (i.e., precipitation plus evaporation).

Step	Mass flow		Time period (Year 1998)				Mean	Total (m^3)
			18 Jun.- 9 Jul.	10 Jul.- 6 Aug.	7 Aug.- 3 Sep.	4 Sep.- 1 Oct.		
T	system in	m^3/week	3.7	2.7	2.0	3.0	2.3	43.9
	out (to M)	m^3/week	2.9	1.9	0.9	1.7	2.1	30.7
	avg. retention time	weeks	3.4	5.2	10.9	5.8	4.8	
M	in 1 (from T)	m^3/week	2.9	1.9	0.9	1.7	2.1	30.7
	in 2 (from S)	m^3/week	8.7	14.6	10.6	20.3	18.6	232.9
	fresh water in	m^3/week	–	0.9	3.5		–	3.5
	out (to A)	m^3/week	10.2	15.1	10.1	19.6	20.3	240.8
	avg. retention time	weeks	2.0	1.3	2.0	1.0	1.0	
A	in (from M)	m^3/week	10.2	15.1	10.1	19.6	20.3	240.8
	fresh water in	m^3/week	0.6	2.8	2.3	7.8	0.5	44.6
	out (to D)	m^3/week	8.6	15.4	8.7	23.0	21.3	246.7
	avg. retention time	weeks	5.8	3.2	5.7	2.2	2.3	
D	in (from A)	m^3/week	8.6	15.4	8.7	23.0	21.3	246.7
	fresh water in	m^3/week	–	1.5	6.0	–	–	6.0
	out 1 (to S)	m^3/week	5.2	9.8	5.7	13.7	14.6	156.2
	out 2 (to P)	m^3/week	2.8	5.3	3.1	7.4	7.8	84.1
	avg. retention time	weeks	7.6	4.0	6.9	2.9	2.7	
F	fresh water in	m^3/week	3.5	4.5	7.2	6.5	1.0	72.4
	out (to S)	m^3/week	2.8	3.8	5.7	4.5	2.1	60.1
	avg. retention time	weeks	21.5	16.0	10.6	13.4	28.9	
S	in 1 (from F)	m^3/week	2.8	3.8	5.7	4.5	2.1	60.1
	in 2 (from D)	m^3/week	5.2	9.8	5.7	13.7	14.6	156.2
	out (back to M)	m^3/week	8.7	14.6	10.6	20.3	18.6	232.9
	avg. retention time	weeks	8.3	4.7	6.1	3.7	3.4	
P	in (from D)	m^3/week	2.8	5.3	3.1	7.4	7.8	84.1
	system out	m^3/week	1.4	3.7	0.3	3.8	9.3	59.0
	avg. retention time	weeks	25.3	9.5	138.4	9.3	3.8	
Summary	waste water in	m^3/week	3.6	2.7	2.0	3.0	2.4	43.9
	fresh water in	m^3/week	4.1	9.7	19.0	14.3	1.5	126.5
	treated water out	m^3/week	1.4	3.7	0.3	3.8	9.3	59.0
	intern recycling	m^3/week	8.7	14.6	10.6	20.3	18.6	232.9

FIGURE 2. Precipitation and estimated evaporation values (weekly average value) for the whole system and resulting net atmospheric water exchange (AWE) rate for outdoor basins.

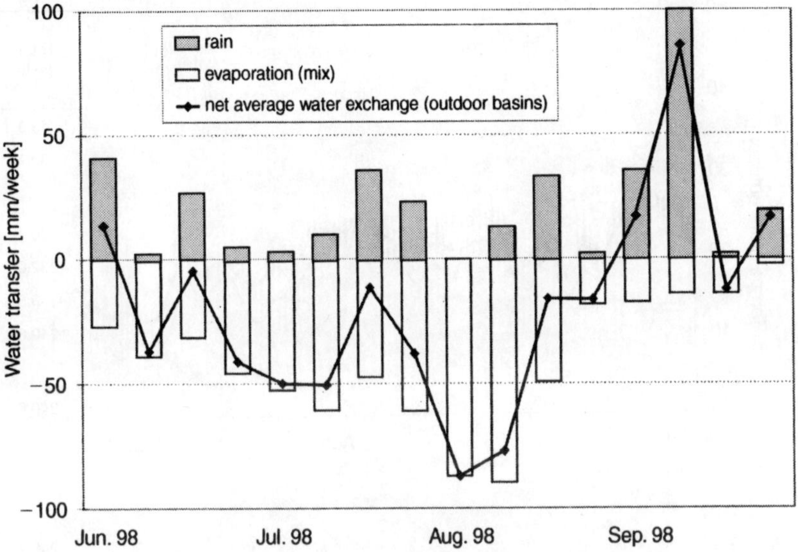

dilution effect. Notwithstanding, the aquaculture outflow was below the value that could be expected due merely to physical effects (95 g/m^3 instead of 1,250 g/m^3, being expected).

pH values in the aquaculture basins and in the outflow ranged generally between 7.5 and 9.0. These values were significantly higher than values in the inflow (6.6-7.5), except for those of step M (6.5-8.0), where the measurements were taken in the last, non-aerated basin. Strong daily fluctuations were observed mainly in basins of step A (and to a lesser extent in D, S, and P), based on the effect of carbon dioxide reduction by activity of primary producers, leading to peak values of pH > 10. Parallel to the pH-peaks, oxygen over-saturation (up to 150%) was often observed. DO values, were low (around 0.6 g O_2/m^3) in the inflowing waste water, but were near the saturation levels in all basins, due to aeration and mixing.

The weekly concentrations of TOC, nitrogen (TAN), and total PO_4-P in every step's outflow, as well as in the system's input (wastewater inflow), are given in Figures 4 to 8.

FIGURE 3. Outdoor and indoor (greenhouse covered) temperature curve of water (representative basin) and air as three-day moving average values with daily minima and maxima measurement points.

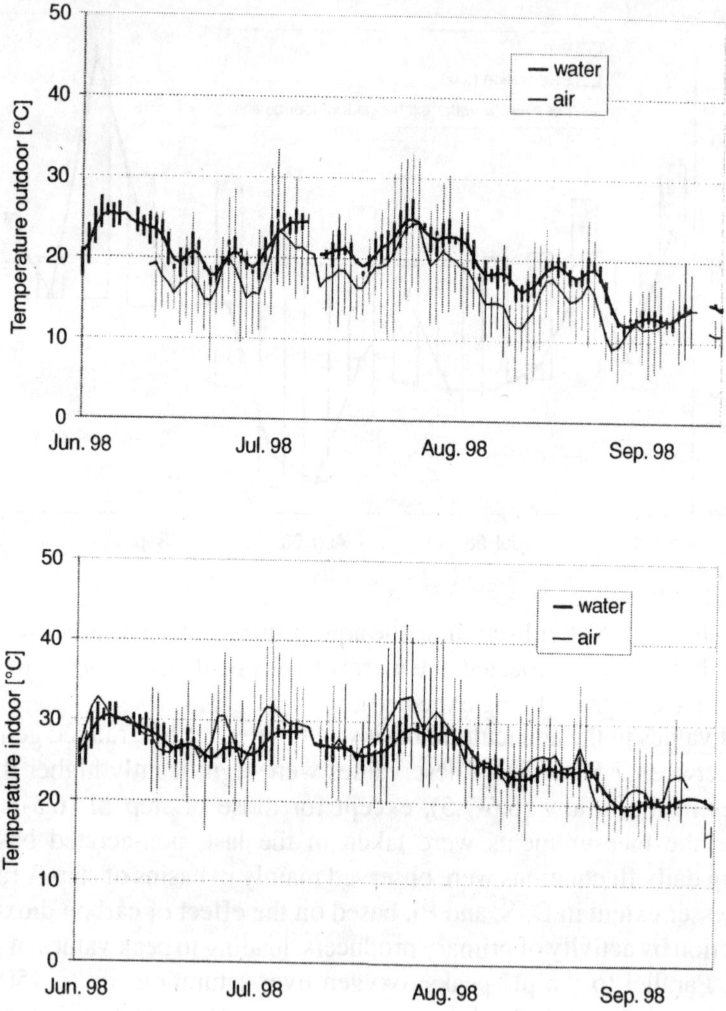

TABLE 3. Monthly average values of main physico-chemical characteristics of system inflow (waste water) and discharge (after step P).

Parameter			Time period (Year 1998)				Mean
			18 Jun.- 9 Jul.	10 Jul.- 6 Aug.	7 Aug.- 3 Sep.	4 Sep.- 1 Oct.	
Total organic carbon (TOC)	inflow	(g/m^3)	594	637	739	711	670
	discharge	(g/m^3)	18	36	38	62	39
Total Nitrogen (TN)	inflow	(g/m^3)	419	260	160	181	255
	discharge	(g/m^3)	6.1	4.8	4.1	4.4	4.9
Ammonium- nitrogen (NH_4-N)	inflow	(g/m^3)	106	77	109	90	95
	discharge	(g/m^3)	0.04	0.04	0.04	0.06	0.05
Nitrate-nitrogen (NO_3-N)	inflow	(g/m^3)	287	169	104	61	155
	discharge	(g/m^3)	4.4	1.4	0.81	1.2	2.0
Total phosphate-phosphorus (PO_4-P)	inflow	(g/m^3)	82	37	48	43	52
	discharge	(g/m^3)	0.25	0.12	0.10	0.16	0.16
Electric conductivity (EC)	inflow	(µS)	4,640	3,830	3,540	4,880	4,220
	discharge	(µS)	573	678	715	863	707
Chemical oxygen demand (COD)	inflow	(g/m^3)	1,380	1,580	1,830	1,960	1,690
	discharge	(g/m^3)	39	100	108	124	93
pH	inflow	(–)	6.8	6.8	7.1	7.5	7.1
	discharge	(–)	8.4	8.4	8.6	8.6	8.5
Dissolved oxygen (DO)	inflow	(g/L)	0.64	1.4	0.33	0.20	0.63
	discharge	(g/L)	9.8	9.9	8.7	8.8	9.3

PRODUCTION, BIOMASS ANALYSIS, AND ELIMINATION RATES

A total of 2,150 kg fresh weight (FW) of biomass was harvested during the 16-week period (137 kg/week); approximately 97% (i.e., 2,080 kg) were floating macrophytes. The rest of the harvested biomass were *Daphnia* (67 kg) and fish (4.4 kg). Compared to total yield, most macrophytes were harvested in steps M, T, D, and P, i.e., 53%, 20%, 14%, and 9%, respectively. A majority (55%) of the plant yield was produced as water hyacinth, 36% as water lettuce, and 6.6% as duckweed. The rest (2.4%) were different macrophytes (tiger grass, *Hygrorhiza* sp.; reussia, *Reussia* sp.; anchored water hyacinth, *Eichhornia azurea*;

FIGURE 4. Total organic carbon (TOC) in system inflow and in the outflow of the different steps.

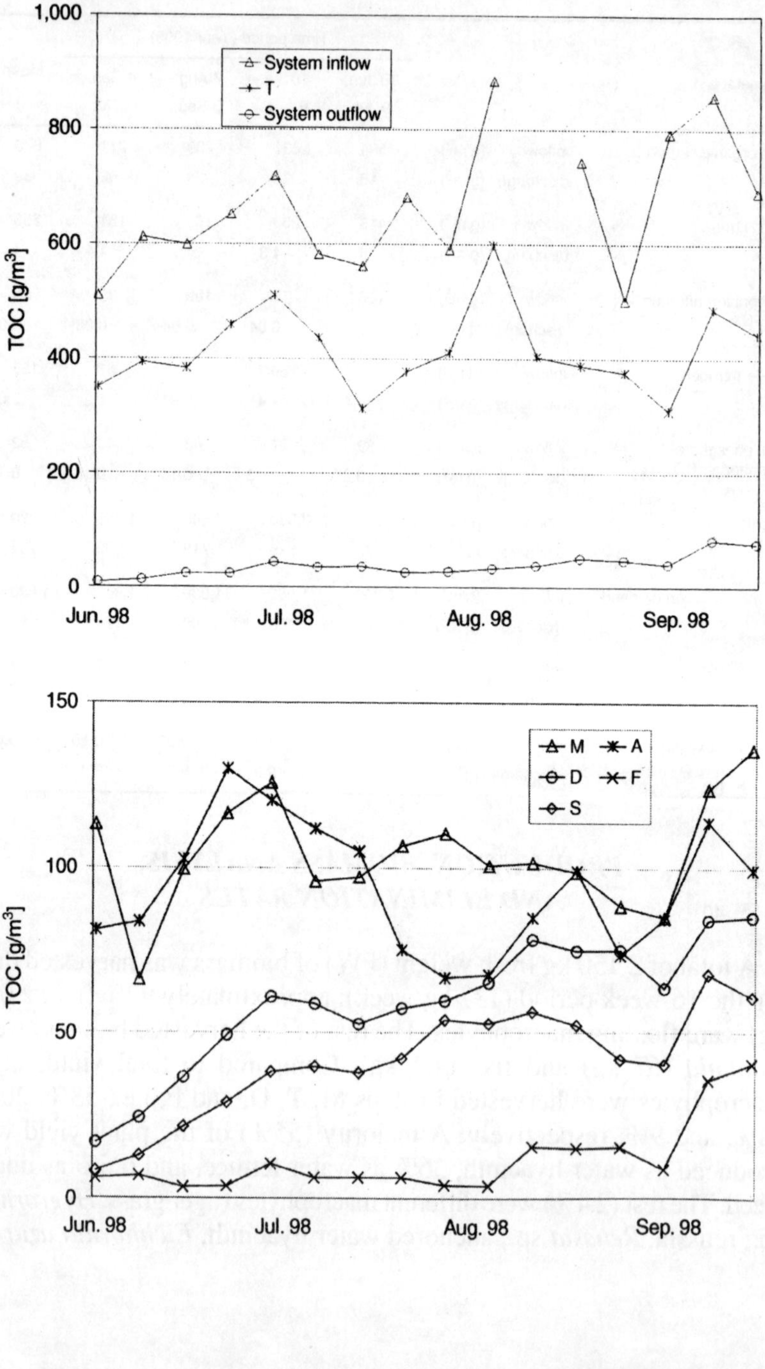

FIGURE 5. Total nitrogen in system inflow and in the outflow of the different steps.

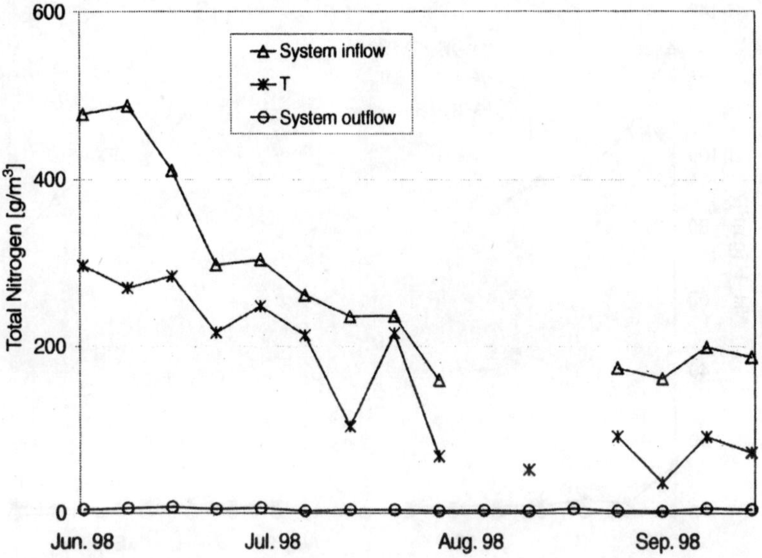

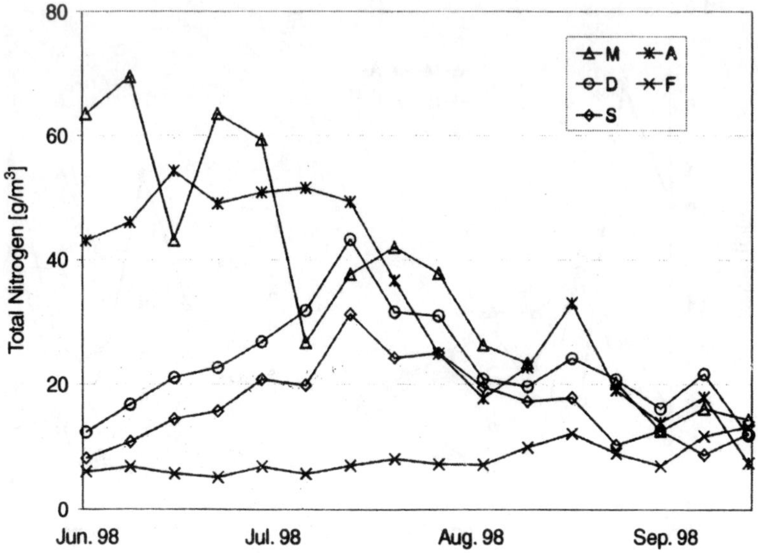

FIGURE 6. Ammonia-nitrogen (NH_4-N) in system inflow and in the outflow of the different steps.

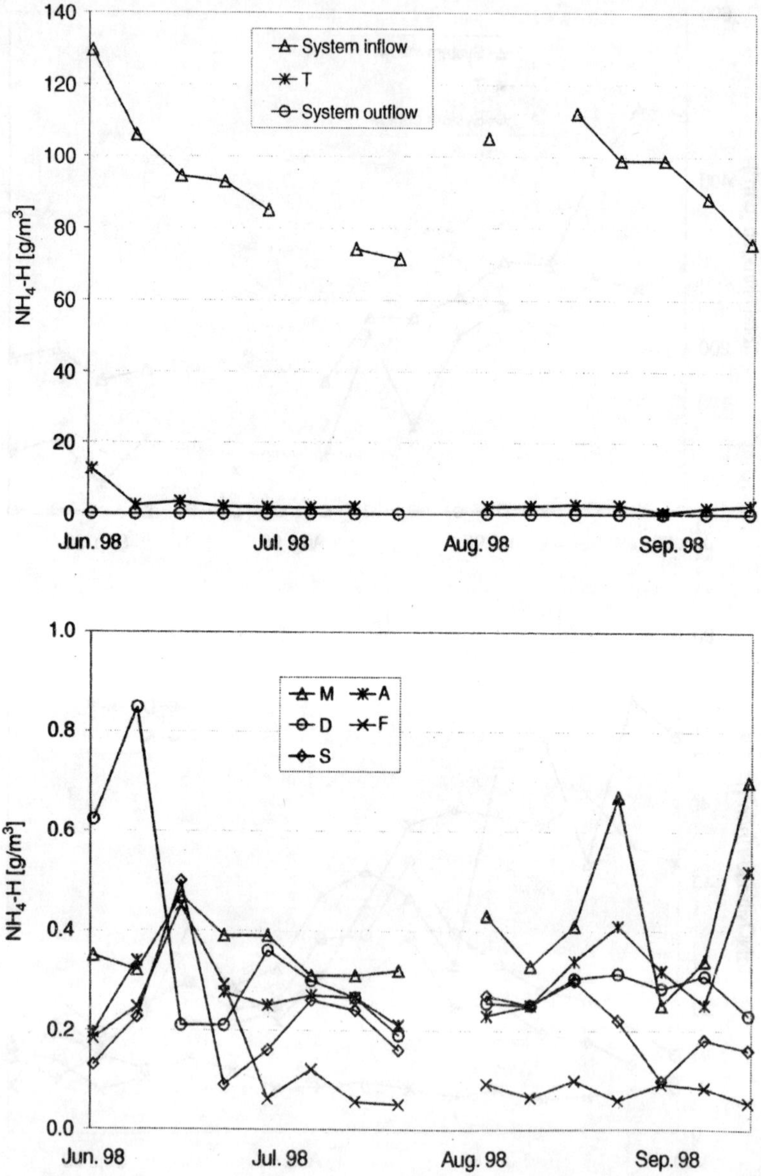

FIGURE 7. Nitrate-nitrogen (NO_3-N) in system inflow and in the outflow of the different steps.

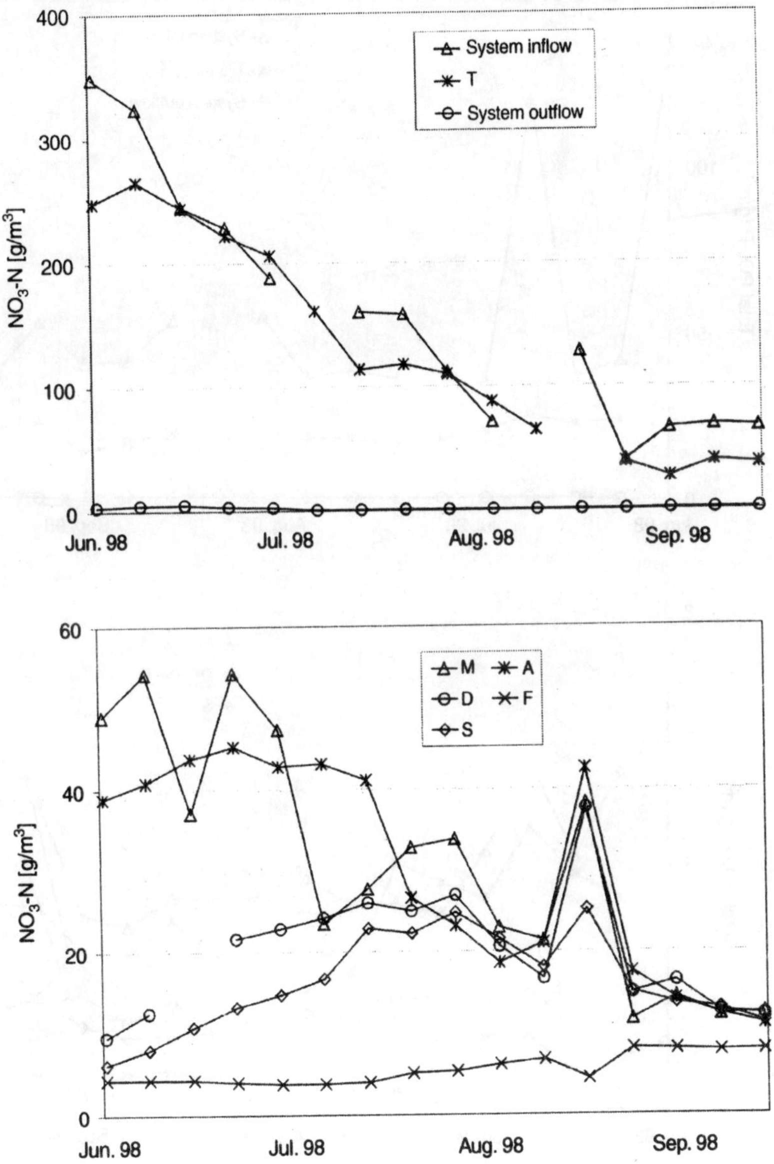

FIGURE 8. Total phosphate-phosphorus concentration in system inflow and in the outflow of the different steps.

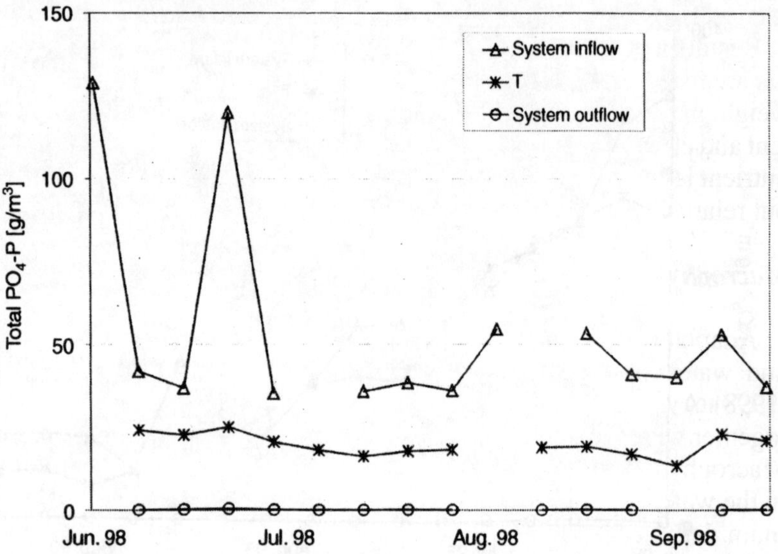

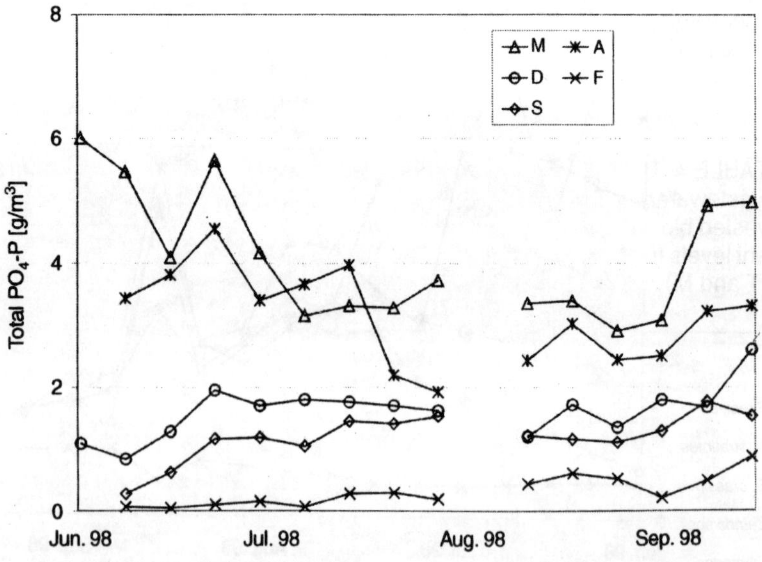

brazilian watermilfoil, *Myriophyllum aquaticum*). The harvested macrophyte biomass (87.9%) was either sold as ornamental plants or recycled to the fermentation process. The rest (6.1%, mainly *Lemna* spp.) was used as fish feed in step F or to start plant culture in different basins.

Results from biomass analysis for harvested macrophytes and *Daphnia* are listed in Table 4. Fresh macrophyte biomass contained more dry weight in basins with rather low nutrient levels (e.g., S and P). Ash content and elemental composition of macrophytes also varied due to water nutrient levels: dry weight contained less ash, nitrogen, and phosphorus but relatively more carbon in basins with low nutrient levels.

Macrophytes

Adaptation, growth, and elimination performance of water hyacinth and water lettuce under different conditions were studied by Todt (1998). Average and maximum specific growth rates are given in Table 5, together with calculated elimination rates for nitrogen and phosphorus. Macrophytes' daily growth rate depended on nutrient levels and ratios in the water on one side and on initial plant density on the other. Maximum daily macrophyte growth was observed in the basins of steps M and T, being between 20-25%/day at initial standing crop of 2,000 g(FW)/m^2 and 3,000 g(FW)/m^2 for water hyacinth and water lettuce, respectively. The populations grew to maximum densities of 15,000 g (FW)/m^2 (water hyacinth) and 16,500 g (FW)/m^2 (water lettuce), although the doubling rate decreased below the indicated maximum val-

TABLE 4. Biomass composition of some macrophytes and of *D. magna* from wastewater-fed aquaculture Otelfingen. For macrophytes, two values for harvested biomass composition are given: for biomass from basins with low nutrient levels (D, F, S, and P), and for biomass from basins with high nutrient levels (T and M).

	Dry weight (DW)		Composition of biomass dry weight							
	(% FW)		ash (% DW)		C (% DW)		N (% DW)		P (% DW)	
Species	low	high	low	high	low	high	low	high	low	High
P. stratiotes	5.32	4.44	21.1	25.6	34.7	32.9	2.00	2.29	0.66	0.77
E. crassipes	5.80	4.36	16.3	22.0	36.7	33.6	1.45	3.50	0.26	0.89
Lemna spp.	4.25	4.98	18.8	20.4	37.4	35.5	2.08	4.27	0.78	1.61
D. magna (average)	4.26		23.4		36.5		6.16		1.17	

ues at such densities (i.e., 10-15%/day in the range of 5,000 g (FW)/m^2 and below 5%/day for densities over 7,000-8,000 g (FW)/m^2).

Zooplankton

Zooplankton (i.e., *Daphnia*) was harvested periodically from steps D, S, P, A, and was used as fish feed in step F. The total harvest was on average 4.2 kg (FW)/week, and had maximum specific growth rate of 0.26 and 3.5 g (DW)/m^2/day in the basins maintained for *Daphnia* production (Table 6) (Staudenmann and Junge-Berberovic 2000).

Fish

During the experimental period in the eight basins of step F, common carp, tench, and Nile tilapia together gained 4.36 kg (FW) (Graber

TABLE 5. Average and maximum specific growth rates of macrophytes observed in different nutrient concentration levels (data adapted from Todt 1998).

		Nutrient level (g/m^3)		Growth rate (g DW/m^{2}d)		N assimilation (g N/m^{2}d)		P assimilation (g P/m^{2}d)	
Step	Species	N	P	average	max.	average	max.	average	max.
T	*E. crassipes*	100-300	15-25	–	–	–	–	–	–
	P. stratiotes	100-300	15-25	12.6	24	0.29	0.55	0.10	0.18
M	*E. crassipes*	20-70	3-6	16.6	42	0.58	1.5	0.15	0.37
	P. stratiotes	20-70	3-6	12.7	14.9	0.29	0.34	0.10	0.11
D/P	*E. crassipes*	3 -30	0.1-2.0	1.4-5.2	7.2	0.02-0.08	0.10	0.01	0.02
	P. stratiotes	3-30	0.1-2.0	2.5-2.6	3.8	0.05	0.08	0.02	0.03

TABLE 6. Total crustacean zooplankton (mainly *Daphnia magna*) harvest during four experimental periods.

	Time period (Year 1998)				Total
Step	18 Jun.-9 Jul.	10 Jul.-6 Aug.	7 Aug.-3 Sep.	4 Sep.-1 Oct.	(16 weeks)
M (kg FW)	40	0	2,892	556	3,488
D (kg FW)	4,682	5,833	2,672	4,609	17,796
S (kg FW)	5,885	6,713	5,207	2,555	20,360
P (kg FW)	6,440	6,150	4,885	7,998	25,473
Total (kg FW)	17,047	18,696	15,656	15,718	67,117

1998). Total diet consumed was 63.4 kg FW of *Daphnia* and 115 kg FW of duckweed. Conversion rates, along with diet composition are indicated in Table 7. The average conversion of the three species thus equaled 2.2. Tench showed a similar conversion efficiency as common carp, whereas Nile tilapia incorporated the given feeds the best. Under favorable conditions, growth rates of Nile tilapia and common carp were comparable, while tench grew only half as fast. When the feeding level was lower, Nile tilapia showed the best growth rate, followed by common carp and tench. Under the climatic conditions given, culturing Nile tilapia can impose severe problems. As a result of a decrease in water temperature from 22°C down to 17°C within five days at the end of June 1998, high mortality occurred (93 out of 105 fish).

Algae

In pilot scale experiments in 200-L basins placed *in situ*, the effect of inoculation of the axenically cultivated strain of green alga, *Scenedesmus acuminatus*, into the water of module A1 was tested. After two weeks of open-air cultivation, there was no difference in the species' composition and abundance between the inoculated and non-inoculated (Figure 9). This means that disproportionately large amounts of inoculum would be needed to overcome the spontaneous colonization of the basins. Therefore, instead of culturing the algae in a batch mode, with peri-

TABLE 7. Fish growth rates, average food conversion rate, and diet composition during experimental period (115 days) in mono- and polyculture.

Species	Culture type	Fish growth (%/day)	Food conversion[1]	*Daphnia* (% BW/day)	Duckweed (% BW/day)
Common carp	monoculture	0.5	2.2	6.9	11.2
	polyculture	0.6- 2.4	1.3-2.3	5.6-5.9	6.9-7.5
Tench	monoculture	0.23	2.3	6.3	2.5
	polyculture	0.1-1.0	1.3-2.3	5.6-5.9	6.9-7.5
Nile tilapia	monoculture	--[2]/1.2	1.6		
	polyculture	1.8-2.3	1.3	5.9	6.9
Silver carp	monoculture	0.5-0.6	--	--	--
	polyculture	0.3-2.3			

[1] Total ingested food (kg DW)/fish growth (kg FW).
[2] Tilapia in one monoculture system died after 17 days because of cold water temperatures.

FIGURE 9. The effect of inoculation with *Scenedesmus acuminatus* (species number 3) on species composition in experimental basin.

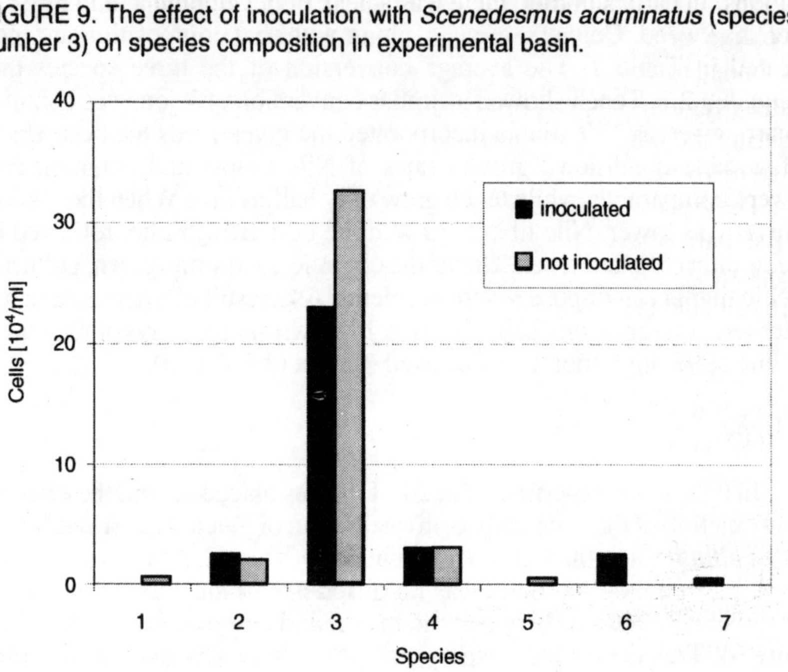

odic inoculation, the algae were cultivated in semi-continual culture with manipulation of retention times, dilution rates, and mixing methods instead.

The indoor basins contained at all times extremely low algal concentrations (under the detection limit of 2,500 cells/mL). The reasons for this were probably reduced light availability (quality and quantity) and periodically very high water temperatures in the greenhouse. After August 1, 1998 the algae production in the four indoor basins of module A1 was discontinued. Approximately 60 distinct forms of microalgae were encountered during the growth season, and 28 could be identified to species level. Approximately 70% of the encountered species had diameters smaller than 50 μm and can therefore be considered ingestible by water flea, especially species *D. magna* (Knisely and Geller 1986). At the beginning of the season a bloom of indigestible filamentous algae was observed. It was possible to avoid this by reduction of water retention times in algae basins.

Seasonal development of algae population in the last basin of module A2, which was used as a charging basin for step D, showed large fluctuations. In early summer, there was a peak of development, and the total volume of algae cells reached over 40 × 10^3 mm^3/L, which corresponded to 100 × 10^4 cells/mL. This was followed by gradual decrease and irregular fluctuations toward the autumn. Average cell numbers during the months of August and September were approximately 50 × 10^4 cells/mL and 10 × 10^4 cells/mL, respectively.

There were 17 algae species that could be termed dominant. In the first half of the season, the prevalent species was single cell alga *Ankistrodesmus falcatus* (up to 80 × 10^4 cells/mL); in late summer 4 to 16 cell-colony forming *Scenedesmus* species (*Sc. platydiscus*, *Sc. arcuatus*, and *Sc. magnus)* predominated (approximately 8 × 10^4 cells/mL).

Table 8 shows some important indices of phytoplankton community structure within steps with different nutrient concentrations and different species composition. It seems that the number of species, as well as average colony size (cells per colony) and total biomass, depended more on the species stocking of a particular step than on the nutrient concentration. In basins with zooplankton filter-feeders the average colony size increased, and in basins with filter-feeding silver carp the average cell size decreased drastically. Standing crops of algae tended to be higher in basins of filter-feeding silver carp, but the species richness tended to be higher in polycultures.

TABLE 8. Some indices describing phytoplankton community on 4 September 1998 in aquaculture modules with different stocking pattern and nutrient levels.

Step	Stocking	Nitrogen (mg/L)	Phosphorus (mg/L)	Algae species (N)	Cells/ colony	Standing crop volume (mL/m^3)
A	Algae	33.1	3.02	10	4.1	32.7
D	*Daphnia*	24.4	1.74	15	11.4	40.7
F2	Silver carp monoculture	8.2	0.17	11	1.0	35.6
F3	Fish polyculture with Nile tilapia	12.2	0.61	12	3.6	42.8
F4	Fish polyculture without Nile tilapia	7.8	0.50	17	15.2	105.6
P	Polyculture: Macrophytes, *Daphnia*, Nile tilapia	17.3	0.87	18	6.7	29.3
S1	Polyculture: *Astacus*, common carp, tench, Nile tilapia. No macrophytes.	13.2	0.50	19	21.3	222.2

BALANCE AND SYSTEM PERFORMANCE

Referring to the mass flows (water), only 30% of total waste- and freshwater input was released as effluent from the system. Together with all the rain input, the rest was lost by evapotranspiration to the atmosphere, corresponding to 90% of total freshwater input. Water exported by biomass harvest could be neglected (approximately 1%). The target of using only as much freshwater as necessary to compensate water loss via atmosphere was mostly achieved. It remains to be answered whether it would make sense to further reduce water outflow rate, eventually targeting a non-effluent system, and to avoid freshwater input altogether. In such a case, a strategy for dealing with precipitation and/or wastewater storage would have to be developed. Regarding the nutrient balance, it would theoretically be conceivable to increase retention time to allow the macrophytes to assimilate more of the nutrient load. On the other hand, it remains unanswered if and how rapidly side effects such as accumulation of toxic and non-degradable compounds, exhaustion of essential substances, and distortion of nutrient ratios would appear, affecting the system's performance or products' quality.

Compared to wastewater inflow, 98% and 99.5% of the total nitrogen and phosphorus load, respectively, was removed from the water (Table 9). Compared to water flowing in, the concentrations in the effluent were lower for total nitrogen (98%), total phosphate (99.7%), TOC (94%), COD (94%), ammonium (> 99.9%), nitrate (99%), and EC (83%). Nitrification was excellent, regarding both the ammonium elimination rate and the value in the plant's effluent. The average ratio between NH_4-N and total nitrogen declined from 33% in the inflow to 1.9% after T, and 1.0% in the final effluent (after P). However, the absolute concentrations in the effluent would probably be in violation of Swiss law for some parameters concerning release into a natural water body (Table 10). The reason for that is found in the much higher loads in the input compared to communal waste water on one hand and in the massive evapotranspiration rates bringing along an additional increase of absolute concentrations by up to factor 3.3 on the other. An additional problem was related to the dark coloring of the effluent due to dissolved persistent organic compounds (mostly humic acids), typical for effluents from fermentation processes and similar to brown water lakes. Further research is necessary to examine the possibilities of enhancing the degradation of these compounds.

TABLE 9. Nutrient pathways and balance (nitrogen and phosphorus) for steps T & M, for the rest of the steps, and summarized for the whole system.

Steps	Path	Nitrogen		Phosphorus	
		g/week	%[1]	g/week	%[1]
	Waste water inflow	700	100	136	100
T & M	Additional input to M via recycling loop	240	34	17	13
	Biomass harvest T and M	136	19	36	26
	Denitrification/sedimentation [2]	300	43	33	24
	Outflow M	510	73	84	61
Others	Biomass harvest	40	6	11	8
	Denitrification/loss[2]	240	34	58	43
	Discharge (after P)	17	2	0.7	0.5
Total system	System inflow	700	100	136	100
	Biomass harvest	176	25	47	35
	Denitrification/sedimentation [2]	508	73	89	65
	System outflow	17	2	0.7	0.5

[1] Compared to the system input.
[2] Unaccounted for.

Assimilation of Nitrogen and Phosphorus

Major nutrient elimination rates occurred in the two first steps of the plant (i.e., pre-treatment step T and macrophyte growing step M). Here, average biomass growth by floating macrophytes (i.e., 12.5 g(DW)/m^2/day) accounted for elimination of 0.32 g N/m^2/day, and 0.086 g P/m^2/day, and denitrification averaged 0.7 g N/m^2/day. Although these two steps together only amounted to 17% of the system's total surface and 15% to the total retention time, their contribution to nutrient elimination averaged 62% for nitrogen and 50% for phosphorus. Regarding moreover the maximum growth and nutrient elimination rates observed, the superior status of steps M and T compared to the rest of the system's elimination performance is obvious.

While approximately 25% and 35% of the N and P elimination, respectively, was found in the harvested biomass, almost all of the rest was unaccounted for. Most likely, a major part of the "lost" nitrogen was eliminated from the system by denitrification as mentioned already. Phosphorus missing in the balance was presumably accumulated, most likely in the form of particulate phosphorus compounds in

TABLE 10. The elimination rates and outflow parameters of the Otelfingen Aquaculture Pilot Plant compared with Swiss law requirements for introducing treated communal waste water released by small treatment plants (i.e., < 10,000 inhabitants) into a natural water body (Anonymous 1998).

Parameter	Swiss law requirements		Otelfingen Aquaculture effluent	
	Elimination rate (% of inflow)	Concentration[1] (g/m^3)	Elimination rate (% of inflow)	Concentration (g/m^3)
SS	> 90	< 20 (−50)	n.n.	n.n.
BOD_5	> 90	< 20 (−40)	n.n.	n.n.
COD	--	--	94.5	93
DOC (in relation to TOC in influent)	> 85[2]	< 10 (−20)[2]	94.2[4]	39[4]
Total N	--[3]	--[3]	98.1	4.9
NO_3-N	--[3]	--[3]	98.7	2.0
NH_4-N	> 90	< 2	99.9	0.05
NO_2-N	--	< 0.3	n.n.	< 0.1
Total PO_4-P (for sensitive water bodies)	> 80	< 0.8	99.7	0.16

[1] Average requirements (maximum concentration not to be exceeded).
[2] Only for plants with more than 2,000 inhabitants connected.
[3] The state (Kanton) can settle individual maximum loads and concentrations for nitrogen release.
[4] TOC instead of DOC was determined.

different basins. The system, having taken up operation in May 1998, was obviously not yet in a steady state regarding both nutrient and TOC concentrations in the steps after A and the missing total phosphate in the nutrient balance. This is understandable, taking into account the rather long average retention time of the aquaculture system (especially steps D and P with high evapotranspiration rates during summer).

Macrophytes

Water lettuce had fastest and most luxuriant growth (leaves' length up to 0.2 m) in step T with highest loads of waste water. Water hyacinth developed best in step M (height up to 0.8 m). In step T, water hyacinth turned light green and white quickly after introduction, although continuing to grow at the same biomass growth level as water lettuce until July. Since preliminary experiments showed no problems for water hyacinth at the given nutrient, EC, and TOC levels and since it grew very well in the following steps, it must be assumed that either some compounds in the water inhibited or limited its development or that some essential nutrients were not available for water hyacinth. On the other hand, the perfect development of water lettuce in the first basins indicated no general plant growth inhibition (toxicity) from the waste water, corresponding also to preliminary labor-scale experiments (Staudenmann and Junge-Berberovic, unpublished data). In the steps D, F, S, and P with low nutrient levels, both water hyacinth and water lettuce developed little height and small leaf diameter, together with large, dense underwater roots. Both roots and swim corps of water hyacinth turned white and violet, whereas the color of the leaves of both plants was generally lighter in lower nutrient concentration basins. More experiments are needed to investigate the factors influencing the morphology and the development of these aquatic macrophytes in temperate climates.

The growth and elimination rates under optimum conditions were comparable to data from previous projects (Heeb 1995) and literature (Reddy and DeBusk 1987; Reddy and Smith 1987, Záková et al. 1995; Staudenmann et al. 1996). Generally it seems that maximum elimination rates that can be achieved by a macrophyte system do not exceed 1 gN/m^2/day, and ca. 0.3 g P/m^2/day, which are also nearly the maximum observed rates in this study (Table 5).

Since the average net elimination rates in relation to the surface of the whole pilot plant were 0.072 g N/m^2/day and 0.019 g P/m^2/day, the nutrient assimilation could theoretically be enhanced up to factor ten, if the whole surface were covered with macrophytes. In this case, the flow scheme should be adjusted so as to obtain higher nutrient levels in the later steps and thus promote plant growth. This, on the other hand, would probably imply fewer consumer organisms in the system. Another possibility to enhance growth rates would be a manipulation of N:P ratios. The P limitation was increased along the system flow. Whereas the N:P ratio was approximately 26:1 in step M, it was 59:1 in step S. It would be conceivable to add phosphate-rich waste water from other sources at this stage in order to improve the nutrient elimination rates of the system.

Tropical macrophytes could be produced successfully during the summer in Swiss environmental conditions. In contrast to tropical countries, where these plants are pests, they achieve substantial market value in Switzerland, between US$ 2.00-3.50 per plant. Their high growth rate, as well as their tolerance to high nutrient levels, put them at an advantage over native crops during summer months. In basins with moderate nutrient levels (such as *Daphnia* or polishing steps) and during transition months (spring, autumn) the growth of tropical species was comparable to that of native plants. In winter, tropical species could survive only in a heated and lighted glasshouse and were clearly at a disadvantage compared to native species.

Zooplankton

Zooplankton population growth and harvest rates depended on algae growth and were therefore influenced by fluctuations and low performance in algae counts. A dense *Daphnia* population developed in several basins of step M. These basins featured high nutrient and TOC concentrations, low oxygen levels, as well as no microalgae population. *Daphnia* living in these basins had intensive red color, which indicated higher hemoglobin production for better oxygen supply, an adaptation to low oxygen levels (Fox et al. 1950; Chandler 1954; Kobayashi and Hoshi 1982). Obviously, *Daphnia* in this step grazed on suspended (organic) particles and bacteria from the waste water, implying an unexpected potential for direct elimination of organic suspended matter.

Fish

Growth rates of some of the fish reached moderate levels. Maximum annual yields in basins of step F (extrapolated as double summer rate) were up to 4,400 kg/ha/year and are roughly comparable to other extensive fish cultures. Yields between 6,400 and 19,000 kg/ha/year were reported in Chinese polyculture by Yao (1993), and yields between 900 and 9,350 kg/ha/year were reported by Jana (1998) for Indian polycultures in Calcutta wetlands. It is important to note that the experimental system in Otelfingen was not designed to optimize fish growth but to allow for investigation of different fish species and communities. Fish growth was partially food-limited due to poor *Daphnia* growth rates. With other fish (grass carp) that made complete use of available *Lemna*, higher yields could be attained.

The effectiveness of fish production in wastewater-fed aquaculture in temperate climate must be questioned both for economic and for energy and resource conversion efficiency reasons. Even if zooplankton, and hence fish production, were improved, nutrient elimination by fish harvest would remain about a factor 10 lower compared to the macrophyte modules (presently approximately 100 times lower). Moreover, the demand for fish protein as a recycling product in Switzerland and in Europe is small, and marketing potential meager. Therefore, we see the ecological role of fish in wastewater-fed aquaculture in temperate climates primarily as an accompanying, system-stabilizing element. Some fish species with clearly defined nutrition demands could play a regulatory role in the system, to support the growth of target organisms, as opposed to non-target organisms, e.g., Nile tilapia, and common carp consuming undesired zooplankton, floating plants, and snail eggs in algae basins; filter feeding carps eliminating large algae colonies and particulate matter. In tropical regions with higher growth rates, low labor costs, and high demand for protein, the carp production could probably reach commercial levels.

Algae

The semi-continual culture of microalgae in step A was only partially successful for several reasons. Algae growth was subject to fluctuations as described in the classical Plankton Ecology Group (PEG) Model of many lakes (Sommer et al. 1986). It is possible that the growth was partly limited by light, as the water was colored with humic substances. For laboratory cultures of *Daphnia*, algae suspensions with 5×10^4 cells/mL are usually used (Ebert 1991). Therefore, in spite of possible light limitation, for most of the season step A provided enough algae cells to sustain growth of *Daphnia* population. The mean and maximum standing crop of algae in the last basin of module A2 were 0.08 and 0.22 kg (FW)/m^2 (calculated from volume). This means that maximal elimination rates for N and P were 0.07 and 0.01 g/m^2/day, respectively (unpublished data).

Algae biocoenosis in any lake is a function of temperature, light availability, nutrient levels, and presence of predators (Sommer 1987). All these also play a role in determining the species composition, colony size, and cell numbers of algae population in basins of wastewater-fed aquaculture. Yao (1993) reported higher phytoplankton primary production in ponds with filter-feeding fish as compared to non-filter-feeding fish. This is to some extent corroborated with our results. Our results show that for successful outdoor culture of dense algae suspension suitable for zooplankton feed, it is important to observe several factors, but water retention times and appropriate stocking (filter-feeding fish) seem to be more sensitive parameters than the nutrient concentration.

GENERAL ASPECTS AND PERSPECTIVE

This study's pilot plant lay-out and running scheme involved much labor and material investment. This was due largely to the research and monitoring requirements. But also methods of both macrophyte harvest (meeting quality standards for selling as ornamental plants) and zooplankton culture have to be optimized. The latter problem was mainly due to of the exclusion of the fish basins from the (waste-) water main course; the *Daphnia* had to be collected with nets manually. All attempts of automated, continual harvest of (living) zooplankton using air-lift pumps failed. Should that be optimized, a great potential for increased commercial application can be expected, since living (or fresh) zooplankton has a high value on the European market.

No investigations have been made concerning the energy consumption for aeration and water pumping, as well as in general about the economic benefit of the plant. These very important aspects for assessing the ecological and economical feasibility will be examined in another experiment and should provide reference parameters for the comparison of wastewater-fed aquaculture with conventional treatment systems or biomass production sites. Such a system can only be a sustainable option in Switzerland if either the profit from the produced goods is relevant compared to the costs (Switzerland's first priorities are labor and space), or it has other advantages over the well-established method on a long-term basis, such as a better degradation or elimination performance for specific parameters not fully in our concern yet (e.g., degradation of estrogenic substances or direct and efficient reuse of nutrients). Other research questions to be examined include winter performance, accumulation of undesired compounds in the biomass, and use of extra heat and carbon dioxide from the fermentation process.

It can be summarized that the approach of wastewater-based biomass production has several potential advantages over conventional systems, especially concerning nutrient assimilation and recycling aspects. When assessing the pros and cons for such a system, it is important to balance the disadvantages, such as increased space and labor demand, with the advantages (recycling aspect and income generation). Wastewater-fed aquaculture should be regarded as an integrated production plant rather than a wastewater disposal site, and the processes should be compared to other biomass production and/or nutrient consuming systems.

ACKNOWLEDGMENTS

The authors thank W. Schmid AG, Glattbrugg Switzerland, particularly Mr. W. Schmid, owner and director of the company, for their trust and investment in the project. Without their colleagues, students, and research assistants from the University of Applied Sciences, Waedenswil and other institutes this project would not have been possible. The authors especially thank Jean Bernard Baechtiger and Andreas Graber for their continual support. They also thank the Swiss Federal Commission for Technology and Innovation for the financial research grant (No. KTI-3487.2).

REFERENCES

Anon. 1998. Gewaesserschutzverordnung (GSchV) vom 28. Oktober 1998 (in Kraft am 1. Januar 1999). EDMZ, Swiss Government, Bern, Switzerland.

Anon. 1992. Agenda 21: The United Nations Programme of Action from Rio. UN Publications, United Nations, New York, New York.

Berberovic, R. 1990. Nutritional state of *Daphnia* in a temperate large lake as measured by elemental composition of biomass. Oecologia 84:340-350.

Costa-Pierce, B.A. 1998. Preliminary investigation of an integrated aquaculture-wetland ecosystem using tertiary-treated municipal wastewater in Los Angeles County, California. Ecological Engineering 10:341-354.

Chandler, A. 1954. Causes of variation in the haemoglobin content of *Daphnia* (Crustacea: Cladocera) in nature. Proceedings of Zoological Society London 124: 625-630.

DeBusk, T.A., and J.H. Ryther. 1987. Biomass Production and Yields of Aquatic Plants. Magnolia Publishing Inc., Orlando, Florida.

DeBusk, T.A., L.D. Williams, and J.H. Ryther. 1983. Removal of nitrogen and phosphorus from wastewater in a water hyacinth-based treatment system. Journal of Environmental Quality 12:257-262.

Ebert, D.M. 1991. Phenotypic Plasticity, Developmental Constraints and Genetics of *Daphnia magna* Straus. Doctoral dissertation, University of Basel, Basel, Switzerland.

Edelmann, W., H. Engeli, M. Glauser, H. Hofer, Y. Membrez, J. Meylan, and J.P. Schwitzguébel. 1993. Vergaerung von haeuslichen Abfaellen und Industrieabwässern. Bundesamt für Konjunkturfragen, Impulsprogramm Erneuerbare Energien, Bern, Switzerland.

Fox, H.M., B.M. Gilchrist, and E.A. Phear 1950. Functions of haemoglobin in *Daphnia*. Nature 166:609-610.

Graber, A. 1998. Fische in einer Aquakultur zur Naehrstoffrueckgewinnung aus Prozesswasser einer Kompogas-Anlage. Diplom thesis, Swiss Federal Institute of Technology, Zürich, Switzerland.

Greenway, M., and A. Woolley. 1999. Constructed wetlands in Queensland: Performance efficiency and nutrient bioaccumulation. Ecological Engineering 12:39-55.

Gumbricht, T. 1993. Nutrient removal capacity in submersed macrophyte pond systems in a temperate climate. Ecological Engineering 2:49-61.

Guterstam, B. 1991. Ecological engineering for wastewater treatment: Theoretical foundations and practical realities. Pages 38-54 *in* C. Etnier and B. Guterstam, eds.

Ecological Engineering for Wastewater Treatment. Conference Proceedings of the 1st International Conference on Ecological Engineering for Wastewater Treatment Bokskogen, Gothenburg, Sweden.

Guterstam, B. 1996. Demonstrating ecological engineering for wastewater treatment in a Nordic climate using aquaculture principles in a greenhouse mesocosm. Ecological Engineering 6:73-97.

Heeb A. 1995. Efficiency of Wastewater Purification by Aquatic Macrophytes. Diplom thesis, University of Zurich, Zurich, Switzerland.

Iqbal, S. 1999. Duckweed Aquaculture. Potential, Possibilities and Limitations for Combined Wastewater Treatment and Animal Feed Production in Developing Countries. SANDEC report 6/99. EAWAG, Duebendorf, Switzerland.

Jana, B.B. 1998. Sewage-fed aquaculture: The Calcutta model. Ecological Engineering 11:73-85.

Junge-Berberovic, R., and J. Staudenmann. 1998. Wastewater-fed aquaculture in temperate climate: Possibilities and limits of nutrient recycling. Page 187 *in* A. Moser, ed. The Green Book of Eco-Technology. Strategy Report from 4th International Ecological Engineering Conference, University of Graz, Austria.

Knisely, K., and W. Geller. 1986. Selective feeding of four zooplankton species on natural lake phytoplankton. Oecologia 69:86-94.

Kobayashi, M., and Hoshi, T. 1982. Relationship between the haemoglobin concentration of *Daphnia magna* and the ambient oxygen concentration. Comparative Biochemistry Physiology 72A:247-279.

Landolt, E. 1996. Duckweeds (Lemnaceae), morphological and ecological characteristics and their potential for recycling of nutrients. Pages 289-296 *in* J. Staudenmann, A. Schoenborn, and C. Etnier, eds. Recycling the Resource. Proceedings of the 2nd International Ecological Engineering Conference, Waedenswil. Transtech Publications, Uetikon, Switzerland.

Lange, J., and R. Otterpohl 1997. Abwasser-Handbuch zu einer zukunftsfähigen Wasserwirtschaft. Mallbeton GmbH, Pfohren, Germany.

Mueller, B. 1999. Eignet sich der Edelkrebs (*Astacus astacus*) für die Aufzucht in Aquakulturen mit hohem Nährstoffeintrag? Diplom thesis, University of Zurich, Zurich, Switzerland.

Reddy, K.R., and W.F. DeBusk. 1987. Nutrient storage capabilities of aquatic and wetland plants. Pages 337-357 *in* K.R. Reddy and W.H. Smith, eds. Aquatic Plants for Water Treatment and Resource Recovery. Magnolia Publishing Inc., Orlando, Florida.

Reddy, K.R., and W.H. Smith. 1987. Aquatic Plants for Water Treatment and Resource Recovery. Magnolia Publishing, Orlando, Florida.

Saxer, P. 1999. Wachstumsverhalten von Helophyten in der Aquakulturanlage Otelfingen. Term thesis, University of Applied Sciences Waedenswil, Waedenswil, Switzerland.

Schmid, W. 1992. Kompogas. Technical and Business Description of the Kompogas System for Methanization of Organic Waste. W. Schmid AG, Glattbrugg, Switzerland.

Schwoerbel, J. 1986. Methoden der Hydrobiologie, Süsswasserbiologie. 3. Edition. Fischer, Stuttgart, Germany.

Soeder, C.J. 1981. Chemical composition of microalgal biomass as compared to some other types of single cell protein (SCP). Pages 73-85 *in* J.U. Grobelaar, C.J. Soeder, and D.T. Toerien, eds. Wastewater for Aquaculture. Proceedings of a Workshop on Biological Production Systems and Waste Treatment, University of the Orange Free State, Bloemfontein, South Africa.

Sommer, U. 1987. Factors controlling the seasonal variation in phytoplankton species composition. A case study for a deep, nutrient rich lake. Progress in Phycological Research 5:123-178.

Sommer, U., M. Gliwicz, W. Lampert, and A. Duncan 1986. The PEG-model of seasonal succession of planktonic events in fresh waters. Archiv Hydrobiologie 106: 433-471.

Staudenmann J., and R. Junge-Berberovic. 2000. Treating biogas plant effluent through aquaculture: first results and experiences from the pilot plant Otelfingen (Switzerland). Pages 51-57 *In*: B.B. Jana, Banerjee, R.D., Guterstam, B., and Heeb, J., eds. Waste Recycling and Resource Management in the Developing World. University of Kalyani, India and International Ecological Engineering Society, Wolhusen, Switzerland.

Staudenmann, J., A. Schoenborn, and C. Etnier. 1996. Recycling the Resource. Proceedings of the 2nd International Ecological Engineering Conference, Waedenswil. Transtech Publications, Uetikon, Switzerland.

Todd, J., and B. Josephson. 1996. The design of living technologies for waste treatment. Ecological Engineering 6:109-136.

Todt, D. 1998. Produktion von tropischen Pflanzen zur Nährstoffelimination von Prozesswasser. Diplom thesis, University of Applied Sciences Waedenswil, Waedenswil, Switzerland.

VSA (Verband Schweizer Abwasser-und Gewaesserschutzfachleute). 1995. Die Klaerschlammsituation in der Schweiz. Verbandsbericht Nr. 495. Zuerich, Switzerland.

Wissing, F. 1995. Wasserreinigung mit Pflanzen. Ulmer, Stuttgart, Germany.

Wyss, P., and B. Zuest. 2000. Sustainable Wastewater Treatment with Soil Filters. Intermediate Technology Publications, 103-105 Southampton Row, London, United Kingdom.

Yao, H. 1993. Phytoplankton production in integrated fish culture high-output ponds and its status in energy flow. Ecological Engineering 2:217-229.

Yan, J., and H. Yao. 1989. Integrated fish culture management in China. Pages 375-408 *in*: J. Mitsch, and S.E. Jorgensen, eds. Ecological Engineering. John Wiley and Sons, New York, New York.

Záková, Z., M. Palát, E. Kocková, and J. Toufar. 1995. Ist der Einsatz von Wasserhyazinthen bei der Abwasserbehandlung und Nährstoffelimination in Mitteleuropa sinnvoll? Mitteilungen der Oswald-Schulze-Stiftung, 18:495-503. Gladbeck, Germany.

Nutrition of Marine Fish Larvae

Akio Kanazawa

SUMMARY. In the hatchery production of aquatic animals for aquaculture, live foods such as diatoms; rotifer, *Brachionus plicatilis* and brine shrimp, *Artemia salina*, have been used throughout the world. However, such production requires large facilities, maintenance expenses, and labor to produce a desired amount of live foods constantly and reliably. Also, the nutritive value of planktonic organisms is occasionally variable, indicating that the dietary quality of these live foods varies with the content of n-3 highly unsàturated fatty acid (n-3 HUFA). Therefore it is necessary to develop microparticulate diets as a substitute for live foods to further increase the productivity of seed for fish culture. The nutritional components of microparticulate diets for fish larvae should be determined on the basis of requirements of the larval fish for proteins and amino acids, lipids and fatty acids, carbohydrates, vitamins, and minerals. Moreover, the efficient development of microparticulate diets for the fish larvae has promoted the improvement of nutritional requirement studies. The present review concerning the nutrition of marine fish larvae focuses on the proteins, amino acids, peptides, fatty acids, phospholipids, depigmentation of flatfish, stress tolerance of lipids, incorporation of HUFA in neural tissues, HUFA in egg and larvae, HUFA enrichment of live food, carbohydrates, vitamins, energy source during embryo and larval stages, enzyme supplement in microparticulate diets, and application of microparticulate diets in aquaculture. *[Article copies available for a fee from The Haworth Document Delivery Service: 1-800-HAWORTH. E-mail address: <getinfo@haworthpressinc.com> Website: <http://www.HaworthPress.com> © 2003 by The Haworth Press, Inc. All rights reserved.]*

Akio Kanazawa, Faculty of Fisheries, Kagoshima University, 4-50-20 Shimoarata, Kagoshima 890-0056, Japan.

[Haworth co-indexing entry note]: "Nutrition of Marine Fish Larvae." Kanazawa, Akio. Co-published simultaneously in *Journal of Applied Aquaculture* (Food Products Press, an imprint of The Haworth Press, Inc.) Vol. 13, No. 1/2, 2003, pp. 103-143; and: *Sustainable Aquaculture: Global Perspectives* (ed: B. B. Jana, and Carl D. Webster) Food Products Press, an imprint of The Haworth Press, Inc., 2003, pp. 103-143. Single or multiple copies of this article are available for a fee from The Haworth Document Delivery Service [1-800-HAWORTH, 9:00 a.m. - 5:00 p.m. (EST). E-mail address: getinfo@haworthpressinc.com].

KEYWORDS. Nutrition, marine fish larvae, microparticulate diet, highly unsaturated fatty acid, phospholipid

INTRODUCTION

An area of recent progress in aquaculture is larval fish nutrition. Much knowledge of the nutritional requirement of marine fish larvae after the transition from endogenous to exogenous feeding is required in order to compose a satisfactory diet to produce improved quality larvae. Although critical factors are involved in the improvement of first-feed larvae, a considerable amount of research has been accumulated to elucidate the nutritional requirement of marine fish larvae in terms of protein, essential fatty acids, phospholipids, vitamins, biochemical composition of eggs and larvae, enrichment of live food, and selected formulation of microparticulate diets.

Reviews on the nutrition of marine fish larvae have been presented by Watanabe and Kiron (1994), Sorgeloos (1994), Sorgeloos et al. (1995) and Izquierdo and Fernandez-Palacios (1997). Fish have been successfully raised to juvenile and adult stages on a purified or semi-purified artificial diet, in order to study nutritional requirements. As for the use of an artificial diet for early stages of marine fish larvae, however, little is known. Microparticulate diets for larval fish require the following: (1) small particles of 5-300 μm diameter that are stable in water until consumed by the larvae; (2) nutritionally well-balanced ingredients that are digestible in the alimentary canal. Research on attempts to culture larval fish on microparticulate diets have been described by many researchers (Kanazawa et al. 1981, 1982a, 1989; Teshima et al. 1982; Kanazawa and Teshima 1988; Kanazawa 1991a, 1993a, 1997; Coutteau et al. 1995; Bergot 1996). Composition of the semi-purified microparticulate diet is shown in Table 1 (Kanazawa 1997).

PROTEINS

Protein Requirement

Proteins are indispensable nutrients for growth and maintenance of living organisms. Little is known, however, about the optimum protein level in microparticulate diets for fish larvae. Generally, the hatchery production of ayu, *Plecoglossus altivelis*, is carried out in seawater.

TABLE 1. Typical formula of microparticulate diet for marine fish larvae.

Ingredient	g/100 g of dry diet
Defatted squid meal	25.0
Defatted krill meal	20.0
Defatted white fish meal	10.0
Milk casein (vitamin free)	15.0
Pollack liver oil	5.0
Soybean lecithin	3.0
Oleic acid	2.0
Vitamin mixture[1]	5.3
Mineral mixture[2]	5.0
Attractants[3]	1.7
Zein (binder)	8.0

[1,2]See Kanazawa (1997).
[3]Taurine, 0.6; Betaine, 0.6; Alanine, 0.4; Inosine-5′-monophosphoric acid, 0.1 g/100 g dry diet.

Microparticulate diets incorporating three different protein levels were compared with live food (control) diets in a feeding trial using 10-day-old ayu larvae. The results of the 30-day feeding trial demonstrated an optimum protein level of 40% in microparticulate diet, when the amino acid profile of the diet simulated that of the body protein of larval ayu fish (Kanazawa 1991a). Also, the optimum protein level in a microparticulate diet for Japanese flounder, *Paralichthys olivaceus*, larvae (initial size = 5 mm) was 60% out of diets with 50, 55, and 60% protein levels incorporating sardine meal, squid meal, and scallop meal (Kanazawa 1988). The optimum protein levels for fish larvae differ, probably due to a variety of factors, such as the differences in food habits, age of larvae, water temperature, protein sources used, and energy level of the diet.

Essential Amino Acid

Because fish have difficulty efficiently utilizing free amino acid in diets, the incorporation of radioactive acetate into individual amino acids was investigated in Japanese flounder in order to determine the qualitative essential amino acids. Six days after an injection of 200 μCi [U-^{14}C]acetate (four times every two days) in Japanese flounder juveniles, the livers were removed. After protein separation and hydrolysis, the radioactivity of each individual amino acid was measured using a high-performance liquid chromatography (HPLC) and a radioanalyzer. Based on the radioactive incorporation data, it was concluded that the essential amino acids for the flounder are as follows: arginine, methionine, valine, lysine, threonine, isoleucine, leucine, histidine, phenylalanine, and tryptophan (Table 2; Kanazawa 1991b).

Although the nutritional requirements of essential amino acids in juvenile and adult fish have been widely studied (Hidalgo et al. 1987; Borlongan and Benitez 1990; Moon and Gatlin 1991; Tibaldi and Lanari 1991; Craig and Gatlin 1992; Boren and Gatlin 1995; Forster

TABLE 2. Specific activities in amino acids of protein residue from liver of Japanese flounder after injection of [U-^{14}C] acetate (adapted from Kanazawa 1991b).

Amino acid	Specific activity (c.p.m./nmol)	Incorporation of radioactivity
Tau	9.71	+
Asp	1.44	+
Ser	0.76	+
Glu	2.96	+
Pro	1.03	+
Gly	0.51	+
Ala	1.33	+
Cys	1.38	+
Thr	0.03	-
Val	-	-
Met	-	-
Ile	-	-
Leu	-	-
Tyr	-	-
Phe	-	-
His	-	-
Lys	-	-
Trp	-	-
Arg	-	-

and Ogata 1998), little research has been done on amino acid requirements in fish larvae. In order to study the effect of different dietary arginine levels on the growth of red seabream, *Pagrus major*, larvae, two types of diets were prepared: one group of five zein microbound diets with increasing levels of arginine and one group of five microcoated diets with the same arginine levels as those in the microbound diets. All diets had a crude protein content (N × 6.25) of 60%. Twenty-day-old red seabream larvae were fed on these diets for 28 days, and at the end of the experiment, the amino acids in the diets and larvae were analyzed. Growth of red seabream larvae was enhanced by increasing arginine levels to 2.5% of the diet. Further increase in arginine level did not improve the growth response. For practical purposes, at least 2.5% arginine in the diet is recommended for larval feeding of red seabream (López-Alvarado and Kanazawa 1994).

To study the nutritional requirements for essential amino acids in fish larvae, a purified diet is needed, and to assess the optimum level of crystalline amino acids in larval diets, red seabream larvae were fed five microbound diets with different levels of crystalline amino acids. The larvae were cultured for 30 days. Assessment of growth, survival, and body amino acid composition after the culture period indicates that replacing casein with low levels of crystalline amino acids is beneficial. Growth and survival were enhanced when 5 g casein per 100 g of diet was replaced with crystalline amino acids. The effect of replacing 10 g casein per 100 g of diet with crystalline amino acids was that the larval performance was similar to that of those on the control diet. However, replacing 15 g or more of casein produced a detrimental effect on larval performance. The amino acid leakage from the diet particles is high, and microencapsulation of crystalline amino acids is therefore recommended (López-Alvarado and Kanazawa 1995).

The ability of larval turbot, *Scophthalmus maximus*, to ingest labeled ^{14}C-alanine seawater during their yolk-sac stages was investigated by means of autoradiographic and microscopic methods. Autoradiograms showed that after 24 hours of exposure to ^{14}C-alanine virtually the entire label was located in the intestinal lumen. Also, the release of labeled CO_2 by the yolk-sac larvae was low, but the amount of tracer incorporated into the TCA-insoluble fraction of the larval tissues was large (Korsgaard 1991).

Oligopeptides

Little is known about feeding attractants for larval fish. An attempt to clarify the growth-promoting effect of oligopeptides on red seabream larvae was carried out by a feeding trial using microparticulate diets. Using casein and fish meal as major protein sources and zein as the binder, diets supplemented with 0.5% of several oligopeptides were prepared. One hundred seventy-three red seabream larvae, 48 days after hatching (total length 18 mm), were stocked in 100 L tanks and were fed with the microparticulate diets (250-750 μm particle size) twelve times a day for 30 days. Seawater flow rate was 0.7 L/minute, and temperature was 24.6-26.0°C. Duplicate tanks were assigned for each test diet. Microparticulate diets supplemented with 0.5% of low molecular weight nitrogen compounds such as alanyl-glycine, alanyl-serine, alanyl-isoleucine, alanyl-alanine, and alanyl-leucine gave higher growth rates than a control group without supplemental oligopeptide (Table 3; unpublished data). The results indicated that fish larvae can utilize oligopeptides as feeding attractants in microparticulate diets, improving the nutritional values of the diet, and/or such diets can trigger some other unknown biological activity. Two fish attractants, a dipeptide (Arcamine, hypotauryl-2-carboxyglycine) and Strombine (C-methyl-imino-diacetic acid), which were isolated from the marine invertebrates mussel, *Arca zebra*, and conch, *Strombus gigas*, at 10^{-8} g/L had been active in inducing fish to display feeding behavior in experimental aquaria (Sangster et al. 1975). All oligopeptides tested on rainbow trout (Hara 1977) re-

TABLE 3. Oligopeptides as feeding attractant of red seabream larvae. Initial total length: 18.4 mm. Feeding period: 30 days. Treatment means with the same letters are not significantly different ($P < 0.05$).[1]

Oligopeptide	Final total length (mm)	Body weight (g)	Survival (%)
Free	42.4 a	1.2a	96
Ala-Pro	43.5ab	1.3ab	98
Ala-Gly	48.0e	1.8d	88
Ala-His	44.5abc	1.5bc	99
Ala-Ser	46.9de	1.7d	100
Gly-Ala	43.7ab	1.4ab	98
Pro-Gly-Gly	43.7ab	1.3ab	97
Ala-Ile	47.3de	1.7d	100
Ala-Ala	46.1cde	1.6cd	95
Ala-Leu	45.5bcd	1.5bcd	98

[1]Kanazawa, unpublished original data.

sulted in only a slight olfactory stimulation at 10^{-4} M and were negligible at 10^{-5} M.

Soybean Peptide as Protein Source

The effect of dietary soybean peptide on growth and survival of Japanese flounder was examined via feeding trials. Soybean peptide was isolated from soybean protein by hydrolysis with proteases (Fuji Seiyu Ltd., Japan[1]). The chain length of soybean peptide was 3.2. Casein was used as a protein source of test diets, and attempts to partially replace proteins in diets with soybean peptide were carried out. Methionine, threonine, lysine, and arginine were used as dietary supplements to adjust to amino acid profiles of Japanese flounder protein. At 56 days after hatching, the Japanese flounder larvae, total length 15 mm; body weight 28 mg, were divided into lots of 200 fish per 100 L tank and then raised on microparticulate diets for 30 days. A diet of 350-1,000 μm size was fed 8 times per day. Duplicate tanks were assigned for each test diet. The inclusion of soybean peptide in the microparticulate diet promoted growth of fish larvae (Table 4; unpublished data). A similar result was also reported on ayu fish larvae (Kanazawa 1995). The effects of peptides on fish larvae were shown by Zambonino-Infante et al. (1997) and Cahu et al. (1999).

To determine whether incorporation of peptides into diets can improve larval development, sea bass, *Dicentrarchus labrax*, larvae were fed for 21 days one of three isonitrogenous, isoenergetic semi-purified diets, in which enzymatic hydrolysate (75% di- and tripeptides) of fish meal proteins was substituted for 0, 20, or 40% of native fish meal pro-

TABLE 4. Effects of soybean peptide on growth of Japanese flounder larvae. Initial total length: 15.2 mm. Initial body weight: 27.9 mg. Feeding period: 30 days (Kanazawa, unpublished original data).

Soybean peptide added in diet (%)	Final totaL length (mm)	Final body weight (mg)	Body weight gain (%)	Survival rate (%)
0	25.5	139.7	401	87.5
10	28.8	193.1	592	92.0
20	28.9	210.3	654	95.0
30	27.7	177.2	535	92.5

1. Use of trade or manufacturer's name does not imply endorsement.

teins. Growth and survival were significantly greater ($P < 0.05$) in larvae fed peptide diets compared to those fed only native protein, with the best performance exhibited by those fed the 20% level of peptides. The better larval performances observed in groups fed diets containing peptides may be related to the enhanced proteolytic capacity of the pancreas and earlier development of intestinal digestion.

Protein Sources in Microparticulate Diet

Microparticulate diets were prepared by using several protein sources and crystalline essential amino acids. The essential amino acid profiles of the diets were similar to those of larval whole body protein. The formulated diets were used in feeding experiments to examine whether microparticulate diets could sustain growth and good survival of the larval puffer fish, *Fugu rubripes*. Four microparticulate diets were formulated with brown fish meal, squid meal, soybean protein, white fish meal, krill meal, and casein. Test diets composed with four different protein sources had similar essential amino acid profiles (% of dry diet) as compared with those of larval body protein. Feeding experiments were conducted using larval puffer fish at 34 days of age and 8 mm in length. Microparticulate diets with particle sizes of 350 to 500 μm were used.

The results showed that growth and survival of larvae fed diets containing 37% brown fish meal or 37% squid meal were superior to those of larvae fed the diet containing 31% soybean protein (Kanazawa 1991c). Similar studies have been carried out on the ayu fish (Kanazawa 1986); Japanese flounder (Kanazawa and Teshima 1988; Kanazawa 1995); red seabream (Kanazawa 1993a); and striped jack, *Caranx delicatissimus*, (Kanazawa 1993b). These studies have demonstrated that the essential amino acid compositions of larval fish bodies closely match the fishes' dietary requirements. The use of two protein sources, krill meal (protein 62% dry matter) and soybean protein (protein 85.2% dry matter), and their effect on growth performance and feeding behavior of red seabream during weaning and metamorphosis was also studied using microbound diets. High levels of dietary soybean protein (25% of the diet) resulted in poor feeding response and marked anorexia, resulting in high mortalities and poor growth. Moderate levels of dietary soybean protein (17-19% of diet) resulted in higher growth. Krill meal contributed to an increase in food consumption (López-Alvarado and Kanazawa 1997).

LIPIDS

Essential Fatty Acids

Since fish are incapable of *de novo* synthesis of linoleic acid (18:2n-6), linolenic acid (18:3n-3), eicosapentaenoic acid (EPA, 20:5n-3), and docosahexaenoic acid (DHA, 22:6n-3), dietary sources of these fatty acids are probably essential for normal growth and survival. In fact, this has been demonstrated in feeding experiments with a variety of fish species (Kanazawa 1985). The most notable difference in essential fatty acid requirements existed between freshwater and marine fish species. *Tilapia zillii* and *Oreochromis niloticus* require fatty acids of the linoleic family (n-6), while rainbow trout, *Oncorhynchus mykiss*, requires linolenic family (n-3) as essential fatty acid. Carp, *Cyprinus carpio*; eel, *Anguilla japonica*; and chum salmon, *Oncorhynchus keta*, require not only linolenic but also linoleic acids for good growth. However, n-3 highly unsaturated fatty acids (HUFAs) such as EPA and DHA were very effective as essential fatty acids for marine fish species. Most of these results are of essential fatty acid requirements in the juvenile stage of marine fish, while little is known about the requirements of the larvae (Kanazawa 1993b, 1995; Watanabe 1993; Rainuzzo et al. 1997; Takeuchi 1997).

EPA and DHA requirements of marine fish larvae and juveniles have been demonstrated by the feeding trials in which fish were fed rotifer and/or *Artemia* enriched with EPA and DHA (Takeuchi et al. 1990, 1992a, 1996; Furuita et al. 1996a, 1996b) and fed microparticulate diets with EPA and DHA (Kanazawa 1993b, 1995, unpublished data). Marine fish larvae requirements of EPA and DHA are estimated at about 1-2% of diet as shown in Table 5. Several workers have shown that DHA is a more efficient essential fatty acid than EPA for red seabream (Watanabe et al. 1989a; Izquierdo et al. 1989; Takeuchi et al. 1990, 1992b); striped jack (Watanabe et al. 1989b); Japanese parrotfish, *Oplegnathus fasciatus* (Kanazawa 1993b); and Japanese flounder (Kanazawa 1995; Table 6). The value of HUFAs in microparticulate diets for red drum, *Scianops ocellatus*, larvae was recently emphasized (Brinkmeyer and Holt 1998). Growth and survival of larvae fed diets with varied DHA-to-EPA ratios were unaffected; however, larvae fed a diet with the highest ratio (3.78) exhibited significantly superior performance in a salinity challenge test, indicating a possible nutritional requirement for DHA over EPA.

TABLE 5. EPA and DHA requirements of marine fish larvae and juveniles.

Species		Requirement (% of dry diet)	References
Larvae			
Japanese parrotfish	EPA	1	Kanazawa (1993b)
	DHA	1-2	
Japanese flounder	EPA	1	Kanazawa (1995)
	DHA	1	
Red seabream	EPA	2.25	Fruita et al. (1996a)
	DHA	0.95-1.62	
Yellowtail	EPA	3.79	Fruita et al. (1996b)
	DHA	1.39	
Striped jack	DHA	1.6-2.2	Takeuchi et al. (1996)
Juveniles			
Red seabream	EPA	1	Takeuchi et al. (1990)
	DHA	0.5	
Striped jack	EPA	0.8	Takeuchi et al. (1992a)
	DHA	1.7	
Puffer fish	EPA	1-1.5	Kanazawa unpubl. data
	DHA	1	

TABLE 6. Comparison of essential EPA and DHA for marine fish.

Species		Essential fatty acid value	References
Red seabream	Larvae	EPA < DHA	Watanabe et al. (1989a)
	Juveniles	EPA < DHA	Takeuchi et al. (1990)
Japanese parrotfish	Larvae	EPA < DHA	Kanazawa (1993b)
Striped jack	Juveniles	EPA < DHA	Watanabe et al. (1989b)
Japanese flounder	Larvae	EPA < DHA	Kanazawa (1995)

The effects of linoleic and linolenic acids on growth, survival, and fatty acid composition of milkfish, *Chanos chanos*, were tested (Bautista and Dela Cruz 1988). Both linoleic and linolenic acids are effective for good growth and survival of juvenile milkfish; however, the effect of linolenic acid on the growth is better than that of linoleic acid. The effect of dietary fatty acids on growth of milkfish larvae in brackish water was shown (Alava and Kanazawa 1996). Five purified microparticulate diets containing 1% of 18:2n-6, 18:3n-3, arachidonic acid (ARA, 20:4n-6), or n-3 HUFA (60% EPA + 40% DHA) in addition to 8% oleic acid (18:1n-9), and a control diet containing 9% 18:1n-9 were fed to milkfish larvae for 30 days. The salinity was 16-17‰, and temperature was 27 ± 1°C during the culture periods. In each trial, 60 fish (5 mg, 6 mm) were stocked per 30 L plastic tank. Specific growth rates, weight, length, and feed conversion ratio of milkfish larvae fed the various diets did not differ significantly ($P > 0.05$). However, incidence of eye abnormality was highest in milkfish larvae fed the 18:1n-9 diet. The essential fatty acid requirements of milkfish in brackish water seem to be different from that of freshwater and marine fish species.

Also, Robin (1994, 1995) and Sargent et al. (1997) presented the importance of n-6 fatty acid for marine fish larvae. The effect of purified diets containing different combinations of ARA and DHA on growth, survival, and fatty acid composition of juvenile turbot was investigated (Castell et al. 1994). In this trial, feeding the diet containing ARA as the only HUFA resulted in higher growth and survival than any of the mixtures of the two fatty acids or DHA alone. Effects of dietary borage oil (18:3n-6 rich) or marine fish oil (EPA-rich) on growth, survival, liver histopatholgy, and lipid composition of juvenile turbot were demonstrated (Bell et al. 1995a). Results exhibited that no differences were observed in weights between dietary treatments; mortalities in the borage oil-fed group were significantly greater than in the marine fish oil-fed group, and the carcass composition of turbot can be altered by means of dietary lipids to contain increased levels of EPA and dihomo-gamma-linolenic acid (20:3n-6).

Furthermore, the competition between n-3 HUFA (EPA and DHA) and n-6 HUFA (ARA) in Japanese flounder larvae was investigated (Kanazawa et al. unpublished data). The microparticulate diets containing 1% DHA, 1% EPA, 1% ARA, 1% DHA + 1% ARA, and 1% EPA + 1% ARA, respectively, were used as test diets. Feeding trials were conducted using larvae 12 days after hatching, with length of 5 mm; 1,000 larvae were kept in each 100-L tank. Flounder larvae fed with diet containing DHA showed higher growth than the larvae fed with diets containing EPA or ARA. The growth of flounder larvae fed with DHA + ARA diet was inferior to that of larvae fed with DHA diet. From these results, it was suggested that Japanese flounder larvae demonstrated competition between n-3 and n-6 HUFAs. The effect of ARA on survival and growth of yellowtail, *Seriola quinqueradiata*, during *Artemia* feeding period was conducted (Ishizaki et al. 1998). In this experiment it was shown that 4% ARA content in *Artemia* has a negative effect on growth performance in the culture of yellowtail larvae and juveniles. Bell et al. (1994) determined the production of prostaglandins E and F of the 1-, 2-, and 3-series in primary cultures of turbot brain astroglial cells after supplementation with 20:3n-6, ARA (20:4n-6), and EPA (20:5n-3).

Fatty Acid Deficiency

In the 1980s, HUFAs as EPA were reported as an essential fatty acid of marine fish; thus, rotifer and *Artemia* enrichment with EPA in *Nannochloropsis* is widely practiced. On the hand, if DHA content in live foods is low, the vitality of fish larvae fed on these foods is reduced, leading to decreased seedling production. From these results, it is suggested that the physiological mechanism of DHA may be different from that of EPA. Dietary deficiency of essential fatty acids resulted in various abnormalities and diseases. In red seabream (Kitajima 1985) and sea bass, scoliosis was observed to be caused by EPA deficiency. Kanazawa et al. (1982b) investigated the fate of exogenous radioactive EPA. Radioactive measurements of tissues and organs showed high incorporation of label into the gall bladder, swim bladder, liver, and pyloric caecae. Therefore, EPA is likely to be utilized as a constituent of cellular membranes of these tissues. The occurrence of abnormal pigmentation in hatchery-raised flatfish was observed when DHA, phospholipid, and vitamin A were deficient in the diet (Kanazawa 1993c). DHA deficiency frequently resulted in hydropsy in red seabream (Watanabe et al. 1989a) and in decreased stress tolerance and vitality of marine fish larvae (Table 7).

Phospholipids

During the course of investigation of microparticulate diets for fish larvae, it was found that dietary sources of phospholipids are essential for normal growth and survival of the ayu (Kanazawa et al. 1981, 1983a, 1985), red seabream (Kanazawa et al. 1983b), and Japanese parrotfish (Kanazawa et al. 1983b) larvae. After that, investigations of

TABLE 7. Essential fatty acid deficiency disease in fish larvae.

Essential fatty acid	Fish	Disease symptoms
EPA (20:5n-3)	Red seabream	Scoliosis
	Sea bass	Scoliosis
DHA (22:6n-3)	Red seabream	Hydropsy
	Flatfish[1]	Depigmentation of skin
	Fish larvae[2]	Lowering of vitality
	Fish larvae[2]	Lowering of stress tolerance
	Fish larvae and juveniles[3]	Lowering of growth and survival

[1] Flounder, sole, and turbot, etc.
[2] Seabream, flounder, and seabass, etc.
[3] Seabream, parrotfish, striped jack, yellowtail, and flounder, etc.

phospholipid requirements of marine fish larvae have been established by many researchers (Kanazawa 1985, 1993d, 1997; Kanazawa and Teshima 1988; Takeuchi et al. 1992a; Koven et al. 1993; Geurden et al. 1995a; Coutteau et al. 1997; Tago et al. 1999). In recent years, Geurden et al. (1995b, 1998a, and 1999) and Fontagne et al. (1998) have also suggested the essentiality of phospholipid addition to diets of common carp larvae.

Dietary phospholipids were not only effective at the larval stage, but also at the juvenile stage of Japanese flounder and Japanese parrotfish (Kanazawa 1993d). The larval fish had high growth and survival rates when given a diet with soybean lecithin including 1.0-3.0% as phosphatidylcholine (PC) plus phosphatidylinositol (PI) (Table 8). In order to clarify which components of soybean lecithin were most effective for the Japanese flounder, fractionation and isolation of each phospholipid class were done by column chromatography using silica gel. The purity of isolated phosphatidylethanolamine (PE), PI, and PC was 92, 98, and 94%, respectively.

Regarding body weight gain of Japanese flounder larvae (initial body weight of 5 mg and total length of 8 mm) raised for 40 days, those fed a diet containing 1% PC had the highest, whereas those receiving 1% PI and 1% PE diets had poor weight gains (Table 9; Kanazawa 1993d). Fish larvae were thought to require the molecular form of phospholipids that have unsaturated fatty acids at the C-2 position and either inositol or choline groups at the C-3 position. These compounds, in addition to n-3 HUFA, are believed to be indispensable for normal growth and sur-

TABLE 8. Phospholipid requirements of fish larvae and juveniles.

Species	Requirement (g/100 g dry diet; as PC + PI in soybean lecithin)	References
Japanese flounder	1.6-2.2	Kanazawa (1993d)
Marbled sole	1	Unpublished data
Red seabream	1.6	Kanazawa et al. (1983b)
Japanese parrotfish	1-1.6	Kanazawa (1993d)
Yellowtail	1	Unpublished data
Striped jack	1-1.5	Takeuchi et al. (1992a)
Puffer fish	1-2	Kanazawa (1991c)
Red spotted grouper	2.3	Unpublished data
European seabass	2	Geurden et al. (1995a)
Ayu	1-1.6	Kanazawa et al. (1983a)

PC: Phosphatidylcholine, PI: Phosphatidylinositol.

TABLE 9. Necessity of phospholipids for growth and survival of Japanese flounder and ayu larvae.

Phospholipid	Japanese flounder	Ayu
Soybean PC	+++	++
Soybean PI	+	+++
Soybean PE	+	+

+++ Most effective, ++ effective, + less effective.
PC: Phosphatidylcholine, PI: Phosphatidylinositol, PE: Phosphatidylethanolamine.

vival. Geurden et al. (1998b) recently showed the effect of the degree of saturation of dietary PC on the enhancement of dietary fatty acid incorporation in lipid of postlarval turbot.

Phospholipid Deficiency

Incidences of scoliosis and twist-of-jaw in larval ayu were reduced by the supplementation of phospholipids (Kanazawa et al. 1981). Also, depigmentation of skin in larval flatfish and lowering of stress tolerance in fish larvae were prevented by phospholipids (Kanazawa 1993c, 1997; Table 10). Why are phospholipids essential nutrients for fish and crustacean larval? Experiments with ^{14}C-labelled lipids to clarify the mechanism by which dietary phospholipids enhance the growth of the fish and crustaceans were conducted. Results demonstrated that phospholipids are required for the transport of dietary lipids, particularly cholesterol and triglycerides, in the body (Teshima et al. 1986a, 1986b). Koven et al. (1993) tested the effect of dietary lecithin and exogenous lipase on the incorporation of labeled oleic acid in the tissue lipids of gilthead seabream, *Sparus aurata*, larvae. A significant effect of dietary lecithin on the incorporation of labeled oleic acid in both larval neutral and phospholipid fractions was demonstrated. Since the necessity of dietary phospholipids for growth of fish larvae has been found, diets containing phospholipids such as commercial soybean lecithin are widely used in hatchery production.

Skin Depigmentation of Flatfish

The occurrence of color abnormality in the ocular side of flatfish has widely been observed during the process of seed production. Many researchers have proposed genetic, environmental, and nutritional factors as possible causes of albinism (Seikai 1985a, 1985b; Seikai et al. 1987a, 1987b; Nakamura and Iida 1986; Nakamura et al. 1986; Fukusho et al. 1986, 1987; Miki et al. 1988, 1989, 1990; Yamamoto et al. 1992; Støttrup and Attramadal 1992; Baker et al. 1998).

Kanazawa (1993c, 1995, 1998a) suggested that the rhodopsin formation of the eye retina was hindered when 10-day-old larvae were fed with phospholipid-, DHA-, and vitamin A-deficient diets, resulting in the interruption of black pigment (melanin) formation. Therefore, when

TABLE 10. Phospholipid deficiency in fish larvae.

Fish	Deficiency
Ayu larvae	Scolisis; Tail deformation
Flatfish larvae[1]	Depigmentation of skin
Fish larvae[2]	Lowering of vitality
Fish larvae[2]	Lowering of stress tolerance
Fish larvae and juveniles[3]	Lowering growth and survival

[1] Flounder and sole, etc.
[2] Seabream, flounder and seabass, etc.
[3] Seabream, flounder, sole, parrotfish, yellowtail, stripedjack, puffer fish, grouper, and seabass, etc.

microparticulate diets or rotifers enriched with phospholipid, DHA, and vitamin A are fed to flatfish at 10 days after hatching, the prevention of albinism is possible. The effects of phospholipids, PC, PI, and PE on the occurrence of albinism in flatfish were studied (Kanazawa, unpublished data).

Marbled sole, *Limanda yokohamae*, larvae at 8 days after hatching (total length 6 mm) were fed with microparticulate diets for 50 days. The microparticulate diets were composed of vitamin-free casein, defatted squid meal, amino acids, vitamin mixture, mineral mixture, and dextrin. Zein was used as a binder of the diet. Lipids, 5% soybean oil, 1% DHA, and 1% phospholipids were added. The experimental treatments were soybean lecithin-, PC-, PI-, and PE-added groups, respectively; phospholipid deficient group; and live food (rotifers and *Artemia*). Albinism (completely abnormal partially abnormal) in marbled sole was 26% in the group fed with the PC-added diet and 37% in the PI-added diet but 75% in the PE diet and 85% in the phospholipid deficient diet (Table 11).

The responses of normal and albino flounder to a visual test were compared (Kanazawa 1993c). Each tank was divided into two sections, one covered and one open. A divider provided gates in the middle of the tank for the fish to pass through freely. About half of the fish were stocked in each covered and each open section of the tank. After four days the distribution of the fish in the dark and light sections was determined. Normal fish preferred to gather in the covered, dark section of

TABLE 11. Effect of the phospholipids phosphatidylcholine (PC), phosphatidylinositol (PI), and phosphatidylethanolamine (PE) on skin depigmentation of marbled sole.[1]

Phospholipid added in diet	Occurrence rate (%)	
	Normal	Abnormal and partially abnormal
Soybean lecithin	69	31
Phosphatidylcholine (PC)	74	26
Phosphatidylinositol (PI)	63	37
Phosphatidylethanolamine (PE)	25	75
Phospholipid deficient	15	85
Live-food (Rotifer and *Artemia*)	6	94

[1] Kanazawa, unpublished original data.

the tanks during daytime, whereas at nighttime, they were found in the uncovered, open section of the tanks. In contrast, abnormal fish were widely distributed in both sections of the tanks, regardless of time of day. These results show that abnormal or albino Japanese flounder have weak visual perception. Feeding behaviors of normally and abnormally pigmented Japanese flounder were compared at twilight (Kanazawa, unpublished data). Normal and abnormal flounder, total length 3-4 cm, were fed on same pellet diet for 7 days. The light intensity of the culture tank alternated between 10 and 40 lux every 24 hours. Amount of food intake of normal flounder was the same at 10 and 40 lux during 7 days, but that of abnormal flounder was low at 10 lux and high at 40 lux. However, total amount of food intake of both groups was almost the same and the growth rates of both groups were similar. These results suggest that feeding behavior of albino flatfish attributed to DHA, phospholipid, and vitamin A deficiency is different from that of normal flatfish.

Larval fish with poor vision have difficulty identifying and capturing food. In the retina of herring, *Clupea harengus*, rods are recruited from about 8 weeks after hatching, and from this time there is a linear relationship between the number of rods in the photoreceptor cell population and the content of di-22:6n-3 molecular species of phospholipid. The effects of DHA on vision at low light intensities in juvenile herring were determined (Bell et al. 1995b). They concluded that a dietary deficiency of DHA during the period early in rod development impairs visual performance so that the fish can no longer feed at low light intensities. Diets reduced in methionine and with oxidized oil were formulated to induce visual deficiencies in Japanese flounder, in an attempt to link vision with pigmentation development (Estévez et al. 1997a). The results suggest that a deficient intake of amino acids and fatty acids produces a change in retinal structure and composition, leading to reduce visual capability and suppression of the development of normal pigmentary pattern in flatfish.

The lipid class and fatty acid composition of brain and eyes of unpigmented and normally pigmented Japanese flounder, raised on live food were analyzed (Estévez and Kanazawa 1996). The neural tissues of normally-pigmented and unpigmented fish showed a similar lipid class composition. However, normally-pigmented individuals had a higher content of polyunsaturated fatty acids in the polar lipid fraction, especially ARA, EPA, and DHA. Considering the importance of DHA in the composition of neural tissues, as well as EPA and ARA in eicosanoid production, a relationship among vision, neural transmission, and pigmentation is suggested.

The dietary effect of ARA (20:4n-6) on the composition of the head and the body and on the pigmentation of Japanese flounder larvae was studied by means of microbound diets (Estévez et al. 1997b). Increasing levels of this essential fatty acid produced higher growth and pigmentation success in juveniles after 45 days of feeding. The role played by ARA in the composition of PI and eicosanoid synthesis seems to be affected in relation to the normal function of the neural system and the production of hormones involved in the metamorphosis and pigmentation of flatfish. Rainuzzo et al. (1997) suggested that turbot larvae tended to exhibit less pigmentation with lower DHA: EPA ratio in the total lipid of larvae, especially when the amounts of EPA were high compared to those of DHA. Takeuchi et al. (1995) investigated whether the occurrence of color abnormality and malformation in larval Japanese flounder could be prevented by enriching *Artemia* nauplii with vitamin A and beta-carotene. Results suggest that vitamin A seems to be effective in preventing the abnormal coloration caused by Tien-tsin *Artemia*, but an excess of this vitamin can lead to ill effects on normal growth of flounder.

Stress Tolerance of DHA and Phospholipid

Marine fish larvae were found to have a requirement for DHA and phospholipids on growth and survival. It has also been demonstrated that DHA and phospholipids affect stress tolerance, in the cases of changes in water temperature and salinity, and exposure to low dissolved oxygen (DO) in mahimahi, *Coryphaena hippurus* (Ako et al. 1991; Kraul et al. 1991, 1993); Asian seabass, *Lates calcarifer* (Dhert et al. 1992); striped mullet, *Mugil cephalus* (Ako et al. 1994); red seabream (Kanazawa 1997); Japanese flounder (Kanazawa 1998a); and milkfish (Gapasin et al. 1998) larvae.

Two lipid nutrients, phospholipids and n-3 HUFAs, are believed to be indispensable for normal growth and survival of marine fish larvae. Recently, the effects of 1,2-di-20:5-PC, 1,2-di-22:6-PC, and 22:6-triglyceride (TG) on growth and stress tolerance of larval Japanese flounder were compared (Tago et al. 1999). Japanese flounder (20 days after hatching, mean total length of 10 mm) were fed microparticulate diets containing 1% phospholipids (purity 99%) for 30 days at 17.2-19.5°C. After the feeding trials, their tolerance to stress factors such as changes in water temperature and salinity and exposure to low dissolved oxygen

(DO) were determined. Results indicated that dietary 1,2-di-22:6-PC was more efficient than 1,2-di-20:5-PC and 22:6-TG for stress tolerance of Japanese flounder larvae towards increased water temperature and reduced DO (Table 12). Lactic acid bacteria isolated from rotifers, which increase the resistance of turbot larvae against pathogenic *Vibrio*, were shown by Gatesoupe (1994).

Incorporation of HUFA in Neural Tissues

Dietary lipids are important sources of energy and essential fatty acids for all animals. The n-3 fatty acid such as EPA and DHA are HUFAs that are commonly found in marine organisms. The useful roles and beneficial effects of these fatty acids have been recognized for marine animals and human health. It has been demonstrated that EPA is generally biosynthesized by phytoplankton, and it then is assimilated by zooplankton, of which a part of EPA is bioconverted into DHA. Both n-3 HUFAs are deposited and accumulated in marine fish. Furthermore, considerable attention has been focused on the n-3 HUFA requirements of marine fish larvae. Studies such as turbot (Witt et al. 1984), gilthead seabream (Koven et al. 1989), dolphin, *Coryphaena hippurus* (Ostrowski and Divakaran 1990), and Japanese parrotfish (Kanazawa 1993b) larvae have shown that DHA is strongly retained.

Many studies have shown that the level of DHA accumulates in the brain and retina of fish larvae. DHA was strongly retained in brain of sea bass (Pagliarani et al. 1986); in brain and retina of cod, *Gadus morhua* (Tocher and Harvie 1988); in brain of turbot (Mourente et al.

TABLE 12. Effect of 20:5-PC and 22:6-PC on growth, survival and stress tolerance of Japanese flounder larvae.[1]

Phospholipid and triglyceride supplemented in diet	Body weight (mg, dried)	Survival (%)	Stress tolerance test (Time × no. of surviving fish)		
			Increased temp.	Reduced salinity	Reduced DO
1,2-di-22:6n-3-Phosphatidylcholine	3.9	62	3,448	6,720	379
1,2-di-20:5n-3-Phosphatidylcholine	3.8	64	2,937	4,102	774
Tri-22:6n-3-glyceride (TG)	3.6	50	2,948	5,057	746
Phospholipid deficient	3.6	50	2,498	3,880	683

[1]Adapted from Tago et al. (1999).

1991; Mourente and Tocher 1992); in heads and eyes of Atlantic herring (Navarro et al. 1993); in brain of gilthead seabream (Mourente and Tocher 1993); in eyes of sea bass (Bell et al. 1996); in brain and eyes of Japanese flounder (Estévez and Kanazawa 1996; Kanazawa 1998a; Furuita et al. 1998); and in brain and eyes of Atlantic salmon, *Salmo salar* (Brodtkorb et al. 1997). The incorporation of [1-^{14}C]18:3n-3 and [1-^{14}C]22:6n-3 (DHA) and the metabolism via the desaturase/elongase pathways of [1-^{14}C] 18:3n-3 and [1-^{14}C]20:5n-3 (EPA) were investigated in brain cells from newly-weaned (1-month-old) and 4-month-old turbot (Tocher et al. 1992).

HUFA Content in Egg and Larvae

Japanese parrotfish; striped jack; and red-spotted grouper, *Epinephelus akaara*, are three important species in Japanese aquaculture. Growth and survival during larval stages, however, are comparatively low, and reliable techniques for seed production have yet to be established. Watanabe (1993) showed that the concentration of DHA was generally high in the egg of marine fishes such as cod, seabream, striped jack, and Japanese flounder and quickly decreased during development.

As an initial step in the development of microparticulate diets for striped jack, biochemical analyses of eggs and larvae were carried out (Kanazawa 1993b). The eggs and larvae at 0, 8, 15, 18, 21, and 24 days post-hatch were used for lipid analyses. Lipid class composition in the egg and larvae after hatching of striped jack indicated that sterylester and triglyceride contents decreased after hatching; however, free fatty acid, free sterol, monoglyceride, PI, and PC increased after hatching. EPA content in the neutral lipid fraction isolated from the eggs and larvae after hatching increased, whereas DHA content decreased. In the phospholipids fraction, DHA level in the eggs was high, and rapidly decreased after hatching; however, EPA levels in both eggs and larvae were almost the same. The apparent depletion of DHA suggests that it plays an important role in the development of striped jack larvae. Also, the fatty acid and amino acid compositions of high-vitality, live larvae and shock-dead larvae during transfer and harvest of red-spotted grouper were analyzed (Kanazawa 1998b). Lipid class composition of the whole body of red-spotted grouper showed that triglyceride content in the shock-dead larvae is low. In the neutral lipid fraction, polyunsaturated fatty acids such as linoleic acid, linolenic acid, EPA, and DHA of live larvae were higher than those of the shock-dead larvae. In the polar lipid fraction, DHA of live larvae was higher than that of the dead larvae. Results suggested that the high vitality of red-spotted grouper larvae is due to a large amount of DHA in the body.

Three different groups of cod larvae fed on natural plankton in large enclosures were analyzed for fatty acid composition through development (Van der Meeren et al. 1993). Rainuzzo et al. (1992) and Rainuzzo (1993) have reported on fatty acid and lipid composition of fish egg and larvae. Rainuzzo et al. (1997) also reviewed the significance of lipids at different early stages of marine fish larvae. Lipids in broodstock nutrition are considered to be important for spawning, egg quality, and quality of larvae. Lipids as a source of energy at the embryonic and larval stages before feeding are evaluated.

HUFA Enrichment of Live-Food Organisms

The seed production of marine fish has widely been conducted using both live foods, such as rotifers and *Artemia*, and microparticulate diet. However, the larvae fed on rotifers and *Artemia* containing a low level of n-3 HUFA showed poor growth, high mortality and high rate of deformity. Enrichment of live food for marine fish larvae with n-3 HUFA has been shown to improve growth and survival by many researchers (Watanabe et al. 1989c; Izquierdo et al. 1989, 1990, 1992; Koven et al. 1989, 1990, 1992; Dhert et al. 1992; Kraul et al. 1992; Takeuchi et al. 1992b; Robin et al. 1993; Craig et al. 1994; Rodriguez et al. 1994; Salhi et al. 1994; Cairrao et al. 1995; Navarro et al. 1995; Evjemo et al. 1997; Reitan et al. 1997; Gapasin et al. 1998; Liu 1998; Robin 1998). These results demonstrated the effect of HUFA enrichment in enhancing marine larval fish.

CARBOHYDRATES

Carbohydrates are considered to be the cheapest form of dietary energy for fish aquaculture, but their utilization by fish varies, and little is known. The effects of dietary carbohydrate: lipid ratio on growth and body composition of hybrid striped bass juveniles (1.5 g) were shown (Nematipour et al. 1992). Results indicate hybrid striped bass are able to efficiently utilize carbohydrate for energy, and lipid could be partially replaced with carbohydrate to improve fish quality and productivity. The effects of dietary carbohydrate on egg quality, larval performance, and broodstock maturation in cod were reported (Mangor-Jensen and Birkeland 1993). Results demonstrated that cod broodstock dietary levels of carbohydrates have little or no impact on the measured egg characteristics.

Diaz et al. (1994) showed the assimilation of dissolved glucose from sea water by sea bass and gilthead seabream larvae. Larvae were transferred to normal or glucose-enriched seawater immediately after mouth opening to assess their ability to absorb and assimilate glucose at the beginning of the larval period. The results showed that glucose absorption resulted in glycogen accumulation in the hepatocytes; glucose delayed the pathological effects of fasting.

VITAMINS

The vitamin requirements of marine fish juveniles are known; however those of larval fish are limited.

An experiment has been conducted to identify signs of hypervitaminosis and establish a safe incorporation level of vitamin A in *Artemia* nauplii in terms of biological performances and uptake of vitamin A by larval Japanese flounder. The high concentrations of different vitamin A compounds added in *Artemia* medium were toxic to flounder, probably due to the high content of retinoic acid, a vitamin A metabolite formed in *Artemia* (Dedi et al. 1995, 1997, 1998; Takeuchi et al. 1995, 1998). Recently, Rønnestad et al. (1998) reported that feeding *Artemia* to larvae of Atlantic halibut, *Hippoglossus hippoglossus*, resulted in lower larval vitamin A content compared with feeding copepods. They suggested that halibut larvae are not able to efficiently convert the available carotenoids or the unknown retinoid component into retinal and retinol during the first period after onset of exogenous feeding.

Japanese parrotfish, and spotted parrotfish, *Oplegnathus punctatus*, larvae were fed with beta-carotene-supplemented rotifers and unsupplemented control rotifers for 24 days after hatchout. Results show that survival rates of beta-carotene-supplemented groups of both species' larvae were higher than those of control groups and that the supplementation of beta-carotene to rotifer might be of benefit in production of healthy, resistive larvae against infectious disease (Tachibana et al. 1997).

Vitamin B_6 content in Atlantic halibut larvae fed natural zooplankton was shown (Rønnestad et al. 1997).

Studies on the role of vitamin C, L-ascorbic acid (AsA), in fish nutrition have focused on requirements; its effects on immature fish; disease deficiency; and body defense mechanism. AsA content of sprat, *Sprattus*

sprattus, larvae was analyzed (Hepette et al. 1991). Also, AsA content in live-food organisms such as algae, rotifers, and *Artemia* was determined (Merchie et al. 1995a). The dietary AsA requirements with first-feeding larvae common carp were determined (Gouillou-Coustans et al. 1998).

Several AsA derivatives that are more stable than AsA–L-ascorbyl-2-sulphate, L-ascorbyl-2-phosphate, L-ascorbyl-2-polyphosphate, and L-ascorbyl-6-palmitate–are used. L-Ascorbyl-2-phosphate Mg (APM) has two characteristics: a high stability to the heat used in feed manufacturing processes and retention of activity for a longer period of time under normal conditions of storage. APM was tested for efficacy as a vitamin C source in Japanese flounder juveniles (Teshima et al. 1993) and red seabream juveniles (Kosutarak et al. 1994). The results indicated that 6 to 10 mg of APM/100 g diet was sufficient to support good growth and survival of Japanese flounder, and 5 mg of APM/100 g diet did the same for red seabream.

Live-food enrichment techniques, using formulated diets and emulsions for improving the nutritional quality of rotifers and *Artemia*, were studied as tools for transferring AsA to sea bass larvae. AsA was well incorporated into the larvae from rotifers (Merchie et al. 1995b). Formulated diets containing variable levels of stable ascorbyl-2-phosphate esters were used for the determination of minimal requirements for AsA in sea bass and turbot larvae. Results indicated that 2 mg AsA/100 g diet is sufficient for normal growth and survival (Merchie et al. 1996a, 1996b, 1997).

Recently, the effects of essential fatty acids and vitamin C-enriched live food on growth, survival, resistance to salinity stress, and incidence of deformity in milkfish larvae were investigated (Gapasin et al. 1998). The effects of dietary vitamin C on maturation and egg quality of cod were tested. Difference in free amino acid profile, egg strength, and neutral buoyancy were found, whereas no effects on fertilization rate and survival rate were detected (Mangor-Jensen et al. 1994).

Interactions of APM and oxidized fish oil on red seabream juveniles were investigated. The analysis of variance revealed that there was significant interaction of APM level and oil type in the tissues on AsA level and condition factor. A depletion of APM from the diets resulted in accelerating lipid peroxidation of muscle (Kosutarak et al. 1995a). Similarly, interactions of APM and n-3 HUFAs on Japanese flounder juveniles were conducted. Analysis of variance on weight gain, feed conversion efficiency and concentration of AsA in the liver revealed that there were significant interactions between APM and n-3 HUFA levels (Kosutarak et al. 1995b).

ENERGY SOURCE DURING EMBRYOS AND LARVAE

Fyhn (1989) has initially proposed that free amino acids are an important energy source during embryonic development of marine fish, from studies of developing eggs and larvae of halibut and cod. The large pool of free amino acid is almost contained in the yolk-sac compartment. The free amino acid pool is depleted during development and reaches low levels at first feeding. These findings suggest that in the early yolk-sac stage, free amino acid enters the embryo from the yolk and is utilized for both energy and protein synthesis (Rønnestad and Fyhn 1993; Rønnestad et al. 1993). Respiration, nitrogen, and energy metabolism of developing yolk-sac larvae of Atlantic halibut were investigated (Finn et al. 1995a). Quantitative data for the changes in the contents of total lipids, lipid classes, and fatty acids, together with the changes in caloric content of developing egg and yolk-sac larvae of Atlantic cod (Finn et al. 1995b) and Atlantic halibut (Rønnestad et al. 1995) were measured. Conceicao et al. (1997) have shown the influence of different food regimens on the free amino acid pool, the rate of protein turnover, the flux of amino acids, and their relation to growth of larval turbot from first feeding until metamorphosis.

Two major endogenous nutrients, lipids and free amino acids, in developing marine fish embryos and larvae were studied to determine the nutritional requirements of these fishes. From the perspective of broodstock and larval nutrition, lipid and amino acid metabolism should not be considered separately (Zhu et al. 1997). Changes in content of free amino acids, lipovitellin, and lipids were examined in developing embryos and larvae of barfin flounder, *Verasper moseri*, to elucidate the sequential utilization of these nutrient stocks before first feeding (Ohkubo and Matsubara 1998). Total free amino acid content showed no change during the first 4 days, then decreased to about 13% of the initial level by the 13th day after fertilization. The lipovitellin content was stable during the 13th day after fertilization, then decreased rapidly until the end of yolk-sac absorption. Phospholipids decreased gradually after hatching. From these results, they considered the following four periods for the sequential nutrient utilization in barfin flounder embryos and larvae: (1) before free amino acid utilization period, 0-4th day; (2) free amino acid utilization period, 4-10th day; (3) switching period, 10-13th day; and (4) lipovitellin and phospholipid utilization period, 13-21st day.

Kanazawa et al. (1991) studied on utilization of free amino acid as a source for energy in developing embryos of Japanese flounder. The availability of amino acids and fatty acids as energy sources during embryonic and larval development of red seabream were examined (Kanazawa 1998b). Newly spawned eggs were kept in a 500-L tank under a slow seawater flow rate. Eggs, from 1 hour after being spawned until hatching, and larvae, from hatching to open-mouth stage, were sampled every 4 hours and analyzed for amino acid and fatty acid contents. With advancing embryonic and larval development of red seabream, free amino acids gradually decreased. Fatty acid contents in the neutral and polar lipids were almost constant during embryonic development; however, after hatching, fatty acid in the neutral lipid gradually decreased. From these results, it was demonstrated that free amino acids are used during embryonic development, while free amino acids and neutral lipids are utilized after hatching, pointing to their importance as energy sources in eggs and larvae.

ENZYME SUPPLEMENT IN DIET

Dabrowski and Culver (1991) investigated the digestive tract and formulation of starter diets in larval fish. A study of changes in total lipids and lipase activity during development of red drum larvae was carried out (Holt and Sun 1991). The success of microparticulate diets presently used in seed production of marine fish larvae is limited. Poor performance of microparticulate diets is considered to be associated with the low digestive enzyme activity due to an undeveloped digestive tract. Enhanced growth in gilthead seabream larvae that were given microparticulate diet with added digestive enzyme has been reported (Tandler and Kolkovski 1991; Kolkovski et al. 1993a, 1993b). The addition of 0.05% pancreatin resulted in a significantly higher proportion of neutral lipid and protein being assimilated by the larvae. Results suggested the inclusion of exogenous digestive enzymes in microparticulate diets to improve digestibility in gilthead seabream larvae.

The addition of algae, *Isochrysis galbana*, in culture water and the activity level of some digestive enzymes in sea bass larvae were tested (Cahu et al. 1998a). Results suggested that the presence of algae facilitates the onset of hydrolytic functions of cell membranes and that the algae act by triggering digestive enzyme production at both the pancreatic and intestinal level. Whether sea bass larvae fed compound diets incorporating different levels of fish protein hydrolysate could normally develop their digestive functions was investigated (Cahu et al. 1999). Results showed that the incorporation of a moderate dietary level of fish protein hydrolysate facilitates the onset of the adult mode of digestion in developing fish.

APPLICATION OF MICROPARTICULATE DIETS IN AQUACULTURE

In the hatchery production of animals for aquaculture, live foods such as rotifer, *Brachionus plicatilis*, and brine shrimp, *Artemia salina*, have been widely used throughout the world. However, they require much equipment, maintenance expense, and manpower to produce a desired amount of live food safely and constantly. Therefore, it is necessary to develop artificial diets as substitutes for live foods to further increase the production of seed for fish culture. Various types of microparticulate diets for marine fish larvae have been studied (Adron et al. 1974; Gatesoupe and Luquet 1977; Gatesoupe et al. 1977; Metailler et al. 1979; Kanazawa et al. 1982a, 1985, 1989; Teshima et al. 1982; Bromley and Howell 1983; Kanazawa 1985, 1986, 1991c; Appelbaum 1985; Bromley and Sykes 1985; Kanazawa and Teshima 1988; Devresse et al. 1991; Juario et al. 1991; Juwana 1991; Leibovitz et al. 1991; Marte and Duray 1991; Parazo 1991; Walford and Lam 1991; Walford et al. 1991; Melotti et al. 1992). Jones et al. (1993) have reviewed detailed achievements in the substitution of artificial diets for conventional live feeds in bivalve, crustacean, and fish larval culture. Further, marine fish larvae feeding based on microparticulate diet has been reported by many researchers (Holt 1993; Person Le Ruyet et al. 1993; Escaffre and Bergot 1994; Hayashi et al. 1995; Yufera et al. 1995; Borlongan et al. 1996; Lee et al. 1996; Kolkovski et al. 1997; Rosenlund et al. 1997; Cahu et al. 1998b).

The microparticulate diets reported are categorized into three groups as shown below. Microencapsulated diets (MED) are defined as microparticulate diets made by encapsulating a solution, colloid, or suspension of diet ingredients within a membrane. Microbound diets (MBD) are powdered diets with a binder. Microcoated diets (MCD) are prepared by coating MBD with some material such as zein and cholesterol-lecithin. The nutritional components of microparticulate diets for fish larvae should be determined on the basis of the requirements of the larval fish for proteins, amino acids, lipids, carbohydrates, vitamins, and minerals. However, because the requirements of larval fish are still undefined, protein sources having high nutritional values are used, such as krill meal, squid meal, scallop meal, short-necked clam extract, chicken egg, skim milk, casein, gelatin, egg albumin, yeast, and fish meal (Teshima et al. 1982).

Commercial microparticulate diets are currently available in many countries. Combination of live food and microparticulate diet to supplement each other the defects of both diets has been widely used for stable and high quality hatchery production in aquaculture.

REFERENCES

Adron, J.W., A. Blair, and C.B. Cowey. 1974. Rearing of plaice (*Pleuronectes platessa*) larvae to metamorphosis using an artificial diet. Fishery Bulletin 72:352-357.

Ako, H., S. Kraul, and C. Tamaru. 1991. Pattern of fatty acid loss in several warmwater fish species during early development. Pages 23-25 *in* P. Lavens, P. Sorgeloos, E. Jaspers, and F. Ollevier, eds. Larvi L91-Fish & Crustacean Larviculture Symposium. European Aquaculture Society, Special Publication No.15, Gent, Belgium.

Ako, H., C.S. Tamaru, P. Bass, and C.S. Lee. 1994. Enhancing the resistance to physical stress in larvae of *Mugil cephalus* by the feeding of enriched *Artemia* nauplii. Aquaculture 122:81-90.

Alava, V.R., and A. Kanazawa. 1996. Effect of dietary fatty acids on growth of milkfish *Chanos chanos* fry in brackish water. Aquaculture 144:363-369.

Appelbaum, S. 1985. Rearing of the Dover sole, *Solea solea* (L), through its larval stages using artificial diets. Aquaculture 49:209-221.

Baker, E.P., D. Alves, and D.A. Bengtson. 1998. Effects of rotifer and *Artemia* fatty acid enrichment on survival, growth and pigmentation of summer flounder *Paralichthys dentatus* larvae. Journal of the World Aquaculture Society 29:494-498.

Bautista, M.N., and M.C. Dela Cruz. 1988. Linoleic (n-6) and linolenic (n-3) acids in the diet of fingerling milkfish (*Chanos chanos* Forsskal). Aquaculture 71:347-358.

Bell, J.G., D.R. Tocher, and J.R. Sargent. 1994. Effect of supplementation with 20:3(n-6), 20:4(n-6) and 20:5(n-3) on the production of prostaglandins E and F of the 1-, 2- and 3-series in turbot (*Scophthalmus maximus*) brain astroglial cells in primary culture. Biochimica et Biophysica Acta 1211:335-342.

Bell, J.G., D.R. Tocher, F.M. MacDonald, and J.R. Sargent. 1995a. Effects of dietary borage oil [enriched in gamma-linolenic acid, 18:3(n-6)] or marine fish oil [enriched in eicosapentaenoic acid, 20:5(n-3)] on growth, mortalities, liver histopathology and lipid composition of juvenile turbot (*Scophthalmus maximus*). Fish Physiology and Biochemistry 14:373-383.

Bell, M.V., R.S. Batty, J.R. Dick, K. Fretwell, J.C. Navarro, and J.R. Sargent. 1995b. Dietary deficiency of docosahexaenoic acid impairs vision at low light intensities in juvenile herrinig (*Clupea harengus* L.). Lipids 30:443-449.

Bell, M.V., L.A. McEvoy, and J.C. Navarro. 1996. Deficit of didocosahexaenoyl phospholipid in the eyes of larval sea bass fed an essential fatty acid deficient diet. Journal of Fish Biology 49:941-952.

Bergot, P. 1996. Purified diets used to study nitritional requirements of larvae. Page 4 (Hydrobiologie) INRA IFREMER Workshop on Fish Nutrition. Collected Papers. INRA Station. Saint Pre-Sur-Nivell, France.

Boren, R.S., and D.M. Gatlin, III. 1995. Dietary threonine requirement of juvenile red drum *Sciaenops ocellatus*. Journal of the World Aquaculture Society 26:279-283.

Borlongan, I.G., and L.V. Benitez. 1990. Quantitative lysine requirement of milkfish (*Chanos chanos*) juveniles. Aquaculture 87:341-347.

Borlongan, I., C. Marte, and J. Nocillado. 1996. Development of artificial diets for milkfish (*Chanos chanos*) larvae. Page 116 *in* C.B. Santiago, R.M. Coloso, O.M. Millamena, and I. Borlongan, eds. Feeds for Small Scale Aquaculture. Proceedings of the National Seminar Workshop on Fish Nutrition Feeds. Iloilo, Philippines.

Brinkmeyer, R.L., and G.J. Holt. 1998. Highly unsaturated fatty acids in diets for red drum (*Sciaenops ocellatus*) larvae. Aquaculture 161:253-268.

Brodtkorb, T., G. Rosenlund, and Ø. Lie. 1997. Effects of dietary levels of 20:5n-3 and 22:6n-3 on tissue lipid composition in juvenile Atlantic salmon, *Salmo salar*, with emphasis on brain and eye. Aquaculture Nutrition 3:175-187.

Bromley, P.J., and B.R. Howell. 1983. Factors influencing the survival and growth of turbot larvae, *Scophthalmus maximus* L. during the change from live to compound feeds. Aquaculture 31:31-40.

Bromley, P.J., and P.A. Sykes. 1985. Weaning diets for turbot (*Scophthalmus maximus* L.), sole (*Solea solea* L.) and cod (*Gadus morhua* L.). Pages 191-212 *in* C.B. Cowey, A.M. Mackie, and J.G. Bell, eds. Nutrition and Feeding in Fish. Academic Press, London, England.

Cahu, C.L., J. Zambonino Infante, A. Péres, P. Quazuguel, and M.M. Le Gall. 1998a. Algal addition in sea bass (*Dicentrarchus labrax*) larvae rearing: effect on digestive enzymes. Aquaculture 161:479-489.

Cahu, C.L., J.L. Zambonino Infante, A.M. Escaffre, P. Bergot, and S. Kaushik. 1998b. Preliminary results on sea bass (*Dicentrarchus labrax*) larvae rearing with compound diet from first feeding. Comparison with carp (*Cyprinus carpio*) larvae. Aquaculture 169:1-7.

Cahu, C.L., J.L. Zambonino Infante, P. Quazuguel, and M.M. Le Gall. 1999. Protein hydrolysate vs. fish meal in compound diets for 10-day old sea bass *Dicentrarchus labrax* larvae. Aquaculture 171:109-119.

Cairrao, F., L. Narcisco, and P. Pousao-Ferreira. 1995. Evaluation of the uptake of fatty acids by sea bream larvae (*Sparus aurata*) through enriched *Artemia* metanauplii. Page 194 *in* K. Pittman, and J. Verreth, eds. Mass Rearing of Juvenile Fish 201. Selected Papers from a Symposium Held in Bergen. Bergen, Norway.

Castell, J.D., J.G. Bell, D.R. Tocher, and J.R. Sargent. 1994. Effects of purified diets containing different combinations of arachidonic and docosahexaenoic acid on survival, growth and fatty acid composition of juvenile turbot (*Scophthalmus maximus*). Aquaculture 128:315-333.

Conceicao, L.E.C., T. Van der Meeren, J.A.J. Verreth, M.S. Evjen, D.F. Houlihan, and H.J. Fyhn. 1997. Amino acid metabolism and protein turnover in larval turbot

(*Scophthalmus maximus*) fed natural zooplankton or *Artemia*. Marine Biology 129:255-265.

Coutteau, P., G. Van-Stappen, and P. Sorgeloos. 1995. A standard experimental diet for the study of fatty acid requirements of weaning and first ongrowing stages of European sea bass (*Dicentrarchus labrax* L.): Selection of the basal diet. Pages 130-137 *in* K. Pittman, and J. Verreth, eds. Mass Rearing of Juvenile Fish 201. Selected Papers from a Symposium Held in Bergen. Bergen, Norway.

Coutteau, P., I. Geurden, M.R. Camara, P. Bergot, and P. Sorgeloos. 1997. Review on the dietary effects of phospholipids in fish and crustacean larviculture. Aquaculture 155:149-164.

Craig, S.R., and D.M. Gatlin, III. 1992. Dietary lysine requirement of juvenile red drum *Sciaenops ocellatus*. Journal of the World Aquaculture Society 23:133-137.

Craig, S.R., C.R. Arnold, and G.J. Holt. 1994. The effects of enriching live foods with highly unsaturated fatty acids on the growth and fatty acid composition of larval red drum *Sciaenops ocellatus*. Journal of the World Aquaculture Society 25:424-431.

Dabrowski, K., and D. Culver. 1991. The physiology of larval fish. Digestive tract and formulation of starter diets. Aquaculture Magazine 17(2):49-61.

Dedi, J., T. Takeuchi, T. Seikai, and T. Watanabe. 1995. Hypervitaminosis and safe levels of vitamin A for larval flounder (*Paralichthys olivaceus*) fed *Artemia* nauplii. Aquaculture 133:135-146.

Dedi, J., T. Takeuchi, T. Seikai, T. Watanabe, and K. Hosoya. 1997. Hypervitaminosis A during vertebral morphogenesis in larval Japanese flounder. Fisheries Science 63:466-473.

Dedi, J., T. Takeuchi, K. Hosoya, T. Watanabe, and T. Seikai. 1998. Effect of vitamin A levels in *Artemia* nauplii on the caudal skeleton formation of Japanese flounder *Paralichthys olivaceus*. Fisheries Science 64:344-345.

Devresse, B., P. Candreva, Ph. L ger, and P. Sorgeloos. 1991. A new artificial diet for the early weaning of sea bass (*Dicentrarchus labrax*) larvae. Pages 178-182 *in* P. Lavens, P. Sorgeloos, E. Jaspers, and F. Ollevier, eds. Larvi '91-Fish & Crustacean Larviculture Symposium. European Aquaculture Society, Special Publication No.15, Gent, Belgium.

Dhert, P., P. Lavens, and P. Sorgeloos. 1992. State of the art of Asian seabass *Lates calcarifer* larviculture. Journal of the World Aquaculture Society 23:317-329.

Diaz, J.P., L. Mani-Ponset, E. Guyot, and R. Connes. 1994. Assimilation of dissolved glucose from sea water by the sea bass *Dicentrarchus labrax* and the sea bream *Sparus aurata*. Marine Biology 120:181-186.

Escaffre, A.M., and P. Bergot. 1994. Artificial feeding and nutritional requirements of fish larvae. Pisciculture Francaise 118:36-40.

Estévez, A., and A. Kanazawa. 1996. Fatty acid composition of neural tissues of normally pigmented and unpigmented juveniles of Japanese flounder using rotifer and *Artemia* enriched in n-3 HUFA. Fisheries Science 62:88-93.

Estévez, A., M. Sameshima, M. Ishikawa, and A. Kanazawa. 1997a. Effect of diets containing low levels of methionine and oxidized oil on body composition, retina structure and pigmentation success of Japanese flounder. Aquaculture Nutrition 3:201-216.

Estévez, A., M. Ishikawa, and A. Kanazawa. 1997b. Effect of arachidonic acid on pigmentation and fatty acid composition of Japanese flounder, *Paralichthys olivaceus* (Temminck and Schlegel). Aquaculture Reserch 28:279-289.

Evjemo, J.O., P. Coutteau, Y. Olsen, and P. Sorgeloos. 1997. The stability of docosahexaenoic acid in two Artemia species following enrichment and subsequent starvation. Aquaculture 155:135-148.

Finn, R.N., I. Rønnestad, and H.J. Fyhn. 1995a. Respiration, nitrogen and energy metabolism of developing yolk-sac larvae of Atlantic halibut (*Hippoglossus hippoglossus* L.). Comparative Biochemistry and Physiology 111A:647-671.

Finn, R.N., J.R. Henderson, and H.J. Fyhn. 1995b. Physiological energetics of developing embryos and yolk-sac larvae of Atlantic cod (*Gadus morhua*). 2. Lipid metabolism and enthalpy balance. Marine Biology 124:371-379.

Fontagne, S., I. Geurden, A.M. Escaffre, and P. Bergot. 1998. Histological changes induced by dietary phospholipids in intestine and liver of common carp (*Cyprinus carpio* L.) larvae. Aquaculture 161:213-223.

Forster, I., and H.Y. Ogata. 1998. Lysine requirement of juvenile Japanese flounder *Paralichthys olivaceus* and juvenile red sea bream *Pagrus major*. Aquaculture 161:131-142.

Fukusho, K., T. Yamamoto, and T. Seikai. 1986. Influence of various amount of aeration during larval development of hatchery-reared flounder *Paralichthys olivaceus* on the appearance of abnormal coloration. Bulletin of the Natural Resources Institute. Aquaculture No.10:53-56.

Fukusho, K., H. Nanba, T. Yamamoto, Y. Yamasaki, M.T. Lee, T. Seikai, and T. Watanabe. 1987. Reduction of albinism in juvenile flounder *Paralichthys olivaceus* hatchery-reared on fertilized eggs of red sea bream *Pagrus major*, and its critical stage for effective feeding. Bulletin of the Natural Resources Institute. Aquaculture No.12:1-7.

Furuita, H., T. Takeuchi, M. Toyota, and T. Watanabe. 1996a. EPA and DHA requirements in early juvenile red sea bream using HUFA enriched *Artemia* nauplii. Fisheries Science 62:246-251.

Furuita, H., T. Takeuchi, T. Watanabe, H. Fujimoto, S. Sekiya, and K. Imaizumi. 1996b. Requirements of larval yellowtail for eicosapentaenoic acid, docosahexaenoic acid, and n-3 highly unsaturated faty acid. Fisheries Science 62:372-379.

Furuita, H., T. Takeuchi, and K. Uematsu. 1998. Effects of eicosapentaenoic and docosahexaenoic acids on growth, survival and brain development of larval Japanese flounder (*Paralichthys olivaceus*). Aquaculture 161:269-279.

Fyhn, H.J. 1989. First feeding of marine fish larvae: Are free amino acids the source of energy? Aquaculture 80:111-120.

Gapasin, R.S.J., R. Bombeo, P. Lavens, P. Sorgeloos, and H. Nelis. 1998. Enrichment of live food with essential fatty acids and vitamin C: effects on milkfish (*Chanos chanos*) larval performance. Aquaculture 162:269-286.

Gatesoupe, F.J. 1994. Lactic acid bacteria increase the resistance of turbot larvae, *Scophthalmus maximus*, against pathogenic *Vibrio*. Aquatic Living Resources 7:277-282.

Gatesoupe, F.J., and P. Luquet. 1977. Recherche d'une alimentation artificielle adaptée a l'élevage des stades larvaires des poisson I. Comparaison de quelques techniques destinées à améliorer la stabilite à l'eau des aliments. Pages 13-20 *in* Third Meeting, Working Group on Mariculture 4. International Council Exploration Sea. Brest, France.

Gatesoupe, F.J., M. Girin, and P. Luquet. 1977. Recherche d'une alimentation artificielle adaptée a l' élevage des stades larvaires des poisson II. Application a l'élevage alvaire due bar et de la sole. Pages 59-66 *in* Third Meeting, Working Group on Mariculture 4. International Council Exploration Sea. Brest, France.

Geurden, I., P. Coutteau, and P. Sorgeloos. 1995a. Dietary phospholipids for European seabass (*Dicentrarchus labrax* L.) during first on growing. Pages 175-178 *in* P. Lavens, E. Jaspers, and I. Roelants, eds. Larvi '95-Fish & Shellfish Larviculture Symposium. European Aquaculture Society, Special Publication No. 24, Gent, Belgium.

Geurden, I., J. Radünz-Neto, and P. Bergot. 1995b. Essentiality of dietary phospholipids for carp (*Cyprinus carpio* L.) larvae. Aquaculture 131:303-314.

Geurden, I., D. Marion, N. Charlon, P. Coutteau, and P. Bergot. 1998a. Comparison of different soybean phospholipidic fractions as dietay supplements for common carp, *Cyprinus carpio*, larvae. Aquaculture 161:225-235.

Geurden, I., O.S. Reyes, P. Bergot, P. Coutteau, and P. Sorgeloos. 1998b. Incorporation of fatty acids from dietary neutral lipid in eye, brain and muscle of postlarval turbot fed diets with different types of phosphatidylcholine. Fish Physiology and Biochemistry 19:365-375.

Geurden, I., P. Bergot, K. Van Ryckeghem, and P. Sorgeloos. 1999. Phospholipid composition of common carp (*Cyprinus carpio*) larvae starved or fed different phospholipid classes. Aquaculture 171:93-107.

Gouillou-Coustans, M.-F., P. Bergot, and S.J. Kaushik. 1998. Dietary ascorbic acid needs of common carp (*Cyprinus carpio*) larvae. Aquaculture 161:453-461.

Hapette, A.M., S. Coombs, R. Williams, and S.A. Poulet. 1991. Variation in vitamin C content of sprat larvae (*Sprattus sprattus*) in the Irish Sea. Marine Biology 108:39-48.

Hara, T.J. 1977. Further studies on the structure-activity relationships of amino acids in fish olfaction. Comparative Biochemistry and Physiology 56A:559-565.

Hayashi, M., K. Yoda, A. Kanazawa, and S. Kitaoka. 1995. Effects of supplementing *Euglena gracilis* to microbound diets for larval red sea bream *Pagrus major* and Japanese flounder *Paralichthys olivaceus*. Saibai Giken 23:103-107.

Hidalgo, F., E. Alliot, and H. Thebault. 1987. Methionine- and cystine-supplemented diets for juvenile sea bass (*Dicentrarchus labrax*). Aquaculture 64:209-217.

Holt, G.J. 1993. Feeding larval red drum on microparticulate diets in a closed recirculating water system. Journal of the World Aquaculture Society 24:225-230.

Holt, G.J., and F. Sun. 1991. Lipase activity and total lipid content during early development of red drum *Sciaenops ocellatus*. Pages 30-33 *in* P. Lavens, P. Sorgeloos, E. Jaspers, and F. Ollevier, eds. Larvi '91-Fish & Crustacean Larviculture Symposium. European Aquaculture Society, Special Publication No.15, Gent, Belgium.

Ishizaki, Y., T. Takeuchi, T. Watanabe, M. Arimoto, and K. Shimizu. 1998. A preliminary experiment on the effect of *Artemia* enriched with arachidonic acid on survival and growth of yellowtail. Fisheries Science 64:295-299.

Izquierdo, M.S., and H. Fernandez-Palacios. 1997. Nutritional requirements of marine fish larvae and broodstock. Pages 243-264 *in* A. Tacon, and B. Basurco, eds. Feeding Tomorrow's Fish. No. 22, Proceedings of the Workshop of the CIHEAM Network on Technology of Aquaculture in the Mediterranean TECAM. Zaragoza, Spain.

Izquierdo, M.S., T. Watanabe, T. Takeuchi, T. Arakawa, and C. Kitajima. 1989. Requirement of larval red seabream *Pagrus major* for essential fatty acids. Nippon Suisan Gakkaishi 55:859-867.

Izquierdo, M.S., T. Watanabe, T. Takeuchi, T. Arakawa, and C. Kitajima. 1990. Optimal EFA levels in Artemia to meet the EFA requirements of red seabream (*Pagrus major*). Pages 221-232 *in* M. Takeda, and T. Watanabe, eds. The Current Status of Fish Nutrition in Aquaculture. Proceedings of the Third International Symposium on Feeding and Nutrition in Fish. Toba, Japan.

Izquierdo, M.S., T. Arakawa, T. Takeuchi, R. Haroun, and T. Watanabe. 1992. Effect of n-3 HUFA levels in *Artemia* on growth of larval Japanese flounder (*Paralichthys olivaceus*). Aquaculture 105:73-82.

Jones, D.A., M.S. Kamarudin, and L. Le Vay. 1993. The potential for replacement of live feeds in larval culture. Journal of the World Aquaculture Society 24:199-210.

Juario, J.V., M.N. Duray, and J. Fuchs. 1991. Weaning of seabass, *Lates calcarifer* B., larvae to artificial diet. Page 183 *in* P. Lavens, P. Sorgeloós, E. Jaspers, and F. Ollevier, eds. Larvi '91-Fish & Crustacean Larviculture Symposium. European Aquaculture Society, Special Publication No.15, Gent, Belgium.

Juwana, S. 1991. Current status of microencapsulated diets for aquaculture. Oseana 16:25-36.

Kanazawa, A. 1985. Essential fatty acid and lipid requirement of fish. Pages 281-298 *in* C.B. Cowey, A.M. Mackle, and J.G. Bell, eds. Nutrition and Feeding in Fish. Academic Press, London, England.

Kanazawa, A. 1986. New developments in fish nutrition. Pages 9-14 *in* J.L. Maclean, L.B. Dizon, and L.V. Hosillos, eds. The First Asian Fisheries Forum. Asian Fisheries Society, Manila, Philippines.

Kanazawa, A. 1988. Nutritional requirements and formulated feed of flounder-I. Larval fish. Fish Culture 25:116-119.

Kanazawa, A. 1991a. Ayu, *Plecoglossum altivelis*. Pages 23-29 *in* R.P. Wilson, ed. Handbook of Nutrient Requirements of Finfish. CRC Press, Boca Raton, Florida.

Kanazawa, A. 1991b. Recent advances in fish nutrition. Pages 1-26, Proceeding of the 16th Roche Feed Seminar. Nippon Roche, Japan.

Kanazawa, A. 1991c. Puffer fish, *Fugu rubripes*. Pages 123-129 *in* R.P. Wilson, ed. Handbook of Nutrient Requirements of Finfish. CRC Press, Boca Raton, Florida.

Kanazawa, A. 1993a. Larval nutrition and seed production of fish by microparticulate diets. Pages 515-536 *in* S.A. Al-Thobaiti, H.M. Al-Hinty, A.Q. Siddiqui, and G. Hussain, eds. Proceedings of the First International Symposium on Aquaculture Technology and Investiment Opportunities. Ministry of Agriculture and Water, King Abdulaziz City for Science and Technology and Riyadh Chamber of Commerce and Industry. Riyadh, Saudi Arabia.

Kanazawa, A. 1993b. Importance of dietary docosahexaenoic acid on growth and survival of fish larvae. Pages 87-95 *in* C.S. Lee, M.S. Su, and I.C. Liao, eds. Finfish

Hatchery in Asia. Proceedings of Finfish Hatchery in Asia '91. TML Conference Proceedings 3, Tungkang Marine Laboratory, Taiwan Fisheries Research Institute, Tungkang, Taiwan.

Kanazawa, A. 1993c. Nutritional mechanisms involved in the occurrence of abnormal pigmentation in hatchery-reared flatfish. Journal of the World Aquaculture Society 24:162-166.

Kanazawa, A. 1993d. Essential phospholipids of fish and crustaceans. Pages 519-530 *in* S.J. Kaushik, and P. Luquet, eds. Fish Nutrition in Practice. Institut National de la Recherche Agronomique Edition, Les Colloques, No. 61. Paris, France.

Kanazawa, A. 1995. Nutrition of larval fish. Pages 50-59 *in* C.E. Lim, and D.J. Sessa, eds. Nutrition and Utilization Technology in Aquaculture. American Oil Chemists' Society Press, Champaign, Illinois.

Kanazawa, A. 1997. Effects of docosahexaenoic acid and phospholipids on stress tolerance of fish. Aquaculture 155:129-134.

Kanazawa, A. 1998a. Importance of dietary lipids in flatfish. Pages 181-186 *in* W.H. Wowell, B.J. Keller, P.K. Park, J.P. McVey, K. Takayanagi, and Y. Uekita, eds. Proceedings of the 26th US-Japan Aquaculture Symposium, Nutrition and Technical Development of Aquaculture. UJNR Technical Report No.26, Durham, New Hampshire.

Kanazawa, A. 1998b. Nutrition of fish larvae. IV International Symposium in Aquatic Nutrition, 15-18 November, 1998, La Paz, Mexico. Manuscripts of Conference, Part 2.

Kanazawa, A., and S. Teshima. 1988. Microparticulate diets for fish larvae. Pages 57-62 *in* A.K. Sparks. ed. Proceedings of the 14th US-Japan Meeting on Aquaculture, New and Innovative Advances in Biology/Engineering with Potential for Use in Aquaculture. NOAA Technical Report NMFS 70, U.S. Department of Commerce, Seattle, Washington.

Kanazawa, A., S. Teshima, S. Inamori, T. Iwashita, and A. Nagao. 1981. Effects of phospholipids on growth, survival rate, and incidence of malformation in the larval ayu. Memoirs of the Faculty of Fisheries of Kagoshima University 30:301-309.

Kanazawa, A., S. Teshima, S. Inamori, S. Sumida, and T. Iwashita. 1982a. Rearing of larval red sea bream and ayu with artificial diets. Memoirs of the Faculty of Fisheries of Kagoshima University 31:185-192.

Kanazawa, A., S. Teshima, N. Imatanaka, O. Imada, and A. Inoue. 1982b. Tissue uptake of radioactive eicosapentaenoic acid in the red sea bream. Bulletin of the Japanese Society of Scientific Fisheries 48:1441-1444.

Kanazawa, A., S. Teshima, T. Kobayashi, M. Takae, T. Iwashita, and R. Uehara. 1983a. Necessity of dietary phospholipid for growth of the larval ayu. Memoirs of the Faculty of Fisheries of Kagoshimia University 32:115-120.

Kanazawa, A., S. Teshima, S. Inamori, and H. Matsubara. 1983b. Effects of dietary phospholipids on growth of the larval red sea bream and knife jaw. Memoirs of the Faculty of Fisheries of Kagoshima University 32:109-114.

Kanazawa, A., S. Teshima, and M. Sakamoto. 1985. Effects of dietary bonito-egg phospholipids and some phospholipids on growth and survival of the larval ayu, *Plecoglossus altivelis*. Zeitschrift für Angewandte Ichthyologie 4:165-170.

Kanazawa, A., S. Koshio, and S. Teshima. 1989. Growth and survival of larval red sea bream *Pagrus major* and Japanese flounder *Paralichths olivaceus* fed microbound diets. Journal of the World Aquaculture Society 20:31-37.

Kanazawa, A.. S. Teshima, T. Hara, and K. Kobayashi. 1991. Utulization of free amino acid as a source for energy in developing embryos of fish. Page 124 *in* Abstracts of the Annual Meeting of Japanese Society of Scientific Fisheries, Autumn, 1991, Sanriku, Japan.

Kitajima, C. 1985. Living feeds. Pages 75-88 *in* Y. Yone, ed. Fish Nutrition and Diets. Koseisha Koseikaku, Tokyo.

Kolkovski, S., A. Tandler, and G.W. Kissil. 1993a. The effect of dietary enzymes with age on protein and lipid assimilation and deposition in *Sparus aurata* larvae. Pages 569-578 *in* S.J. Kaushik, and P. Luquet, eds. Fish Nutrition in Practice. Institut National de la Recherche Agronomique Edition. Les Colloques, No.61, Paris, France.

Kolkovski, S., A. Tandler, G.W. Kissil, and A. Gertler. 1993b. The effect of dietary exogenous digestive enzymes on ingestion, assimilation, growth and survival of gilthead seabream (*Sparus aurata*, Sparidae, Linnaeus) larvae. Fish Physiology and Biochemistry 12:203-209.

Kolkovski, S., W. Koven, and A. Tandler. 1997. The mode of action of Artemia in enhancing utilization of microdiet by gilthead seabream *Sparus aurata* larvae. Aquaculture 155:193-205.

Korsgaard, B. 1991. Metabolism of larval turbot *Scophthalmus maximus* (L.) and uptake of amino acids from seawater studied by autoradiographic and radiochemical methods. Journal of Experimental Marine Biology and Ecology 148:1-10.

Kosutarak, P., A. Kanazawa, S. Teshima, S. Koshio, and S. Itoh. 1994. L-Ascorbyl-2-phosphate Mg as a vitamin C source for red seabream (*Pagrus major*) juveniles. Pages 729-732 *in* L.M. Chou, A.D. Munro, T.J. Lam, T.W. Chen, L.K.K. Cheong, J.K. Ding, K.K. Hooi, H.W. Khoo, V.P.E. Phang, K.F. Shim, and C.H. Tan, eds. The Third Asian Fisheries Forum. Asian Fisheries Society, Manila, Philippines.

Kosutarak, P., A. Kanazawa, S. Teshima, and S. Koshio. 1995a. Interactions of L-ascorbyl-2-phosphate Mg and oxidized fish oil on red seabream juveniles. Fisheries Science 61:696-702.

Kosutarak, P., A. Kanazawa, S. Teshima, and S. Koshio. 1995b. Interactions of L-ascorbyl-2-phosphate Mg and n-3 highly unsaturated fatty acids on Japanese flounder juveniles. Fisheries Science 61:860-866.

Koven, W.M., G.M. Kissil, and A. Tandler. 1989. Lipid and n-3 requirement of *Sparus aurata* larvae during starvation and feeding. Aquaculture 79:185-191.

Koven, W.M., A. Tandler, G.W. Kissil, D. Sklan, O. Friezlander, and M. Harel. 1990. The effect of dietary (n-3) polyunsaturated fatty acids on growth, survival and swim bladder development in *Sparus aurata* larvae. Aquaculture 91:131-141.

Koven, W.M., A. Tandler, G.M. Kissil, and D. Sklan. 1992. The importance of n-3 highly unsaturated fatty acids for growth in larval *Sparus aurata* and their effect on survival, lipid composition and size distribution. Aquaculture 104:91-104.

Koven, W.M., S. Kolkovski, A. Tandler, G.W. Kissil, and D. Sklan. 1993. The effect of dietary lecithin and lipase, as a function of age, on n-9 fatty acid incorporation in the tissue lipids of *Sparus aurata* larvae. Fish Physiology and Biochemistry 10:357-364.

Kraul, S., H. Ako, K. Brittain, A. Ogasawara, R. Cantrell, and T. Nagao. 1991. Comparison of copepods and enriched *Artemia* as feeds for larval mahimahi, *Coryphaena hippurus*. Pages 45-47 *in* P. Lavens, P. Sorgeloos, E. Jaspers, and F. Ollevier, eds. Larvi '91-Fish & Crustacean Larviculture Symposium. European Aquaculture Society, Special Publication No. 15, Gent, Bilgium.

Kraul, S., A. Nelson, K. Brittain, H. Ako, and A. Ogasawara. 1992. Evaluation of live feeds for larval and postlarval mahimahi *Coryphaena hippurus*. Journal of the World Aquaculture Society 23:299-306.

Kraul, S., K. Brittain, R. Cantrell, T. Nagao, H. Ako, A. Ogasawara, and H. Kitagawa. 1993. Nutritional factors affecting stress resistance in the larval mahimahi *Coryphaena hippurus*. Journal of the World Aquaculture Society 24:186-193.

Lee, P.S., P.C. Southgate, and D.S. Fielder. 1996. Assessment of two microbound artificial diets for weaning Asian sea bass (*Lates calcarifer*, Bloch). Asian Fisheries Science 9:115-120.

Leibovitz, H.E., D.A. Bengtson, and K.L. Simpson. 1991. Evaluation of microcapsule performance on delivering diets to silverside (*Menidia beryllina*) larvae. Pages 173-174 *in* P. Lavens, P. Sorgeloos, E. Jaspers, and F. Ollevier eds. Larvi '91-Fish & Crustacean larviculture Symposium. European Aquaculture Society, Special Publication No. 15, Gent, Belgium.

Liu, J.K. 1998. Effect of artificial regulations of Artemia n-3 HUFA content on growth and survival of black seabream (*Sparus macrocephalus*) larvae. Chinese Journal of Oceanology and Limnology 16:173-176.

López-Alvarado, J., and A. Kanazawa. 1994. Effect of dietary arginine levels on growth of red sea bream larvae fed diets supplemented with crystalline amino acids. Fisheries Science 60:435-439.

López-Alvarado, J., and A. Kanazawa. 1995. Optimum levels of crystalline amino acids in diets for larval red sea bream (*Pagrus major*). ICES Marine Science Symposia 201:100-105.

López-Alvarado, J., and A. Kanazawa. 1997. Effect of dietary protein sources in microdiets on feeding behavior and growth of red sea bream, *Pagrus major*, during weaning and metamorphosis. Journal of Applied Aquaculture 7(3):53-66.

Mangor-Jensen, A., and R.N. Birkeland. 1993. Effects of dietary carbohydrate on broodstock maturation and egg quality in cod. Page 14 Milestone report 1993, 9. Bergen, Norway.

Mangor-Jensen, A., J.C. Holm, G. Rosenlund, Ø. Lie, and K. Sandnes. 1994. Effects of dietary vitamin C on maturation and egg quality of cod *Gadus morhua* L. Journal of the World Aquaculture Society 25:30-40.

Marte, C.L., and M.N. Duray. 1991. Microbound larval feed as supplement to live food for milkfish (*Chanos chanos* Forskal) larvae. Pages 175-177 *in* P. Lavens, P. Sorgeloos, E. Jaspers, and F. Ollevier, eds. Larvi '91-Fish & Crustacean Larviculture Symposium. European Aquaculture Society, Special Publication No. 15, Gent, Belgium.

Melotti, P., M. Amerio, L. Gennari, and A. Roncarati. 1992. Comparative evaluation of some microdiets for seabass (*Dicentrarchus labrax* L.) weaning. Zootecnica Nutrizione Animale 18:191-200.

Merchie, G., P. Lavens, P. Dhert, M. Dehasque, H. Nelis, A. De Leenheer, and P. Sorgeloos. 1995a. Variation of ascorbic acid content in different live food organisms. Aquaculture 134:325-337.

Merchie, G., P. Lavens, R. Pector, A.F. Mai-Soni, H. Nelis, A. De Jeenheer, and P. Sorgeloos. 1995b. Live food mediated vitamin C transfer to *Dicentrarchus labrax* and *Clarias gariepinus*. Journal of Applied Ichthyology 11:336-341.

Merchie, G., P. Lavens, P. Dhert, M. García Ulloa Gómez, H. Nelis, A. De Leenheer, and P. Sorgeloos. 1996a. Dietary ascorbic acid requirements during the hatchery production of turbot larvae. Journal of Fish Biology 49:573-583.

Merchie, G., P. Lavens, V. Storch, U. Übel, H. Nelis, A. De Leenheer, and P. Sorgeloos. 1996b. Influence of dietary vitamin C dosage on turbot (*Scophthalmus maximus*) and European sea bass (*Dicentrarchus labrax*) nursery stages. Comparative Biochemistry and Physiology 114A:123-133.

Merchie, G., P. Lavens, and P. Sorgeloos. 1997. Optimization of dietary vitamin C in fish and crustacean larvae: a review. Aquaculture 155:165-181.

Metailler, R., C. Manant, and C. Depierre. 1979. Microparticules alimentaires inertes destinées à l'élevage larvaire des poissons marinis. Utilizationdes alginates. Pages 181-190. *in* Proceedings of the World Symposium on Finfish Nutrition and Fishfeed Technology. Hamburg, Germany.

Miki, N., T. Taniguchi, and H. Hamakawa. 1988. Effect of rotifer enriched with fat-soluble vitamins on occurrence of albinism in hatchery-reared "hirame" *Paralichthys olivaceus* (Preliminary report). Suisan-zoshoku 36:91-96.

Miki, N., T. Taniguchi, and H. Hamakawa. 1989. Adequate vitamin level for reduction of albinism in hatchery-reared "hirame" *Paralichthys olivaceus* fed on rotifer enriched with fat-soluble vitamins. Suisan-zoshoku 37:109-114.

Miki, N., T. Taniguchi, and H. Hamakawa. 1990. Reduction of albinism in hatchery-reared flounder "hirame" *Paralichthys olivaceus* by feeding on rotifer enriched with vitamin A. Suisan-zoshoku 38:147-155.

Moon, H.Y., and D.M. Gatlin, III. 1991. Total sulfur amino acid requirement of juvenile red drum, *Sciaenops ocellatus*. Aquaculture 95:97-106.

Mourente, G., and D.R. Tocher. 1992. Effects of weaning onto a pelleted diet on docosahexaenoic acid (22:6n-3) levels in brain of developing turbot (*Scophthalmus maximus* L.). Aquaculture 105:363-377.

Mourente, G., and D.R. Tocher. 1993. The effects of weaning onto a dry pellet diet on brain lipid and fatty acid compositions in post-larval gilthead sea bream (*Sparus surata* L.). Comparative Biochemistry and Physilogy 104A:605-611.

Mourente. G., D.R. Tocher, and J.R. Sargent. 1991. Specific accumulation of docosahexaenoic acid (22:6n-3) in brain lipids during development of juvenile turbot *Scophthalmus maximus* L. Lipids 26:871-877.

Nakamura, K., and H. Iida. 1986. Relationship between albinism and riboflavin amount in flounders *Paralichthys olivaceus*. Nippon Suisan Gakkaishi 52:1275-1279.

Nakamura, K., H. Iida, and H. Nakano. 1986. Riboflavin in the skin of albinic flatfish *Liopsetta obscura*. Nippon Suisan Gakkaishi 52:2207.

Navarro, J.C., R.S. Batty, M.V. Bell, and J.R. Sargent. 1993. Effects of two *Artemia* diets with different contents of polyunsaturated fatty acids on the lipid composition of larvae of Atlantic herring (*Clupea harengus*). Journal of Fish Biology 43:503-515.

Navarro, J.C., L.A. McEvoy, F. Amat, and J.R. Sargent. 1995. Effects of diet on fatty acid composition of body sones in larvae of the sea bass *Dicentrarchus labrax*: A chemometric study. Marine Biology 124:177-183.

Nematipour, G.R., M.L. Brown, and D.M. Gatlin, III. 1992. Effects of dietary carbohydrate: lipid ratio on growth and body composition of hybrid striped bass. Journal of the World Aquaculture Society 23:128-132.

Ohkubo, N., and T. Matsubara. 1998. Sequential utilization of free amino acids, yolk protein, and lipids by developing embryos and larvae in barfin flounder *Verasper moseri*. Pages 181-186 *in* W.H. Wowell, B.J. Keller, P.K. Park, J.P. McVey, K. Takayanagi, and Y. Uekita, eds. Proceedings of the 26th US-Japan Aquaculture Symposium, Nutrition and Technical Development of Aquaculture. UJNR Technical Report No. 26, Durham, New Hampshire.

Ostrowski, A.C., and S. Divakaran. 1990. Survival and bioconversion of n-3 fatty acids during early development of dolphin (*Coryphaena hippurus*) larvae fed oil-enriched rotifers. Aquaculture 89:273-285.

Pagliarani, A., M. Pirini, G. Trigari, and V. Ventrella. 1986. Effects of diets containing different oils on brain fatty acid composition in sea bass (*Dicentrarchus labrax* L.). Comparative Biochemistry and Physiology 83B: 277-282.

Parazo, M.M. 1991. An artificial diet for larval rabbitfish, *Siganus guttatus* Bloch. Pages 43-48 *in* S. De Silva, ed. Fish Nutrition Research in Asia, Vol. 5. Proceedings of the Fourth Asian Fish Nutrition Workshop, Manila, Philippines.

Person Le Ruyet, J., J.C. Alexandre, L. Th baud, and C. Mugnier. 1993. Marine fish larvae feeding: formulated diets or live prey? Journal of the World Aquaculture Society 24:211-224.

Rainuzzo, J.R. 1993. Fatty acid and lipid composition of fish egg and larvae. Pages 43-49 *in* L.A. Reinertsen, L. Joergensen, and K. Tvinnereim, eds. Fish Farming Technology. Proceedings of the First International Conference on Fish Farming Technology, Trondheim, Norway.

Rainuzzo, J.R., K.I. Reitan, and L. Joergensen. 1992. Comparative study on the fatty acid and lipid composition of four marine fish larvae. Comparative Biochemistry and Physiology 103B:21-26.

Rainuzzo, J.R., K.I. Reitan, and Y. Olsen. 1997. The significance of lipids at early stages of marine fish: a review. Aquaculture 155:103-115.

Reitan, K.I., J.R. Rainuzzo, G. Øie, and Y. Olsen. 1997. A review of the nutritional effects of algae in marine fish larvae. Aquaculture 155:207-221.

Robin, J.H. 1994. Importance of n-3 and n-6 fatty acids for marine fish larvae feeding. Pisciculture Francaise 118:21-28.

Robin, J.H. 1995. The importance of n-6 fatty acids in the culture of marine fish larvae. Pages 106-111 *in* K. Pittman and J. Verreth, eds. Mass Rearing of Juvenile Fish. Vol. 201, Selected Papers from a Symposium Held in Bergen. Bergen, Norway.

Robin, J.H. 1998. Use of borage oil in rotifer production and *Artemia* enrichment: effect on growth and survival of turbot (*Scophthalmus maximus*) larvae. Aquaculture 161:323-331.

Robin, J.H., M.M. Le Gall, and H. Le Delliou. 1993. Comparison of three kinds of rotifer enrichments for turbot larval culture. Pages 619-622 *in* S.J. Kaushik, and P. Luquet. eds. Fish Nutrition in Practice. Les Colloques, No. 61, Paris, France.

Rodriguez, C., J.A. Perez, A. Lorenzo, M.S. Izquierdo, and J.R. Cejas. 1994. n-3 HUFA requirement of larval gilthead seabream *Sparus aurata* when using high levels of eicosapentaenoic acid. Comparative Biochemistry and Physiology 107A: 693-698.

Rønnestad, I., and H.J. Fyhn. 1993. Metabolic aspects of free amino acids in developing marine fish eggs and larvae. Reviews in Fisheries Science 1:239-259.

Rønnestad, I., E.P. Groot, and H.J. Fyhn. 1993. Compartmental distribution of frees amino acids and protein in developing yolk-sac larvae of Atlantic halibut (*Hippoglossus hippoglossus*). Marine Biology 116:349-354.

Rønnestad, I., R.N. Finn, I. Lein, and Ø. Lie. 1995. Compartmental changes in the contents of total lipid, lipid classes and their associated fatty acids in developing yolk-sac larvae of Atlantic halibut, *Hippoglossus hippoglossus* (L.). Aquaculture Nutrition 1:119-130.

Rønnestad, I., Ø. Lie, and R. Waagb. 1997. Vitamin B_6 in Atlantic halibut, *Hippoglossus hippoglossus*–endogenous utilization and retention in larvae fed natural zooplankton. Aquaculture 157:337-345.

Rønnestad, I., S. Helland, and Ø. Lie. 1998. Feeding Artemia to larvae of Atlantic halibut (*Hippoglossus hippoglossus* L.) results in lower larval vitamin A content compared with feeding copepods. Aquaculture 165:159-164.

Rosenlund, G., J. Stoss, and C. Talbot. 1997. Co-feeding marine fish larvae with inert and live diets. Aquaculture 155:183-191.

Salhi, M., M.S. Izquierdo, C.M. Hernández-Cruz, M. González, and H. Fernández-Palacios. 1994. Effect of lipid and n-3 HUFA levels in microdiets on growth, survival and fatty acid composition of larval gilthead seabream (*Sparus aurata*). Aquaculture 124:275-282.

Sangster, A.W., S.E. Thomas, and N.L. Tingling. 1975. Fish attractants from marine invertebrates. Arcamine from *Arca zebra* and Strombine from *Strombus gigas*. Tetrahedron 31:1135-1137.

Sargent, J.R., L.A. McEvoy, and J.G. Bell. 1997. Requirements, presentation and sources of polyunsaturated fatty acids in marine fish larval feeds. Aquaculture 155:117-127.

Seikai, T. 1985a. Influence of feeding periods of Brazilian *Artemia* during larval development of hatchery-reared flounder *Paralichthys olivaceus* on the appearance of albinism. Nippon Suisan Gakkaishi 51:521-527.

Seikai, T. 1985b. Reduction in occurrence frequency of albinism in juvenile flounder *Paralichthys olivaceus* hatchery-reared on wild zooplankton. Nippon Suisan Gakkaishi 51:1261-1267.

Seikai, T., M. Shimozaki, and T. Watanabe. 1987a. Estimation of larval stage determining the appearance of albinism in hatchery-reared juvenile flounder *Paralichthys olivaceus*. Nippon Suisan Gakkaishi 53:1107-1114.

Seikai, T., T. Watanabe, and M. Shimozaki. 1987b. Influence of three geographically different strains of *Artemia* nauplii on occurrence of albinism in hatchery-reared flounder *Paralichthys olivaceus*. Nippon Suisan Gakkaishi 53:195-200.

Sorgeloos, P. 1994. State of the art in marine fish larviculture. World Aquaculture 25(3):34-37.

Sorgeloos, P., M. Dehasque, P. Dhert, and P. Lavens. 1995. Review of some aspects of marine fish larviculture. Pages 138-142 *in* K. Pittman, and J. Verreth, eds. Mass Rearing of Juvenile Fish. Vol. 201, Papers from a Symposium Held in Bergen. Bergen, Norway.

Støttrup, J.G., and Y. Attramadal. 1992. The influence of different rotifer and *Artemia* enrichment diets on growth, survival and pigmentation in turbot (*Scophthalmus maximus* L.) larvae. Journal of the World Aquaculture Society 23:307-316.

Tachibana, K., M. Yagi, K. Hara, T. Mishima, and M. Tsuchimoto. 1997. Effects of feeding of beta-carotene-supplemented rotifers on survival and lymphocyte proliferation reaction of fish larvae (Japanese parrotfish (*Oplegnathus fasciatus*) and spotted parrotfish (*Oplegnathus punctatus*): preliminary trials. Hydrobiologia 358: 313-316.

Tago, A., Y. Yamamoto, S. Teshima, and A. Kanazawa. 1999. Effects of 1,2-di-20: 5-phosphatidylcholine (PC) and 1,2-di-22:6-PC on growth and stress tolerance of Japanese flounder (*Paralichthys olivaceus*) larvae. Aquaculture 179:231-239.

Takeuchi, T. 1997. Essential fatty acid requirements of aquatic animals with emphasis on fish larvae and fingerlings. Reviews in Fisheries Science 5:1-25.

Takeuchi, T., M. Toyota, S. Satoh, and T. Watanabe. 1990. Requirement of juvenile red sea bream for eicosoapentaenoic and docosahexaenoic acids. Nippon Suisan Gakkaishi 56:1263-1269.

Takeuchi, T., T. Arakawa, S. Satoh, and T. Watanabe. 1992a. Supplemental effect of phospholipids and requirement of eicosapentaenoic acid and docosahexaenoic acid of juvenile striped jack. Nippon Suisan Gakkaishi 58:707-713.

Takeuchi, T., M. Toyota, and T. Watanabe. 1992b. Dietary value of *Artemia* enriched with various types of oil for larval striped knifejaw and red sea bream. Nippon Suisan Gakkaishi 58:283-289.

Takeuchi, T., J. Dedi, C. Ebisawa, T. Watanabe, T. Seikai, K. Hosoya, and J.-I. Nakazoe. 1995. The effect of β-carotene and vitamin A enriched *Artemia* nauplii on the malformation and color abnormality of larval Japanese flounder. Fisheries Science 61:141-148.

Takeuchi, T., R. Masuda, Y. Ishizaki, T. Watanabe, M. Kanematsu, K. Imaizumi, and K. Tsukamoto. 1996. Determination of the requirement of larval striped jack for eicosapentaenoic acid and docosahexaenoic acid using enriched *Artemia* nauplii. Fisheries Science 62:760-765.

Takeuchi, T., J. Dedi, Y. Haga, T. Seikai, and T. Watanabe. 1998. Effect of vitamin A compounds on bone deformity in larval Japanese flounder (*Paralichthys olivaceus*). Aquaculture 169:155-165.

Tandler, A., and S. Kolkovski. 1991. Rates of ingestion and digestibility as limiting factors in the successful use of microdiets in *Sparus aurata* larval rearing. Pages 169-171 *in* P. Lavens, P. Sorgeloos, E. Jaspers, and F. Ollevier, eds. Larvi '91-Fish & Crustacean Larviculture Symposium. European Aquaculture Society, Special Publication No. 15, Gent, Belgium.

Teshima, S., A. Kanazawa, and M. Sakamoto. 1982. Microparticulate diets for the larvae of aquatic animals. Mini Review and Data file of Fisheries Research 2:67-86.

Teshima, S., A. Kanazawa, and Y. Kakuta. 1986a. Role of dietary phospholipids in the transport of [^{14}C]tripalmitin in the prawn. Nippon Suisan Gakkaishi 52:519-524.

Teshima, S., A. Kanazawa, and Y. Kakuta. 1986b. Role of dietary phospholipids in the transport of [^{14}C]cholesterol in the prawn. Nippon Suisan Gakkaishi 52:717-723.

Teshima, S., A. Kanazawa, S. Koshio, and S. Itoh. 1993. L-Ascorbyl-2-phosphate-Mg as vitamin C source for the Japanese flounder (*Paralichthys olivaceus*). Pages 157-166 *in* S.J. Kaushik, and P. Luquet, eds. Fish Nutrition in Practice. Les Colloques No.61, Paris, France.

Tibaldi, E., and D. Lanari. 1991. Optimal dietary lysine for growth and protein utilization of fingerling sea bass (*Dicentrarchus labrax* L.) fed semipurified diets. Aquaculture 95:297-304.

Tocher, D.R., and D.G. Harvie. 1988. Fatty acid compositions of the major phosphoacylglycerols from fish neural tissues: n-3 and n-6 polyunsaturated fatty acids in rainbow trout (*Salmo gairdneri*) and cod (*Gadus morhua*) brains and retinas. Fish Physiology and Biochemistry 5:229-239.

Tocher, D.R., G. Mourente, and J.R. Sargent. 1992. Metabolism of [1-^{14}C]docosahexaenoate (22:6n-3), [1-^{14}C]eicosapentaenoate (20:5n-3) and [1-^{14}C]linolenate (18:3n-3) in brain cells from juvenile turbot *Scophthalmus maximus*. Lipids 27: 494-499.

Van Der Meeren, T., J. Klungsøyr, S. Wilhelmsen, and P.G. Kvenseth. 1993. Fatty acid composition of unfed cod larvae *Gadus morhua* L. and cod larvae feeding on natural plankton in large enclosures. Journal of the World Aquaculture Society 24:167-185.

Walford, J., and T.J. Lam. 1991. Growth and survival of seabass (*Lates calcarifer*) larvae fed microencapsulated diets alone or together with live food. Pages 184-187 *in* P. Lavens, P. Sorgeloos, E. Jaspers, and F. Ollevier, eds. Larvi'91-Fish & Crustacean Larviculture Symposium. European Aquaculture Society, Special Publication No.15, Gent, Belgium.

Walford, J., T.M. Lim, and T.J. Lam. 1991. Replacing live foods with microencapsulated diets in the rearing of seabass *Lates calcarifer* larvae: Do the larvae ingest and digest protein-membrane microcapsules? Aquaculture 92:225-235.

Watanabe, T. 1993. Importance of docosahexaenoic acid in marine larval fish. Journal of the World Aquaculture Society 24:152-161.

Watanabe, T., and V. Kiron. 1994. Review: prospects in larval fish dietetics. Aquaculture 124:223-251.

Watanabe, T., M.S. Izquierdo, T. Takeuchi, S. Satoh, and C. Kitajima. 1989a. Comparison between eicosapentaenoic and docosahexaenoic acids in terms of essential fatty acid efficiency in larval red sea bream. Nippon Suisan Gakkaishi 55:1635-1640.

Watanabe, T., T. Arakawa, T. Takeuchi, and S. Satoh. 1989b. Comparison between eicosapentaenoic and docosahexaenoic acids in terms of essential fatty acid efficiency in juvenile striped jack *Pseudocaranx dentex*. Nippon Suisan Gakkaishi 55:1989-1995.

Watanabe, T., T. Takeuchi, T. Arakawa, K. Imaizumi, and S. Sekiya. 1989c. Requirement of juvenile striped jack *Longirostris delicatissimus* for n-3 highly unsaturated fatty acids. Nippon Suisan Gakkaishi 55:1111-1117.

Witt, U., C. Quantz, D. Kuhlman, and G. Kattner. 1984. Survival and growth of turbot larvae *Scophthalmus maximus* L. reared on different food organisms with special regard to organisms polyunsaturated fatty acids. Aquacultural Engineering 3:177-190.

Yamamoto, T., K. Fukusho, M. Okauchi, H. Tanaka, W.D. Nagata, T. Seikai, and T. Watanabe. 1992. Effect of various foods during metamorphosis on albinism in juvenile of flounder. Nippon Suisan Gakkaishi 58:499-508.

Yúfera, M., C. Fernández-Díaz, and E. Pascual. 1995. Feeding rates of gilthead seabream (*Sparus aurata*), larvae on microcapsules. Aquaculture 134:257-268.

Zambonino-Infante, J.L., C.L. Cahu, and A. Peres. 1997. Partial substitution of di- and tripeptides for native proteins in sea bass diet improves *Dicentrarchus labrax* larval development. Journal of Nutrition 127:608-614.

Zhu, P., R.P. Evans, C.C. Parrish, J.A. Brown, and P.J. Davis. 1997. Is there a direct connection between amino acid and lipid metabolism in marine fish embryos and larvae? Pages 48-50 *in* S.L. Waddy, and M. Frechette, eds. Proceedings of the Contributed Papers No. 97-2. Aquaculture Canada '97, Canada.

Ecological and Ethological Perspectives in Larval Fish Feeding

T. Ramakrishna Rao

SUMMARY. Low survival rates during larval stages constitute a major bottleneck in the successful culture of many marine and some freshwater fish. The availability of live food is recognized as a critical factor influencing larval survival. Live food is still superior to the best larval diets in terms of larval survival and growth. This paper reviews important ecological and ethological aspects of feeding, from hatching to the weaning stage, and relates them to problems in larval culture. In general, freshwater fish larvae are easier to raise than marine fish larvae, because at hatching they are larger and endowed with more yolk reserves, are less sensitive to starvation, and can be weaned to artificial diets sooner. The feeding behavior of the larvae can be analyzed in terms of the sequential components of predation: search, encounter, pursuit, attack, capture, and ingestion. The searching efficiency and encounter rates of the visual predator are influenced by prey parameters such as body size, conspicuousness, and evasiveness. Turbidity of the water and light intensity also affect prey detection. To changing prey densities, the larvae show typical Type II functional responses, which are influenced by prey handling time, which in turn is largely a function of prey size. Knowledge of larval functional responses is helpful in providing the right concentrations of live food for larval culture. The larvae are initially gape-limited and exhibit prey size selectivity but gradually widen their prey size range as they grow. An aquacultural application of this is the commonly em-

---

T. Ramakrishna Rao, Department of Zoology, University of Delhi, Delhi-110 007, India.

[Haworth co-indexing entry note]: "Ecological and Ethological Perspectives in Larval Fish Feeding." Rao, T. Ramakrishna. Co-published simultaneously in *Journal of Applied Aquaculture* (Food Products Press, an imprint of The Haworth Press, Inc.) Vol. 13, No. 1/2, 2003, pp. 145-178; and: *Sustainable Aquaculture: Global Perspectives* (ed: B. B. Jana, and Carl D. Webster) Food Products Press, an imprint of The Haworth Press, Inc., 2003, pp. 145-178. Single or multiple copies of this article are available for a fee from The Haworth Document Delivery Service [1-800-HAWORTH, 9:00 a.m. - 5:00 p.m. (EST). E-mail address: getinfo@haworthpressinc.com].

ployed feeding protocol, prey size sequencing, in which progressively larger live food items are offered as the larvae grow. A thorough knowledge of the feeding behavior is also essential in the formulation of acceptable larval diets.

KEYWORDS. Larval culture, live food, feeding, optimal foraging, prey selection, behavior

INTRODUCTION

The rapidly growing contribution of aquaculture to global fisheries production is reflected convincingly in the latest statistics compiled by the Food and Agricultural Organization of the United Nations (FAO), 1999. Nearly 22% of the total world production of 120 million tonnes was from aquaculture in 1996, compared to 13% in 1990. During the same period, the proportionate contribution of capture fisheries declined by about 8.5%. Going by these trends, aquaculture can be expected to contribute increasingly more to aquatic food production, more so as capture fisheries the world over continues to face problems of overfishing and pollution.

Although the present state-of-the-art aquaculture is still a far cry from that of modern land-based agriculture, there have been significant advances in the last three decades in our efforts to domesticate selected aquatic species including seaweeds, mollusks, crustaceans, and fishes. Domestication of a wild species is considered to be complete and successful only when its entire life cycle is brought under control (Nash 1975). The traditional dependence on nature for large quantities of fry required for economically viable aquaculture has largely been overcome in recent years, due to notable successes in induced breeding of marine and freshwater species of aquacultural importance. The next major bottleneck impeding successful culture of some species is the poor survival rates during the transition from fry to fingerling stage, which results in low fingerling production. This bottleneck in aquaculture is seen as a manifestation of the heavy mortalities suffered in nature by pelagic marine fishes during the early stages of their life (Type III survivorship) (Heath 1992). In contrast, the majority of freshwater fish are demersal spawners, and their habitat is more "offspring-beneficial," and consequently the so-called critical stage is less critical than in marine species and larval survival significantly higher (Houde 1994).

Many extrinsic and intrinsic factors are believed to be responsible for heavy larval mortalities in nature, but in aquaculture, the availability of the right kind of live food in appropriate densities is probably the most critical factor. Further, it is also possible that inbreeding and induced breeding, so commonly practiced in aquaculture, are additional factors contributing to poor larval survival rates. Unlike the juvenile and adult stages, which show a wide spectrum of feeding modes and food niches (Nikolsky 1963; Jana and Chakrabarti 1996), the early larval stages of many fishes are predominantly raptorial because the morphological features necessary for filter feeding develop only gradually (Lazzaro 1987). The larvae, unlike their adult stages, can capture and ingest only limited types and size ranges of food organisms because of morphological and behavioral limitations. The importance of knowledge of ecology and ethology of feeding in fishes in both capture and culture fisheries has been emphasized in recent studies (Bardach et al. 1980; Stouder et al. 1994). Planas and Cunha (1999) pointed out in a recent review that the development of techniques for successful larval culture of many fish species is hampered by the lack of sufficient knowledge of optimal environmental conditions and feeding behavior during the early life stages.

Contributing in a major way to the larval stage bottlenecks in aquaculture are the uncertainties associated with the supply of live food organisms of the right size and in appropriate concentrations. The high costs of live food (generally rotifers and *Artemia* nauplii) production and inconsistency in the nutritional quality of live food organisms for the early larval stages have provided the impetus for the development of artificial diets. Many marine fish larvae at first feeding (2-5 days after hatching) are not able to utilize artificial diets because they lack functional stomachs or gastric glands and the activity levels of the digestive enzymes are very low (Dabrowski 1984, 1986; Govoni et al. 1986). In nature and in larval culture using live food, ingestion of live foods, which come with the enzymes necessary for digestion, helps the larvae to overcome this problem (Lauff and Hoffer 1984; Munilla-Moran et al. 1990). Larval culture with artificial diets has been less problematic with species that have a more advanced or developed digestive system at first feeding (Verreth et al. 1991).

There have been in recent years significant advances in the development of artificial diets for larval fish (Person-Le Ruyet et al. 1993;

Southgate and Partridge 1998). Microencapsulated diets ("artificial plankton") developed for different marine fish larvae (Yufera et al. 1999) have yielded encouraging results in larval culture (Jana and Dutta 1998), but only after the period of dependence on live food had passed. At present it seems unlikely that the ultimate goal of liberating hatcheries from dependence on live food for larval culture will be achieved in the very near future (Jones et al. 1993). There are still some unresolved problems regarding the acceptability of formulated diets, however nutritionally balanced, by larvae of marine fish such as turbot, *Scophthalmus maximus*; milkfish, *Chanos chanos*; red drum, *Sciaenops ocellatus*; and barramundi, *Lates calcarifer* (Kolkovski and Tandler 1995; Tucker 1998). Southgate and Partridge (1998) reviewed the state-of-the-art formulated diets for marine larval culture and concluded that at present there is no artificial diet for any marine finfish larvae that could provide growth and survival comparable to those produced by live food. Even for the culture of larval cyprinids, for many of which satisfactory artificial diets have been developed, neither growth nor survival rates of larvae raised on artificial diets do match those achieved with live food (Kamler 1992; Abi-Ayad and Kestemont 1994; Kolkovski and Tandler 1995; Cahu et al. 1998; Planas and Cunha 1999).

Accumulation of unconsumed microencapsulated diets generally results in rapid deterioration of the water quality due to microbial and metabolite build-up (Muir and Sutton 1994). Another problem with formulated diets is that they are evacuated too rapidly through the alimentary system, without adequate time for digestive enzymes to act on the food (Jobling 1986). It has been reported that amino acids in live foods are catabolized at a lower rate and therefore used to a greater extent in protein synthesis than amino acids in dry diets (Dabrowski et al. 1987). In view of these apparently insurmountable problems, development of artificial diets for larvae may be an extremely difficult process (De Silva and Anderson 1995). Larvae of some species such as yellowtail, *Seriola quinqueradiata*, and sea bream, *Sparus aurata*, do accept artificial diets but do well only when the diets are supplemented with live food (De Silva and Anderson 1995). Co-feeding, now a method frequently adopted in growing some marine species, is a compromise approach for ensuring acceptable rates of survival and growth while cutting down on the prohibitive costs of live food production (Rosenlund et al. 1997). While efforts continue to develop ideal larval feeds, sustained worldwide research on live food production, particularly of rotifers (*Brachionus plicatilis*, *B. rotundiformis*, *B. calyciflorus*) and crustaceans (*Artemia* nauplii and *Moina*) (Hagiwara et al. 1997) clearly points to the undiminished importance of live food in larval culture.

The acceptable level of success presently achieved in raising some fish larvae in aquaculture has been possible only because of the basic knowledge available about the ecology and ethology of feeding in the larval stages of those species. One such species is the exotic African catfish, *Clarias gariepinus*, which appears currently to be a highly favored species for culture in India. In such cases, the short-term benefits are obvious, but the potential long-term consequences of introducing exotic species to natural aquatic ecosystems need to be assessed scientifically (Fernando and Holcik 1990). Ironically, commercial-scale culture success is yet to be achieved for the Indian catfish, known as "singhi," *Heteropneustes fossilis*, because of insufficient knowledge about the feeding ecology and nutritional requirements of its larval stages (Thakur and Das 1986; Mookerji 1992; Datta Munshi and Choudhary 1996). In many south- and southeast Asian countries, it is the natural culture ponds rather than technology-intensive hatcheries and culture facilities that dominate aquaculture. For larval culture, these traditional practices still depend to a large extent on live food organisms developing naturally in culture ponds. The role of zooplankton in the survival of larval carps in these nursery ponds has been documented convincingly (Alikunhi et al. 1955; Mitra and Mohapatra 1956; Jhingran 1991).

Artificial larval diets, however well-balanced nutritionally, should be acceptable to the larvae. Again, an understanding of the feeding ecology and behavior of the larvae is essential for developing feeds that are readily accepted by them. This review will briefly discuss the ecological and ethological aspects of feeding by the early larval stages of fish, the stage during which they are totally or partially dependent on live food, and their relevance to and application in larval culture in aquaculture.

BODY SIZE AND YOLK RESERVES AT HATCHING

In nature, the larvae of a majority of fish have at hatching sufficient yolk reserves to serve their energy requirements until they start feeding on natural food organisms present in their habitat. This period of transition, however short, from endogenous (yolk) nutrition to exogenous feeding, is a critical stage in the life of many fish and often a bottleneck in larval culture in aquaculture. The adequacy of yolk reserves in newly hatched larvae in overcoming the uncertainties associated with first feeding often determines larval survival rates under natural and aquacultural conditions (Nash 1975). The larvae of Indian major carps at hatching measure 3.8 for rohu, *Labeo rohita*, to 4.7 mm for catla, *Catla catla*, and are well endowed with yolk reserves. By the end of the third day, they increase in length to 6.7 mm in rohu and 7.3 mm in catla. Nearly all the yolk is utilized by 96 hours after hatching. For stocking purposes, carp spawn is often classified as desirable or undesirable based on the amount of yolk remaining in the larvae (Jhingran 1991).

Egg Size and Larval Size at Hatching

In many teleosts there exists an inverse relationship between fecundity and egg size (Roff 1984; Winemuller and Rose 1992), reflecting the alternative reproductive strategies of large numbers of small offspring versus small numbers of large offspring. Besides fecundity, age at first reproduction and frequency of reproduction are also known to influence egg size (Winemuller and Rose 1992). Trade-offs among these three reproductive parameters have evolved in different fishes to optimize their overall reproductive potential for a given habitat (Roff 1984; Wootton 1984). Intraspecific phenotypic variation in egg size is also not uncommon, being related to factors such as nutritional status of the female during gonadal maturation (Beacham and Murray 1985; Springate et al. 1985), season, and temperature (Ware 1975). In aquaculture, it is a common practice to selectively breed varieties that produce relatively larger eggs. In fact, the potential success of aquaculture of certain fish species, particularly marine, could be predicted from their egg size (Nash 1975; Rana 1985). The most widely cultured trout produces eggs measuring 4 mm, while the rabbitfish, whose aquaculture is yet to take off on a commercial scale, has eggs measuring no more than 0.5 mm (Nash 1975).

Larger eggs, besides providing greater yolk reserves at hatching, also produce larger-sized larvae. There are many correlates of body size that influence larval survival and growth (Miller et al. 1988): larger larvae can swim faster, search a greater volume of water, capture larger prey, avoid predators more successfully, and withstand a longer duration of food deprivation (Knutsen and Tilseth 1985). As a consequence of trade-offs between offspring size and offspring number (fecundity) dictated by the vicissitudes of their habitat, freshwater fishes in general produce relatively larger larvae at hatching than do marine fishes (Duarte and Alcaraz 1989; Winemuller and Rose 1992; Letcher et al.

1996). There is a tenfold difference between marine and freshwater larvae in their mean dry-weight at hatching: marine: 37.6 ± 6.4 μm, freshwater: 359.7 ± 72.8 μm (Houde 1994). Consequently, the marine fish larvae are relatively more difficult to raise. The larvae of the orange spotted grouper, *Epinepheles coioides*, at hatching measure 1.7 mm, and their survival rates under best culture conditions rarely exceed 10% (Tucker 1998).

PROCESSES AND EVENTS BEFORE FIRST FEEDING

Yolk Utilization

The yolk reserves remaining in fish larvae at the time of hatching are a direct measure of how long the larvae can support itself if deprived of exogenous sources of nutrition. The major factors influencing the rate of yolk utilization by larvae are their activity level and ambient temperature. Depending on these two factors, the yolk may last as long as 30 days in Atlantic halibut, *Hippoglossus hippoglossus* (Olsen 1997), or as short as 94 hours in the spotted grouper (Kohno 1998). Barramundi with a yolk reserve of 1.4×10^{-7} mm^3 at hatching, exhausts its yolk 145 hours after hatching, but milkfish, *Chanos chanos,* with much higher reserves (3.6×10^{-7} mm^3) exhausts its reserves in only 125 hours (Kohno 1998). The temperature dependence of yolk utilization rate has been demonstrated for many species (Wang et al. 1987; Yin and Blaxter 1987; Rana 1990). The breeding seasons of many fish in nature must have evolved to provide optimal temperature conditions for newly hatched larvae. In an aquaculture context, it may be possible, when the supply of live food organisms is erratic, to slow down the rate of yolk utilization by setting where possible, a slightly lower temperature in indoor hatcheries.

Concurrent Exogenous and Endogenous Nutrition

The larvae of many fish may start feeding on live food organisms in water long before their yolk reserves are totally exhausted (Malhotra and Munshi 1985; Balon 1986; Yin and Blaxter 1987; Mookerji and Rao 1999). It was also observed that fed larvae depleted their yolk at a slower rate than unfed larvae. There is thus a period in the early life of larvae during which its energy requirements are met by a combination of exogenous and endogenous sources. This concurrent dependence on yolk and external food is seen as being adaptive in that the larvae are still sustained by its yolk supply while they gradually develop skills to capture live food in the medium and the physiological abilities to digest it.

The transition period during which fish larvae switch from yolk energy to exogenous sources of energy is considered a critical stage in the life history of many pelagic marine fishes (May 1974; Elliott 1989; Chambers and Trippel 1997). The larvae will face starvation if, at the initiation of feeding, they cannot find food organisms that they are physically capable of capturing and that occur in sufficient concentrations to ensure optimal capture success. The vagaries of recruitment in many commercially important marine species have been attributed to variable survival rates of the larval stages during this period (Houde 1987; Bailey and Houde 1989; Letcher et al. 1996; Houde 1997). Although starvation could be a very important factor, other factors such as predation and oceanographic conditions during larval stages have also been implicated in the variability in recruitment.

In an individual-based model incorporating both intrinsic (foraging ability, starvation resistance, and growth capacity of the larva) and extrinsic (prey density and predator size) components, Letcher et al. (1996) found that growth capacity among the intrinsic factors and predator size among the extrinsic factors explained the maximum amount of variance in larval survival rates. Their model, in contrast to earlier studies, did not consider prey density to be a critical variable in larval survival. It should be noted, however, that continued food deprivation, besides leading to irreversible starvation effects, could act indirectly by increasing the predation risk (Rice et al. 1987) because weak larvae take longer to grow out of the "size bottlenecks" of high predation susceptibility (Werner et al. 1983). Since vulnerability to predation was shown in many species to be size-selective (Paradis et al. 1996), larvae which are larger at hatching or grow faster to a large size face less predation risk.

In the absence of suitable food in the water, larvae that have exhausted their yolk reserves start experiencing the physiological effects of starvation and eventually reach a stage beyond which they have no chance of surviving, even if food were made available (Yin and Blaxter 1987). This stage at which 50% of the starving larvae fail to feed and survive if offered food has been called the "point of no return" (PNR) (Blaxter and Hempel 1963). PNR values for the larvae of some marine and freshwater fishes are given in Table 1. They are in general posi-

TABLE 1. Body size at hatching and resistance to starvation (as reflected in the point of no return [PNR]), in the larvae of different marine and freshwater fishes raised at various temperatures.

Species	Culture temperature (°C)	Length at hatching (mm)	PNR (days after hatching)	Reference
Marine				
Turbot, *Scophthalmus maximus*	18		7.0	Jones (1972)
Milk fish, *Chanos chanos*	28	3.46*	3.25	Bagarinao (1986)
Seabass, *Lates calcarifer*	28	1.72*	2.5	Bagarinao (1986)
Rabbitfish, *Siganus guttatus*	28	1.59*	2.5	Bagarinao (1986)
Baltic herring, *Clupea harengus*	9.2	6.87	8	Yin and Blaxter (1987)
Cod, *Gadus morhua*	6.9	4.5	11	Yin and Blaxter (1987)
Flounder, *Platichthys flesus*	9.5	2.6	10	Yin and Blaxter (1987)
Gilthead sea bream, *Sparus aurata*	19.5	3.5	8	Yufera et al. (1993)
Freshwater				
Mozambique Tilapia, *Oreochromis mossambicus*	28	3.85	15	Rana (1985)
	28	4.2	21	Rana (1985)
Common carp, *Cyprinus carpio*	26	4	9	Khadka (1986)
Rohu, *Labeo rohita*	26	4.58	14	Mookerji and Rao (1999)
Singhi, *Heteropneustes fossilis*	26	3.32	8	Mookerji and Rao (1999)

* Standard length, others total length.

tively correlated with the size of larvae at hatching and negatively with temperature and level of activity. Habitat-related differences are also evident.

Typically, marine fish larvae weigh less at hatching, have higher metabolic requirements and longer larval stage duration, are more sensitive to starvation, have lower PNR, and experience higher mortality than do freshwater fish larvae (Houde 1994). Of these, the difference in larval weight at hatching (nearly tenfold) is the most important, as it directly or indirectly accounts for all the other differences. These differences have been attributed to contrasting reproductive strategies of freshwater fishes (which being mostly demersal spawners tend to produce a relatively small number of large-sized eggs) and marine fishes (which being largely pelagic spawners, produce a large number of small-sized eggs [Roff 1984; Duarte and Alcaraz 1989]). The available data on freshwater fish larvae, not as extensive as those for marine fish larvae, indicate that the transition stage from endogenous to exogenous nutrition is probably not as critical as in the early life of marine fish and that its significance may be much less for cultured freshwater fishes.

PNR values nevertheless provide an estimate of the tolerance to food deprivation in the larval stages, information that might be useful under hatchery conditions with unpredictable or inadequate live food supply or during transportation of fry from the hatchery to distant fish farms.

The ability of fish larvae to withstand starvation generally improves with age, as metabolism shifts to lipid deposition and efficiency of catabolic pathways improves (Ehrlich 1974). Concomitantly improving with age are swimming ability, food searching ability, prey capture success, and range of acceptable food items, all of which collectively bring down the chances of a given larva facing starvation (Blaxter 1969; Table 2). With the initiation of behavioral activities relating to foraging, the feeding efficiency of the larva is influenced by various factors relating to the larva itself, prey organisms in the medium, and certain physical conditions around it (Schiemer et al. 1989). This phase of larval life can be analyzed as a typical example of prey-predator interaction.

PREDATION

The act of predation can be broken down into a set of sequentially executed behavioral activities involving decision making at each stage (Figure 1), which can be assessed in terms of the probability of success. The product of all these probabilities gives an estimate of the net probability of a larva finally ingesting the prey (Greene 1983). Various parameters relating to the larva (size, gape, swimming speed, visual acuity, hunger level, and prey selectivity), prey (density, body size, evasiveness, and conspicuousness), and the environment (light, turbidity, other predators, alternate prey) influence each phase of the predation process.

Search and Encounter

The food-searching ability of fish larvae is a function of its swimming ability and to some extent its hunger level. The time spent by a foraging fish in search of prey decreases with increasing prey encounter rate, since higher encounter rates result in a higher capture rate and earlier satiation time (Werner and Hall 1974). In the initial stages of exogenous feeding, the larva is naïve, and has incompletely developed a behavioral repertoire, and hence, its hunting ability is limited and capture success relatively low.

TABLE 2. Important biological parameters of selected marine fish larvae (adapted from Kohno 1998).

Species	Total length at hatching (mm)	Mouth width at first feeding (mm)	Age at first feeding (hours)	Time from first feeding to yolk exhaustion (hours)	Larval survival rates (%) at selected age (days) in	
					Laboratory culture trials	Mass production trials
Seabass, *Lates calcarifer*	1.40	224	55	91	4-24 (5)	2.5-63.0 (30)
Milkfish, *Chanos chanos*	3.72	258	77	48	8-25 (5)	7.5-63.5 (21-24)
Grouper, *Epinephelus coioides*	1.69	268	70	28	0-42 (10)	0.3-12.1 (41-46)
Rabbitfish, *Siganus guttatus*	1.90	187	56	72	0.4-6.0 (5)	0.7-25 (35-45)
Red snapper, *Lutjanus argentimaculatus*	1.72	220	71	44	3.4 (3-8)	0-10.3 (15)

FIGURE 1. Predation cycle of a fish larva involving decision making at each of the four sequential stages: search, attack, capture, and ingestion.

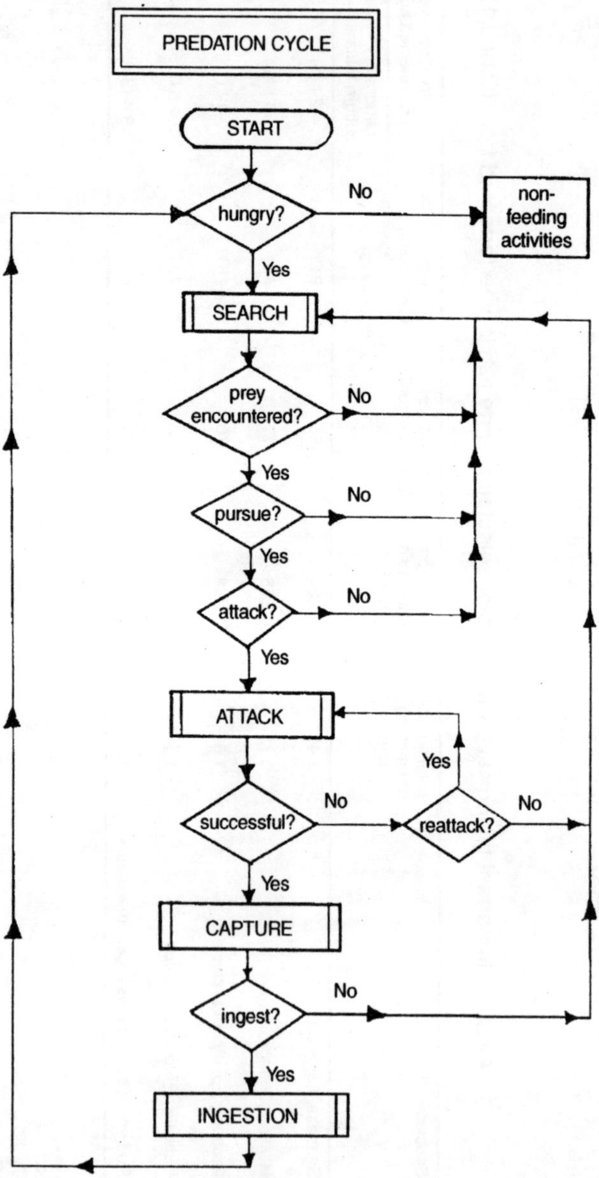

Prey encounter rate is the product of the prey concentration and the volume of medium searched by the fish, the latter calculated as a product of the cross sectional area of its visual field and its swimming speed (Werner and Hall 1974). The visual field of the larva is in turn estimated by measuring two parameters: the reactive distance (RD) and the vertical and horizontal angles at which a specific prey item is detected. These angles are not easily measured, and one alternate method involves a frame-by-frame analysis of high-speed cinematography or video of the actively foraging fish (Hunter 1972). RD refers to the maximum distance from which an individual prey can be detected by the predator; it is often measured using the commonly observed behavioral pattern of a fish showing a sudden acceleration in swimming upon detecting a prey (Werner and Hall 1974; Confer et al. 1978; Khadka 1986). Since reactive distance as well as swimming speed of a fish increases with body size, the volume of medium (V) searched by a larva increases with body size (L) in an exponential fashion ($V = L^3$) (Blaxter and Staines 1971; Hunter 1972).

The RD of a fish is influenced by its visual acuity. Due to the slow development of optic nerves and incomplete development of binocular vision, the eyes of an early stage larva provides only a coarse perception of the prey organism and a limited ability to recognize absolute prey sizes (Blaxter and Staines 1971; Blaxter 1980; Wanzenböck and Schiemer 1989). Visual acuity increases sharply as the larva grows, and the food organisms can be recognized from farther away than in the early stages and pursued.

Body size, conspicuousness, and movements are prey's important attributes that influence RD (Wanzenböck and Schiemer 1989). The larger the prey, the larger is the RD and consequently the higher the probability of being pursued and attacked by the larva (Eggers 1977, 1982). Size-selective predation by planktivorous fishes has been shown to be of evolutionary significance for body size and community organization of zooplankton (Kerfoot and Sih 1987). Most zooplankton species have also evolved transparent or translucent bodies to reduce detection by visual predators (Zaret 1980). Therefore, any feature that makes the prey organism more conspicuous or enhances the prey contrast raises the probability of it being pursued and attacked by the predator. A moving prey is more likely to be detected by a foraging larva from a greater distance. For instance, a continually moving *Daphnia* can be located from farther away than a stationary copepod of comparable size (Wright and O'Brien 1984). Even among copepods, the cyclopoids with their erratic jumping motion attract more attention than the calanoids with their characteristic gliding motion (Confer and Blades 1975).

Prey Concentration: Functional Response

In nature, the concentration of organisms (zooplankton) available as food for larval fish is highly variable, patchy, and unpredictable (Letcher and Rice 1997). In contrast, under culture conditions live food concentrations can be set and maintained at desirable levels in the culture tanks. The important question then is: What is a desirable concentration? Live food densities that are too low might result in a lower percentage of successfully feeding larvae and low individual ingestion rates (Houde and Schekter 1980) and consequently poor larval growth and survival. On the other hand, very high food concentrations, besides being expensive and difficult to provide (Southgate and Partridge 1998), may result in the loss of nutritional quality of the prey before they are eaten and cause deterioration in water quality (lowered dissolved oxygen, higher levels of excretory products), which is detrimental to the larvae (Tucker 1998). It should be noted, however, that stocking densities as well as optimum prey density levels for larval culture are invariably linked to the carrying capacity of the culture system.

The number of prey a predator can capture and ingest in a given foraging period is generally a function of predator size as well as the size and concentration of the prey in the medium. The change in prey consumption rates with change in prey density is called a functional response of the predator and is an important component of predation in nature (Holling 1966). The larvae of many marine and freshwater fish feeding on zooplankton show a type II functional response, a rectangular hyperbola, reflecting progressively declining rates of prey capture by the predator until an asymptote is reached (Houde and Schekter 1980; Lazzaro 1987; Gulbrandsen 1991; Miller et al. 1992; Johnston and Mathias 1994; Rao and Mookerji 1995); however, the changing foraging tactics of older larvae and juveniles of some species might lead to a sigmoidal, type III response, in which the predator shows a lag phase and an accelerating phase at low prey densities (Winkler and Orellana 1992). An exclusively Type III functional response is rarely observed in planktivorous fish larvae. Holling's (1966) model predicts that with increasing prey abundance, search time declines and prey-handling time increases and that the predator's entire foraging time is used up in handling prey when prey abundance is very high. Prey concentrations higher than that at which the asymptote is achieved do not contribute to any higher ingestion rates.

Two important parameters describe this functional response: the instantaneous rate of prey discovery or attack constant (which is the product of the larva's searching rate and probability of finding a prey) and prey handling time. Both these parameters are influenced by the size of the predator as well as prey (Abrams 1990; Johnston and Mathias 1994). The maximum consumption rates generally increase with larval size (Houde and Schekter 1980; Mills et al. 1986; Johnston and Mathias 1994), which is due to an increase concurrent with body size in swimming speed, reactive distance, search rate, and efficiency of prey handling. Although prey size also has a strong effect on the functional response of the larva, the effect is dependent on the prey type; between a copepod and a cladoceran prey of the same size for example, the larva has higher capture success with the latter, since cladocerans are more vulnerable to the suction-type feeding typical of larval fish (Drenner et al. 1978).

Prey-handling time is an important parameter affecting the net energy gained by the larva during its foraging bout. In optimal foraging models, which use cost-benefit analysis to solve the optimal diet breadth, prey-handling time is taken as a cost parameter and energy gained by ingesting the prey as the benefit parameter (Stephens and Krebs 1986). Handling time for a given prey type decreases with the age (or size) of the predator, and for a predator of a given size, it increases with the size of the prey (Mookerji 1992). For gape-limited larvae, prey absolutely or effectively larger than the size commensurate with their gape and those possessing various morphological (helmets, spines) and behavioral (evasiveness) predator-deterrent defenses involve greater handling time. This is the reason why the net profitability of any prey item to the fish is actually taken as the ratio of energy content/handling time (Werner and Hall 1974). Thus, the energy content of the spiny and non-spiny forms of the rotifer *Brachionus calyciflorus* being the same, the spiny morphs, because of the greater handling time, would be less profitable to fish larvae. Similarly, *Moina* sp. (probably the most extensively cultured cladoceran in tropical larval culture), because of its soft carapace has a shorter handling time and is therefore a relatively more profitable prey for older larvae. The profitability of a prey item to the larvae tends to increase as the larvae grow, due to decreasing handling time (Houde and Schekter 1980; Miller et al. 1992).

In larval culture, functional response analysis helps in providing estimates of the minimal prey abundance necessary to obtain maximal feeding rates in the larvae. Based on such an analysis on walleye, *Stizostedion vitreum*, larvae, Johnston and Mathias (1994) suggested a prey abundance of 200-800/L for early (9.5-10.5 mm) larvae and 100/L for older (13-15 mm) larvae, to achieve maximal consumption rates.

For visual predators, light is obviously an important factor (Sibkin 1974; Blaxter 1980, 1986). The spectral quality, as well as the intensity of light reaching a particular depth, influences prey search and detection (Blaxter 1980) and also the pigmentation of the larvae (Boeuf and Le Bail 1999). Generally for locating prey, many fish species appear to have a threshold light intensity, below which the reactive distance is reduced, leading to lower levels of prey detection or to a passive, non-selective prey capture (Vinyard and O'Brien 1976; Wright and O'Brien 1984).

Prey visibility and consequently the reactive distance increases with prey size as well as prey contrast. Since nearly all zooplankton species are translucent or nearly transparent, any thing that increases prey contrast will result ultimately in better feeding rates. This obviously has implications for larval culture; in tanks with light-coloured walls, Dover sole, *Solea solea*, larvae fed artificially stained *Artemia* nauplii were found to survive and grow better than those fed plain *Artemia* (Dendrinos et al. 1984). Prey contrast to the foraging larvae is more easily achieved by providing a dark background. The survival and growth of yellow perch (Hinshaw 1985) and white bass (Denson and Smith 1996) larvae feeding on *Artemia*, and milkfish larvae feeding on rotifers (Duray 1995) were better in dark-walled than in light-walled tanks. Ostrowski (1989) found that the survival of early larval stages of dolphin, *Coryphaena hippurus*, feeding on rotifers was 67% in black-colored tanks but only 29% in light-colored tanks. In indoor larval culture systems, the need to design culture tanks so as to enhance prey contrast cannot be overemphasized (Naas et al. 1996; Planas and Cunha 1999).

Many fish which otherwise are visual predators, have been found to be able to feed under extremely low light intensities with thresholds around 0.1 lux (Mills et al. 1986; Batty 1987; Batty et al. 1990), and even in total darkness (Unger et al. 1984; Batty et al. 1986; Khadka 1986; Mookerji and Rao 1993). Larvae collected during night hours from nursery ponds or hapas revealed guts filled with zooplankton, indicating nocturnal feeding (Jana and Chakrabarti 1988; Mookerji 1992). Although it should be possible for filter feeders to continue to filter water and strain food organisms even in darkness, the ability of particulate feeders to feed in darkness indicates the presence of prey detection mechanisms other than light. The larvae of Indian catfish (singhi) and rohu were not only capable of feeding in total darkness (Mookerji and Rao 1993), but even showed weak prey selectivity and functional responses to prey density (Rao and Mookerji 1995). In a recent experimental study, Janssen et al. (1994) demonstrated that juvenile alewife,

Alosa pseudoharengus, were able to feed in darkness by using their lateral-line sense organ for mechano-detection of prey. They suggested that many fishes that regularly use vision for feeding might simply switch to mechano-detection of prey when ambient light intensity falls below a threshold. There is evidence indicating that the lateral-line organ is fully developed and functional very early in the life of many fishes.

Turbidity of water has an effect similar to that of low light intensity, in that it reduces the RD for foraging larvae (Gardner 1981). In ponds with high concentrations of suspended clay particles. larvae may face problems in efficient prey detection (Gregory 1994). Turbidity may also have indirect effects; in nursery ponds where both cladocerans (such as *Daphnia*) and rotifers (such as *Brachionus*) (both food for larvae) are present, the adverse effect of high suspended clay particle concentrations is generally greater on cladocerans than on rotifers. This differential effect may ultimately lead to the suppression of cladocerans (Kirk and Gilbert 1990).

In the predation cycle, the pursuit phase starts soon after prey location and stops when the fish is close enough to the prey to attack. At this point, a larva might choose to attack a prey or move away in search of another prey item (O'Brien 1979, 1987). Pursuit patterns of the fish are influenced by its hunger level (Ware 1972), its prior experience with the prey type (Vinyard 1980), and prey size and concentration (Werner and Hall 1974). Once within striking distance, the larva assumes a characteristic 'S' or 'C' type striking posture; its actual strike on the prey is initiated in response to prey's escape behavior. In larval culture, one of the reasons that artificial diets are not readily accepted by larvae of some marine fishes, even when they are most suitable in terms of particle size, color, flavor, and nutritional adequacy, is that they show no such escape behavior, which is necessary for fish larvae to act.

At first exogenous feeding, the larva is totally naïve, and its behavioral repertoire may not include any more than a few innate patterns. A critical component of the ontogeny of feeding behavior is the ability of larvae to learn from experience (Drost 1987; Coughlin 1991). The high degree of variability that has been often observed in the feeding success of fish larvae under experimental conditions has been attributed to individual variability present among the larvae in such factors as rate of learning (Magurran 1986; Marcotte and Browman 1986). Through experience larvae gradually improve their accuracy of aiming at the prey and also increase the capture distance (Drost et al. 1988). Their visual resolution also improves rapidly as the larvae grow (Walton et al.

1994). Alevins of Atlantic salmon, *Salmo salar*, initially rely on swimming quickly to the prey, lunging, and ram-feeding capture but gradually switch to a suction mode, where the prey is sucked into the mouth of the fish while it is still some distance from the prey's position (Coughlin 1991). At first feeding, larvae make errors in both perpendicular and parallel aiming at their prey. In the latter case, if the feeding strikes are too early or too late, they miss capturing the prey because suction would have been completed before they reached the prey. With experience, the alevins improve their aim and consistently strike the prey from a relatively constant capture distance. Species-specific differences in the feeding behavior and its ontogeny in the larvae have also been reported. Atlantic salmon alevins tend to aim 'low' (Coughlin 1991), while larval common carp, *Cyprinus carpio*, tend to aim above their prey (Drost 1987).

Not all planktivorous fishes are particulate feeders; filter feeding is probably the major mode of feeding in many species, particularly in the adult stage (Lazzaro 1987). These however, are known to adopt a particulate mode of feeding during larval stages, since gill rakers are formed much later during ontogeny (Kohno 1998). Many species appear to be able to switch from one feeding mode to another, depending on the size and concentration of the prey. Small prey is captured more efficiently by filtering the water through the gill rakers (Drenner et al. 1982), and particulate mode of feeding is more efficient at low prey densities in contrast to filter feeding at high densities (Janssen 1978).

In the initial stages of exogenous feeding, larvae show poor capture success ([number captured/number attacked] × 100), due to low volume of buccal cavity unable to generate enough suction, small gape, and poor aiming at prey (Dabrowski and Bardega 1984). But during ontogeny, these improve rapidly, resulting in higher capture rate (Drost 1987; Wanzenböck 1992; Mookerji and Rao 1995). Kohno (1998) traced the development of bony elements associated with feeding and of fins in important marine fish larvae. He distinguished two types of ontogeny: the "milkfish type" and the "seabass type." In the milkfish type, food organisms are gathered by straining very early in life, and their fin development pattern helps to achieve the right type of swimming and maneuverability. In the seabass type (which includes seabass, *Lates calcarifer*; rabbitfish, *Siganus guttatus*; and red snapper, *Lutjanus argentimaculatus*), the development of oral cavity elements is delayed, and food capture is initially by suction only. The larvae switch to grasping mode of feeding only 4-5 days after hatching. These differences in ontogeny account for the differential prey capture success of the larval stages of the species investigated.

Optimal Foraging Theory and Application

Considering the vagaries of availability of the right type of food in adequate quantities in the habitat of larval fishes and the problems associated with food procurement, natural selection in the larval fish can be expected to favor those foraging parameters that maximize foraging efficiency. Although neo-Darwinian fitness is measured ultimately in terms of the production of viable and fertile offspring, the optimization of energy intake (through feeding) may be considered here as its proximate criterion. The foraging activity of a hungry larva includes different components such as patch choice, foraging time, and diet breadth, and associated with each are costs and benefits (Pyke 1984; Stephens and Krebs 1986). For instance, the energy gained by larvae by ingesting *Daphnia* is the benefit function, while the time and energy spent by larvae in pursuing, attacking, and capturing are the cost function. A larger *Daphnia* certainly has a higher benefit (more biomass or calories), but the costs of capturing a larger prey are also high. Therefore, prey profitability to the predator is more meaningfully measured as E/h, where E is the energy gained and h the handling time for a given prey item.

Optimal foraging models attempt to predict the best possible tradeoffs between costs and benefits in foraging, which maximize the net energy gain to the foraging animal (Stephens and Krebs 1986; Rao 1991). Consider, for instance, the following question: Given a choice of prey types (taxa and sizes), which prey types should the predator include in its diet? In indoor larval culture systems where the larvae are offered only one specific prey type of a defined size range at a time, this question has admittedly no relevance. However, it is not uncommon for an Indian carp fry grown in a traditional nursery pond to encounter concurrently a choice of diet items including 3 to 5 species of rotifers and 2 to 3 species of cladocerans. The predator's search costs will be high when the overall prey concentration is low or when it chooses only the most profitable items (because in either case the prey encounter frequency is less). These search costs decline, and the net energy gains increase exponentially as progressively less profitable prey types are included in its diet.

The optimal diet model (Stephens and Krebs 1986) gives some testable predictions; the most important among them are as follows: the inclusion of a particular prey item in the optimal diet is independent of its own abundance and depends only on the overall abundance of the higher ranked prey types; less profitable prey types outside the optimal diet breadth should be ignored regardless of their abundance; and with increasing prey type abundance the predator should progressively exclude the less profitable prey types from its diet and specialize on the more profitable ones. Thus, with decreasing food abundance in the medium, the predator's diet becomes less and less specialized. These predictions have been tested using planktivores such as sunfish (Werner and Mittelbach 1981) and common carp larvae (Khadka and Rao 1986).

Optimal foraging models have been applied profitably to elucidate food selectivity in fishes (Townsend and Winfield 1985) and are particularly relevant for larval culture in traditional nursery ponds.

Prey Selection

Selectivity for a specific prey type at a given age must be an adaptive feature in larval fish to optimize energy intake (Greene 1986). Mittelbach (1983) found that bluegill growth rates in nature were closer to those predicted from optimal prey selection rather than from random prey selection. When feeding on species of prey selected, larval fish were found to survive and grow better (Mayer and Wahl 1997). Mookerji and Rao (1994) related the differences they observed in the growth rates of rohu and singhi larvae on different diet regimens to ontogenetic changes in prey selection.

In selecting a particular prey type or size, the larval fish may not always behave according to the predictions of optimal foraging theory, because of certain morphological, physiological, and behavioral constraints (Walton et al. 1992). Further, because of fish's incomplete development of gut and digestive enzymes, prey selection patterns may often reflect the nutritional and digestive factors overriding optimal foraging considerations. One of the important morphological constraints in the early life of fishes is the gape of their mouths. Unlike the adult planktivores, larval fish are 'gape-limited' (Lazzaro 1987; Schael et al. 1991; Bremigan and Stein 1994). That the gape size of larvae impose a primary limitation on the largest prey it could capture is clearly indicated by the direct relationship observed in many larval fish between gape size and the average prey size ingested (Shirota 1970; Hunter 1981; Mills et al. 1984; Dabrowski and Bardega 1984; Schael et al. 1991; Mookerji and Rao 1993; Bremigan and Stein 1994; Boubee and Ward 1997).

As the larvae grow, the progressively increasing gape allows them to take progressively larger prey. More than 80% of the variation in mean prey size taken by larvae over a period of time in rohu and singhi was explained by the gape (Mookerji and Rao 1994). However, in many field and laboratory studies larvae were observed to take prey consistently smaller than the size dictated by their gape (Bence and Murdoch 1986; Schael et al. 1991; Bremigan and Stein 1994). Laboratory studies suggest that prey size selected is generally about 25% of the gape size (Hunter 1980). In nature, competitive interactions with other planktivorous species might force some fishes to select smaller prey, even when they are no longer gape-limited (DeVries et al. 1998). Obviously, gape size sets a limit only to the maximum capturable prey size but does not adequately explain prey size selected by the larvae within that limit. The question of whether or not a particular size of prey selected by a larva is optimal can be more realistically addressed only with reference to its mouth size. Consequently, the ratio of prey size to mouth size (PS/MS) has been suggested as a more meaningful measure than prey size alone for assessing optimal prey size selected by gape-limited larvae (Werner 1974). Theoretically, the fish should not accept any prey whose size results in PS/MS > 1.0 (where mouth size is taken as the 45° angle gape); Werner (1974) estimated optimal prey sizes for bluegill, *Lepomis macrochirus*, to be close to PS/MS value of 0.59. Rohu and singhi were observed to maintain nearly a constant PS/MS ratio during the first 3 to 4 weeks of their life (Mookerji and Rao 1994).

Prey profitability, as explained earlier, is a major criterion in prey selection by larvae. When prey-handling times are negligible, as the case would be, for instance, for larvae beyond the gape-limiting stage, it is only prey size that matters. Given a choice of three rotifer prey species–small (*B. angularis*), medium (*B. patulus*), and large (*B. calyciflorus*)–for which the handling times were not significantly different, older rohu larvae strongly selected the large brachionid (Mookerji and Rao 1993).

Prey Size Sequencing in Larval Feeding

From the foregoing account of the ontogenetic changes in prey selection patterns of larval fishes, it will be clear that in nature larvae start with small prey items and select progressively larger prey as they grow. In larval culture also, prey size sequencing ensures optimal survival and growth rates (Tucker 1998). For the first feeding of many marine fish larvae, even the universally used rotifer *B. plicatilis* may be larger than the optimal size, as can be predicted from the PS/MS ratio. The two distinct size groups of this rotifer species, considered until recently as two separate strains, the S-strain (100-210 μm, Y = 160 μm) and the L-strain (130-340 μm, Y = 239 μm), are now recognized as two distinct species, *B. rotundiformis* and *B. plicatilis*, respectively. Feeding incidence among marine fish larvae is generally much higher when the first food offered is the smaller *B. rotundiformis* (Planas and Cunha 1999). Now for many marine species, *B. rotundiformis* is the most preferred first food for larval culture (Tucker 1998).

In the culture of larval turbot, best survival and growth rates were obtained when live food organisms were offered in a sequence of increasing body size as the larvae grew–small (< 150 μ m) rotifers when the larvae were < 4 mm, large (125-200 μm) rotifers at 4.0-4.5 mm length, mixture of rotifers and *Artemia* nauplii (400-450 μm) at 4.5-5.5 mm, and *Artemia* nauplii only when the larvae were > 5.5 mm (Cunha and Planas 1995). In culture of yellowfin porgy, *Acanthopagrus latus*, larvae, trochophore larvae (< 50 μm) are offered as the first food, followed by the rotifer *B. plicatilis*, then *Artemia* nauplii, and finally copepods (Leu and Chou 1996). Gilthead sea bream larvae weighing < 45 μg are fed the small rotifer *B. rotundiformis* and, as they grow beyond 90 μg, the larger *B. plicatilis* (Yufera et al. 1991) are fed.

Whereas in indoor larval culture tanks prey size sequencing is relatively easy and more controlled, in natural larval culture ponds it is often fortuitous. In traditional culture methods, earthen ponds, first drained, dried, and properly prepared by the farmer for larval culture, are known to allow generally a desirable succession of live food organisms (rotifers followed by cladocerans followed by copepods), but a natural succession of live food organisms is not so easily obtained if the ponds cannot be dried completely (Opuszynski and Shireman 1991). It is also common that in nursery ponds, initially copepods and cladocerans might be present concurrently along with rotifers, the desired first food for the larvae. Cyclopoid copepods such as *Mesocyclops* are able to attack and kill very early stages of the larvae and constitute a significant source of larval mortality in certain culture ponds (Jhingran 1991). Besides, even when larvae are big enough not only to avoid predation by cyclopoid copepods but also actually to prey on them, their capture success with these copepods is relatively poor and therefore unprofitable to include in their diet. When dried ponds are filled in preparation for larval culture, the resting eggs or diapause stages of zooplankton present in the bottom soil form the inoculum. Normally, rotifers should be the earliest dominant group of food organisms to develop in the culture pond, since copepods have to go through naupliar and copepodid stages before reaching adult stage, when they become the potential food or predators of growing larvae. If, however, copepodids or adult copepods are concurrently present with the rotifers, some culture practices exclude copepods in the nursery pond by applying selected pesticides such as Sumithion, which kill only crustaceans, without harming rotifers (Jhingran and Pullin 1985).

ARTIFICIAL DIETS

Because of the incomplete development of the digestive system, there is in the life of the larvae of nearly all cultured species, a minimum period of dependence on live food organisms. The earliest age (weaning period) at which the larvae could be shifted from live food to dry feeds varies with the species but is generally longer for marine species–5 days in rohu (Mookerji 1992), 10-15 days in the common sole, 25-30 days in European sea bass, and 40-50 days in Asian red snapper (Tucker 1998).

Two important factors responsible for the sustained efforts to develop artificial diets for culturing fish larvae, notably of marine species, are the high costs of live food production and inconsistencies in the nutritional quality of the live food organisms (Southgate and Partridge 1998). In some marine fish production systems, live food production costs could be as high as 80% (Person-Le Ruyet et al. 1993). Depending upon the type of algal food they have been feeding upon, live food organisms (like the most widely used rotifer *B. plicatilis*) have been shown to be deficient in certain fatty acids essential for the healthy growth of larval fish (Watanabe et al. 1983). Nutritional enrichment of such food organisms before feeding them to the larvae results in a significant improvement in survival and growth (Lubzens et al. 1989).

Considering the costs of providing live food organisms, the earliest age at which larvae can be weaned to artificial diets should be known. It has been suggested that body weight of larvae rather than their age is a better indicator of the readiness for weaning (Bryant and Matty 1981; Kamler et al. 1990). The weight at the attainment of which larvae can be switched to artificial diets without adversely affecting their subsequent survival or growth has been termed the "adaptation weight" (Bryant and Matty 1981). This weight was found to be 0.31 mg (dry weight) in the case of rohu and 0.54 mg in singhi (Mookerji 1992). The age at which the adaptation weight is reached depends mostly upon the quality and quantity of live food received by larvae before switching to dry diet.

The development of artificial diets that can replace live food organisms in larval culture requires a multidisciplinary approach (Bengtson 1993). A sound knowledge of the feeding behavior and ecology of fish larvae is perhaps as important as that of their nutritional requirements for developing artificial diets. Among the most desirable attributes of formulated diets for larval culture (acceptability, digestibility, stability, nutritional quality, and shelf life) (Southgate and Partridge 1998), acceptability is obviously the most important, for if in the first place the diet is not accepted and ingested by the fish, all the other attributes become totally irrelevant. It has been noted that ingestion rates with artificial diets are generally considerably lower than those with live food organisms, even when their nutritional qualities were not significantly different (Person-Le Ruyet et al. 1993). This appears to be particularly true of carnivorous fish larvae, whose prey attack behavior is elicited only by the visual stimulus of a moving prey about to escape (Kamler 1992). This peculiar, species-specific escape movement of a prey facing a larva poised to attack is something that cannot obviously be expected from an inert food particle. Use of certain mechanical devices in the culture tanks (to vibrate the feed particles) and of refractory chemicals in the diets (to create illusion of movement) has been attempted with variable results. Another visual stimulus eliciting feeding response in the foraging larvae may be the color of the prey; dyes could be added to diet formulations to impart the desired color.

Larval feeding responses in nature might also be initiated or enhanced by olfactory stimuli. Live food organisms are known to release into the medium certain free aminoacids which might act as attractants to larvae (Hara 1994). If such feeding stimulants are identified, they can be incorporated into artificial diets for enhancing ingestion rates (Koven et al. 2001). However, it appears from the present literature (Tucker 1998) that in the larvae of many marine fish species the olfactory and gustatory senses are not fully developed in the initial stages of feeding. For them, the primary stimulants are probably the size and movement of the prey organism.

We have seen that optimal feeding protocols for raising larvae invariably incorporate a progressive increase in food particle size commensurate with the increasing gape and improving prey capture ability of the growing larvae. This particle size progression is essential with artificial diets also and is more easily and more precisely achievable perhaps, than with live food organisms.

Whether artificial microdiets should be formulated so as to float to the surface, stay suspended in the water column, or sink to the bottom would depend on the feeding habits of the larvae. A great majority of marine and freshwater fish larvae, particularly in the early stages, are column feeders. Dry diets, unless specially formulated to achieve near neutral buoyancy, tend to sink rapidly to the bottom, leaving a very low particle concentration in the open water, which results in low ingestion rates. Water quality also deteriorates due to excessive accumulation of unconsumed food at the bottom (Southgate and Partridge 1998).

In the foregoing review, those basic aspects of the feeding ecology and ethology of larval fishes that have significant implications for larval culture have been discussed. Experimental studies using fish larvae and their live food as a predator-prey system have proved to be valuable in selecting optimal species, sizes, and densities of live food for raising larvae of aquaculturally important marine and freshwater fish. Insights gained from such basic studies on larval feeding behavior have also been helpful in designing larval culture tanks and in developing artificial larval diets.

ACKNOWLEDGMENTS

The author thanks B. B. Jana for many helpful comments and suggestions and Nandita Mookerji, Rina Chakrabarti, and Ram Kumar for bibliographic assistance.

REFERENCES

Abi-Ayad, A., and P. Kestemont. 1994. Comparison of the nutritional status of goldfish (*Carassius auratus*) larvae fed with live, mixed or dry diet. Aquaculture 128:163-176.

Abrams, P.A. 1990. The effects of adaptive behavior on the type-2 functional response. Ecology 71:877-885.

Alikunhi, K.H., H. Chaudhuri, and V. Ramachandran. 1955. On the mortality of carp fry in nursery ponds and the role of plankton in their survival and growth. Indian Journal of Fisheries 2:257-313.

Bagarinao, T. 1986. Yolk absorption, onset of feeding and survival potential of larvae of three tropical marine fish species reared in the hatchery. Marine Biology 91:449-459.

Bailey, K.M., and E.D. Houde. 1989. Predation on eggs and larvae of marine fishes and the recruitment problem. Advances in Marine Biology 25:1-83.

Balon, E.K. 1986. Types of feeding in the ontogeny of fishes and the life-history model. Environmental Biology of Fishes 16:11-24.

Bardach, J.E., J. Magnuson, R.C. May, and J.M. Reinhart. 1980. Fish Behavior and Its Use in the Capture and Culture of Fishes. ICLARM, Manila, The Philippines.

Batty, R.S. 1987. Effect of light intensity on activity and food-searching of larval herring, *Clupea harengus*: A laboratory study. Marine Biology 94:323-327.

Batty, R.S., J.H.S. Blaxter, and D.A. Libby. 1986. Herring (*Clupea harengus*) filter-feeding in the dark. Marine Biology 91: 371-375.

Batty, R.S., J.H.S. Blaxter, and J.M. Richard. 1990. Light intensity and the feeding behavior of the herring, *Clupea harengus*. Marine Biology 107: 383-388.

Beacham, T.D., and C.B. Murray. 1985. Effect of female size, egg size, and water temperature on developmental biology of chum salmon (*Oncorhynchus keta*) from Nitinat River, British Columbia. Canadian Journal of Fisheries and Aquatic Sciences 42:1755-1765.

Bence, J.R., and W.W. Murdoch. 1986. Prey size selection by the mosquito fish: Relation to optimal diet theory. Ecology 67:324-336.

Bengtson, D.A. 1993. A comprehensive program for the evaluation of artificial diets. Journal of the World Aquaculture Society 24:199-210.

Blaxter, J.H.S. 1969. Development: Eggs and larvae. Pages 177-252 *in* W.S. Hoar, and D.J. Randall, eds. Fish Physiology. Vol. III, Academic Press, New York, New York.

Blaxter, J.H.S. 1980. Vision and the feeding of fishes. Pages 32-56 *in* J.E. Bardach, J.E. Magnuson, R.C. May, and J.M. Reinhart, eds. Fish Behavior and Its Use in the Capture and Culture of Fishes. ICLARM, Manila, Philippines.

Blaxter, J.H.S. 1986. Development of sense organs and behavior of teleost larvae with special reference to feeding and predator avoidance. Transactions of the American Fisheries Society 115:98-114.

Blaxter, J.H.S., and G. Hempel. 1963. The influence of egg size on herring larvae (*Clupea harengus* L.). Journal du Conseil International pour l'Exploration de la Mer., 28:211-240.

Blaxter, J.H.S., and M. Staines. 1971. Food searching potential in marine fish larvae. Pages 467-485 *in* D.J. Crisp, ed. Fourth European Marine Biology Symposium. Cambridge University Press, Cambridge, United Kingdom.

Boeuf, G., and P.-Y. Le Bail. 1999. Does light have an influence on fish growth? Aquaculture 177:129-152.

Boubee, J.A.T., and F.J. Ward. 1997. Mouth gape, food size, and diet of the common smelt *Retropinna retropinna* (Richardson) in the Waikato River System, North Island, New Zealand. New Zealand Journal of Marine and Freshwater Research 31:147-154.

Bremigan, M.T., and R.A. Stein. 1994. Gape-dependent larval foraging and zooplankton size: Implications for fish recruitment across systems. Canadian Journal of Fisheries and Aquatic Sciences 51:913-922.

Bryant, P.L., and A.J. Matty. 1981. Adaptation of carp (*Cyprinus carpio*) larvae to artificial diets. I. Optimum feeding rate and adaptation age for a commercial diet. Aquaculture 23:275-286.

Cahu, C., A.-M. Zanbonino Infante, P. Bergot, and S. Kaushik. 1998. Preliminary results on sea bass (*Dicentrarchus labrax*) larvae rearing with compound diet from first feeding. Comparison with carp (*Cyprinus carpio*) larvae. Aquaculture 169:1-7.

Chambers, R.C., and E.A. Trippel. 1997. Early Life History and Recruitment in Fish Populations. Chapman & Hall, London, England.

Confer, J.L., and P.I. Blades. 1975. Omnivorous zooplankton and planktivorous fish. Limnology and Oceanography 20:571-579.

Confer, J.L., G.L. Howick, M.H. Corzette, S.L. Kramer, S. Fitzgibbon, and R. Landesberg. 1978. Visual predation by planktivores. Oikos 31:27-37.

Coughlin, D.J. 1991. Ontogeny of feeding behavior of first-feeding Atlantic salmon (*Salmo salar*). Canadian Journal of Fisheries and Aquatic Sciences 48:1896-1904.

Cunha, I., and M. Planas. 1995. Ingestion rates of turbot larvae (*Scophthalmus maximus*) using different-sized live prey. ICES Marine Science Symposia 201:16-20.

Dabrowski, K. 1984. The feeding of fish larvae: present 'state of the art' and perspectives. Reproduction, Nutrition and Development 24:807-833.

Dabrowski, K. 1986. Mini-review on ontogenetical aspects of nutritional requirements in fish. Comparative Biochemistry and Physiology 85A: 639-655.

Dabrowski, K., and R. Bardega. 1984. Mouth size and predicted food size differences of larvae of three cyprinid fish species. Aquaculture 40:41-46.

Dabrowski, K., S. J. Kaushik, and B. Fauconneau. 1987. Rearing of sturgeon (*Acipenser baeri* Brandt) larvae. 3. Nitrogen and energy metabolism and aminoacid absorption. Aquaculture 65: 31-41.

Datta Munshi, J.S., and S. Choudhary. 1996. Ecology of *Heteropneustes fossilis* (Bloch): An air-breathing catfish of South-East Asia. Narendra Publishing House, Delhi, India.

Dendrinos, P., S. Dewan, and J.P. Thorpe. 1984. Improvement in the feeding efficiency of larval, postlarval and juvenile Dover sole (*Solea solea* L.) by use of staining to improve the visibility of *Artemia* used as food. Aquaculture 38:137-144.

Denson, M.R., and T.I.J. Smith. 1996. Larval rearing and weaning techniques for white bass *Morone chrysops*. Journal of the World Aquaculture Society 27:194-201.

De Silva, S.S., and A.A. Anderson. 1995. Fish Nutrition in Aquaculture. Chapman and Hall, London, England.

DeVries, D.R., M.T. Bremigan, and R.A. Stein. 1998. Prey selection by larval fishes as influenced by available zooplankton and gape limitation. Transactions of the American Fisheries Society 127:1040-1050.

Drenner, R.W., J.R. Strickler, and W.J. O'Brien. 1978. Capture probability: The role of zooplankter escape in the selective feeding of planktivorous fish. Journal of the Fisheries Research Board of Canada 35:1370-1373.

Drenner, R.W., G.L. Vinyard, M. Gphen, and S.R. McComas. 1982. Feeding behavior of the cichlids, Sarotherodon galialaeum: Selective predation on Lake Kinneret zooplankton. Hydrobiologia 87:17-20.

Drost, M.R. 1987. Relation between aiming and catch success in larval fishes. Canadian Journal of Fisheries and Aquatic Sciences 44:304-315.

Drost, M.R., J.W.M. Osse, and M. Muller. 1988. Prey capture by fish larvae, water flow patterns and the effect of escape movements of prey. Netherlands Journal of Zoology 38:23-45.

Duarte, C.M., and M. Alcaraz. 1989. To produce many small or few large eggs: a size-independent reproductive tactic of fish. Oecologia 80:401-404.

Duray, M.N. 1995. The effect of tank color and rotifer density on rotifer ingestion, growth and survival of milkfish (*Chanos chanos*) larvae. Philippine Scientist 32:18-26.

Eggers, D.M. 1977. The nature of prey selection by planktivorous fish. Ecology 58:46-59.

Eggers, D.M. 1982. Planktivore preference by prey size. Ecology 63:381-390.

Ehrlich, K.F. 1974. Chemical changes during growth and starvation of larval *Pleuronectes platessa*. Marine Biology 24:39-48.

Elliott, J.M. 1989. The critical-period concept for juvenile survival and its relevance for population regulation in young sea trout, *Salmo trutta*. Journal of Fish Biology 35 (Supplement): 91-98.

FAO 1999. The State of World Fisheries and Aquaculture 1998. Food and Agricultural Organization of the United Nations, Rome, Italy.

Fernando, C.H., and J. Holcik. 1990. The impact of fish introductions into tropical freshwaters. Pages. 103-130 *in* P.S. Ramakrishnan, ed. Ecology of Biological Invasions in the Tropics. Special Volume, International Journal of Ecology and Environmental Sciences, New Delhi, India.

Gardner, M.B. 1981. Mechanisms of size selectivity by planktivorous fish: A test of hypotheses. Ecology 62:571-578.

Govoni, J.J., G.W. Boehlert, and Y. Watanabe. 1986. The physiology of digestion in fish larvae. Environmental Biology of Fishes 16:59-77.

Greene, C.H. 1983. Selective predation in freshwater zooplankton communities. Internationale Revue der gesamten Hydrobiologie 68:297-315.

Greene, C.H. 1986. Patterns of prey selection: Implications of predator foraging tactics. American Naturalist 128:824-839.

Gregory, R.S. 1994. The influence of ontogeny, perceived risk of predation, and visual ability on the foraging behavior of juvenile chinook salmon. Pages 271-284 *in* D.J. Stouder, K.L. Fresh, and R.J. Feller, eds. Theory and Application of Fish Feeding Ecology. University of South Carolina Press, Columbia, South Carolina.

Gulbrandsen, J. 1991. Function response of Atlantic halibut larvae related to prey density and distribution. Aquaculture 94:89-98.

Hagiwara, A., T.W. Snell, E. Lubzens, and C.S. Tamaru. 1997. Live Food in Aquaculture. Developments in Hydrobiology Series, Kluwer Publishers, Dodrecht, The Netherlands.

Hara, T.J. 1994. The diversity of chemical stimulation in fish olfaction and gustation. Reviews in Fish Biology and Fisheries 4:1-35.

Heath, M.R. 1992. Field investigations of the early life stages of marine fish. Advances in Marine Biology 28:1-174.

Hinshaw, J.M. 1985. Effects of illumination and prey contrast on survival and growth of larval yellow perch *Perca flavescens*. Transactions of the American Fisheries Society 114:540-545.

Holling, C.S. 1966. The functional response of invertebrate predators to prey density. Memoirs of the Entomological Society of Canada 48:1-87.

Houde, E.D. 1987. Fish early life dynamics and recruitment variability. American Fisheries Society Symposia 2:17-29.

Houde, E.D. 1994. Differences between marine and freshwater fish larvae: Implications for recruitment. ICES Journal of Marine Sciences 51:91-97.

Houde, E.D. 1997. Patterns and consequences of selective processes in teleost early life histories. Pages 173-196 *in* R.C. Chambers, and E.A. Trippel, eds. Early Life History and Recruitment in Fish Populations. Chapman & Hall, London, England.

Houde, E.D., and R.C. Schekter. 1980. Feeding by marine fish larvae: Developmental and functional responses. Environmental Biology of Fishes 5:315-334.

Hunter, J.R. 1972. Swimming and feeding behavior of larval anchovy *Engraulis mordax*. Fishery Bulletin 70:821-839.

Hunter, J.R. 1980. The feeding behavior and ecology of marine fish larvae. Pages 287-330 *in* J.E. Bardach, J.J. Magnuson, R.C. May, and M. Reinhart, eds. Fish Behavior and Its Use in Capture and Culture of Fishes. ICLARM, Manila, Philippines.

Hunter, J.R. 1981. Feeding ecology and predation of marine fish larvae. Pages 37-71 *in* R. Lasker, ed. Marine Fish Larvae: Morphology, Ecology and Relation to Fisheries. Washington Sea Grant Program, University of Washington, Seattle, Washington.

Jana, B.B., and R. Chakrabarti. 1988. Some factors determining the feeding optima in carp fingerlings. Archiv für Hydrobiologie 113:121-131.

Jana, B.B., and R. Chakrabarti. 1996. Feeding spectrum in the rearing of fish larvae: A brief review. Pages 257-284 *in* B.R. Singh, ed. Advances in Fish Research. Vol. 2, Narendra Publishers, Delhi, India.

Jana, B.B., and S. Dutta. 1998. Efficiency of artificial plankton diet on growth of common carp (*Cyprinus carpio*) fry. Journal of Aquaculture in Tropics 13:255-260.

Janssen, J. 1978. Feeding behaviour repertoire of the alewife, *Alosa pseudoharengus*, and the ciscos, *Coregonus hoyi* and *C. artedii*. Journal of the Fisheries Research Board of Canada 35: 249-253.

Janssen, J., W.R. Jones, A. Whang, and P.E. Oshel. 1994. Use of the lateral line in particulate feeding in the dark by juvenile alewife (*Alosa psuedoharengus*). Canadian Journal of Fisheries and Aquatic Sciences 52:358-363.

Jhingran, V.G. 1991. Fish and Fisheries of India, 3rd ed. Hindustan Publishing Corporation, New Delhi, India.

Jhingran, V.G., and R.S.V. Pullin. 1985. A Hatchery Manual for the Common, Chinese and Indian Major Carps. Asian Development Bank, ICLARM, Manila, Philippines.

Jobling, M. 1986. Gastrointestinal overload–a problem with formulated feeds? Aquaculture 51:257-263.

Johnston, T.A., and J.A. Mathias. 1994. Feeding ecology of walleye, *Stizostedion vitreum*, larvae: Effects of body size, zooplankton abundance and zooplankton community composition. Canadian Journal of Fisheries and Aquatic Sciences 51:2077-2089.

Jones, A. 1972. Studies on egg development and larval rearing of turbot, *Scophthalmus maximus* L., and brill, *Scophthalmus rhombus* L., in the laboratory. Journal of the Marine Biological Association, U.K. 52: 965-986.

Jones, D.A., M.S. Kamarudi, and L. Le Vay. 1993. The potential for replacement of live feeds in larval culture. Journal of the World Aquaculture Society 24:199-210.

Kamler, E. 1992. Early life history of fish: An energetics approach. Chapman & Hall, London, England.

Kamler, E., M. Szlamińska, A. Prazybyl, B. Barska, M. Jakubas, M. Kuczylski, and K. Raciborski. 1990. Developmental response of carp, *Cyprinus carpio*, larvae fed different foods or starved. Environmental Biology of Fishes 29:303-313.

Kerfoot, W.C., and A. Sih. 1987. Predation: Direct and indirect impacts on aquatic communities. University Press of New England, Hanover, New Hampshire.

Khadka, R.B. 1986. The Feeding Ecology of Common Carp (*Cyprinus carpio* var. *communis)* Larvae–An Optimal Foraging Approach. Doctoral dissertation, University of Delhi, Delhi, India.

Khadka, R.B., and T.R. Rao. 1986. Prey size selection by common carp (*Cyprinus carpio* var. *communis*) in relation to age and prey density. Aquaculture 54:89-96.

Kirk, K.L., and J.J. Gilbert. 1990. Suspended clay and the population dynamics of planktonic rotifers and cladocerans. Ecology 71:1741-1755.

Knutsen, G.M., and S. Tilseth. 1985. Growth, development, and feeding success of Atlantic cod larvae *Gadus morhua* related to egg size. Transactions of the American Fisheries Society 114:507-511.

Kohno, H. 1998. Early life history features influencing larval survival of cultivated tropical marine finfish. Pages 71-109 *in* S.S. De Selva, ed. Tropical Mariculture. Academic Press, London, England.

Kolkovski, S., and A. Tandler. 1995. Why microdiets are still inadequate as viable alternatives to live zooplankton for developing marine fish larvae. Pages 265-266 *in* P. Lavens, E. Jaspers, and I. Roelants, eds. Larvi'95 Special Publication No.24, European Aquaculture Society, Gent, Belgium.

Koven, W., S. Kolkovski, E. Hadas, K. Gamsiz, and A. Tandler. 2001. Advances in the development of microdiets for gilthead sea bream, *Sparus aurata*: A review. Aquaculture 194: 107-121.

Lauff, M., and R. Hoffer. 1984. Proteolytic enzymes in fish development and the importance of dietary enzymes. Aquaculture 37:335-346.

Lazzaro, X. 1987. A review of planktivorous fishes: Their evolution, feeding behaviors, selectivities, and impacts. Hydrobiologia 146:97-167.

Letcher, B.H., and J.A. Rice. 1997. Prey patchiness and larval fish growth and survival: Inferences from a spatially explicit, individual-based model. Ecological Modeling 95:29-43.

Letcher, B.H., J.A. Rice, L.B. Crowder, and K.A. Rose. 1996. Variability in survival of larval fish: Disentangling components with a generalized individual-based model. Canadian Journal of Fisheries and Aquatic Sciences 53:787-801.

Leu, M.-Y., and Y.-H. Chou. 1996. Induced spawning and larval rearing of captive yellowfin porgy, *Acanthopagrus latus* (Houttuyn). Aquaculture 143:155-166.

Lubzens, E., A. Tandler, and G. Minkoff. 1989. Rotifers as food in aquaculture. Hydrobiologia 186/187: 387-400.

Magurran, A.E. 1986. Individual differences in fish behavior. Pages 338-365 *in* T.J. Pitcher, ed. The Behavior of Teleost Fishes. Croom Helm, Kent, England.

Malhotra, Y.R., and S. Munshi. 1985. First feeding and survival of *Aspidoparia morar* larvae (Cyprinidae). Transactions of the Fisheries Society 114:286-290.

Marcotte, B.M., and H.I. Browman. 1986. Foraging behavior in fishes: Perspectives on variance. Environmental Biology of Fishes 16:25-33.

May, R.C. 1974. Larval mortality in marine fishes and the critical period concept. Pages 3-19 *in* J.H.S. Blaxter, ed. The Early Life History of Fish. Springer-Verlag, New York, New York.

Mayer, C.M., and D.W. Wahl. 1997. The relationship between prey selectivity and growth and survival in a larval fish. Canadian Journal of Fisheries and Aquatic Sciences 54:1504-1512.

Miller, T.J., L.B. Crowder, and J.A. Rice. 1992. Body size and the ontogeny of the functional response in fishes. Canadian Journal of Fisheries and Aquatic Sciences 49;805-812.

Miller, T.J., L.B. Crowder, J.A. Rice, and E.A. Marschall. 1988. Larval size and recruitment mechanisms in fishes: Toward a conceptual framework. Canadian Journal of Fisheries and Aquatic Sciences 45:1657-1670.

Mills, E.L., J.L. Confer, and R.C. Ready. 1984. Prey selection by young yellow perch: The influence of capture success, visual acuity, and prey choice. Transactions of the American Fisheries Society 113:579-587.

Mills, E.L., J.L. Confer, and D.W. Ktretchmer. 1986. Zooplankton selection by young yellow perch: The influence of light, prey density, and predator size. Transactions of the American Fisheries Society 115:716-725.

Mitra, G.N., and P. Mohapatra. 1956. On the role of zooplankton in the nutrition of carp fry. Indian Journal of Fisheries 3:299-310.

Mittelbach, G.G. 1983. Optimal foraging and growth in bluegills. Oecologia 59: 157-162.

Mookerji, N. 1992. Comparative Experimental Studies on the Feeding Ecology of the Larvae of Rohu (*Labeo rohita*) and Singhi (*Heteropneustes fossilis*). Doctoral dissertation, University of Delhi, Delhi, India.

Mookerji, N., and T.R. Rao. 1993. Patterns of prey selection in rohu (*Labeo rohita*) and singhi (*Heteropneustes fossilis*) larvae under light and dark conditions. Aquaculture 118:85-104.

Mookerji, N., and T.R. Rao. 1994. Influence of ontogenetic changes in prey selection on the survival and growth of rohu, *Labeo rohita* and singhi, *Heteropneustes fossilis* larvae. Journal of Fish Biology 44:479-490.

Mookerji, N., and T.R. Rao. 1995. Prey capture success, feeding frequency and daily food intake rates in rohu, *Labeo rohita* (Ham.) and singhi, *Heteropneustes fossilis* (Bloch) larvae. Journal of Applied Ichthyology 11:37-49.

Mookerji, N., and T.R. Rao. 1999. Rates of yolk utilization and effects of delayed initial feeding in the larvae of the freshwater fishes rohu and singhi. Aquaculture International 7:1-12.

Munilla-Moran, R., J.R. Stark, and A. Barbour. 1990. The role of exogenous enzymes in digestion in early turbot larvae, *Scophthalmus maximus* L. Aquaculture 88: 337-350.

Muir, P.R., and D.C. Sutton. 1994. Bacterial degradation of microencapsulated foods used in larval culture. Journal of the World Aquaculture Society 25:371-378.

Naas, K., I. Huse, and J. Iglesias. 1996. Illumination in first feeding tanks for marine fish larvae. Aquacultural Engineering 15:291-300.

Nash, C.E. 1975. Crop selection issues. Pages 183-210 *in* J.A. Hanson, ed. Open Sea Mariculture: Perspectives, Problems and Prospects. Dowden, Hutchinson and Ross, Stroudsburg, Pennsylvania.

Nikolsky, G.V. 1963. The Ecology of Fishes. Academic Press, London. England.

O'Brien, W.J. 1979. The predator-prey interaction of planktivorous fish and zooplankton. American Scientist 67:572-581.

O'Brien, W.J. 1987. Planktivory by freshwater fish. Pages 3-16 *in* W.C. Kerfoot, and A. Sih, eds. Predation: Direct and Indirect Impacts on Aquatic Communities. University Press of New England. Hanover, New Hampshire.

Olsen, Y. 1997. Larval-rearing technology of marine species in Norway. Hydrobiologia 358:27-36.

Opuszynski, K., and J.V. Shireman. 1991. Larviculture of the most important cyprinid species: Ecological, biotechnical and regional aspects. Page 278 *in* P. Lavens, P. Sorgeloos, E. Jaspers, and F. Ollevier, eds. Larvi '91, Special Publication No. 15, European Aquaculture Society No. 15, Gent, Belgium.

Ostrowski, A.C. 1989. Effect of rearing tank background color on early survival of dolphin larvae. Progressive Fish-Culturist 51:161-163.

Paradis, A.R., P. Pepin, and J.A. Brown. 1996. Vulnerability of fish eggs and larvae to predation: Review of the influence of the relative size of prey and predator. Canadian Journal of Fisheries and Aquatic Sciences 53:1226-1235.

Person-Le Ruyet, J., J.C. Alexander, L. Thebaud, and C. Mugnier. 1993. Marine fish larvae feeding: Formulated diets or live preys? Journal of World Aquaculture Society 24:211-224.

Planas, M., and I. Cunha. 1999. Larviculture of marine fish: Problems and perspectives. Aquaculture 177:171-190.

Pyke, G.H. 1984. Optimal foraging theory: A critical review. Annual Review of Ecology and Systematics 15:523-575.

Rana, K.J. 1985. Influence of egg size on growth, onset of feeding, point-of-no-return, and survival of unfed *Oreochromis mossambicus* fry. Aquaculture 46:119-131.

Rana, K.J. 1990. Influence of incubation temperature on *Oreochromis niloticus* (L.) eggs and fry. II. Survival, growth and feeding of fry developing solely on their yolk reserves. Aquaculture 87:183-195.

Rao, T. R. 1991. Optimality models for prey-predator interactions. Pages 342-353 *in* P. Narain, O.P. Kathuria, V. K. Sharma, and Prajneshu, eds. Recent Advances in Agricultural Statistics Research. Wiley-Eastern, Delhi, India.

Rao, T.R., and N. Mookerji. 1995. Functional responses of the larvae of rohu (*Labeo rohita*) and singhi (*Heteropneustes fossilis*) in the presence of alternate prey under dark and light conditions. Pages 409-413 *in* P. Lavens, E. Jaspers, and I. Roelants, eds. Larvi '95, Special Publication No. 24, European Aquaculture Society, Gent, Belgium.

Rice, J.A., L.B. Crowder, and F.P. Binkowski. 1987. Evaluating potential sources of mortality for larval bloater (*Coregonus hoyi*): Starvation and vulnerability to predation. Canadian Journal of Fisheries and Aquatic Sciences 44:467-472.

Roff, D.A. 1984. The evolution of life history parameters in teleosts. Canadian Journal of Fisheries and Aquatic Sciences 41:989-1000.

Rosenlund, G., J. Stoss, and C. Talbot. 1997. Co-feeding marine fish larvae with inert and live diets. Aquaculture 155:183-191.

Schael, D.M., L.G. Rudstam, and J.R. Post. 1991. Gape limitation and prey selection in larval yellow perch (*Perca flavescens*), freshwater drum (*Aplodinotus grunniens*) and black crappie (*Pomoxis nigromaculatus*). Canadian Journal of Fisheries and Aquatic Sciences 38:1919-1925.

Schiemer, F., H. Kockeis, and J. Wanzenböck. 1989. Foraging in Cyprinidae during early development. Polskie Archiwum Hydrobiologii 36:467-474.

Shirota, A. 1970. Studies on the mouth size of fish larvae. Bulletin of the Japanese Society for Scientific Fisheries 36:353-368.

Sibkin Y.N. 1974. Age related changes in the role of vision in the feeding of various fishes. Journal of Ichthyology 14:133-139.

Southgate, P.C., and G.J. Partridge. 1998. Development of artificial diets for marine fish larvae: Problems and prospects. Pages 151-169 *in* S.S. De Selva, ed. Tropical Mariculture. Academic Press, London, England.

Springate, J.R.C., N.R. Bromage, and P.R.T. Cumaranatunga. 1985. The effects of different rations on fecundity and egg quality in the rainbow trout (*Salmo gairdneri*). Pages 371-393 *in* C.B. Cowey, A.M. Mackie, and J.G. Bell, eds. Nutrition and Feeding in Fish. Academic Press, London, England.

Stephens, D., and J.R. Krebs. 1986. Foraging Theory. Princeton University Press, Princeton, New Jersey.

Stouder, D.J., K.L. Fresh, and R.J. Feller. 1994. Theory and Application of Fish Feeding Ecology. University of South Carolina Press, Columbia, South Carolina.

Thakur, N.K., and P. Das. 1986. Synopsis of biological data on singhi *Heteropneustes fossilis* (Bloch 1794). Bulletin of Central Inland Fisheries Research Institute 39:1-32.

Townsend, C.R., and I.J. Winfield. 1985. The application of the optimal foraging theory to feeding behavior in fish. Pages 67-98 *in* P. Tytler, and P. Calow, eds. Fish Energetics: New Perspectives. Croom Helm, London, England.

Tucker, J. W. 1998. Marine Fish Culture. Kluwer Academic Publishers, Boston, Massachusetts.

Unger, P.A., W.M. Lewis, Jr., and D.H. McClearn. 1984. Nonvisual feeding in a visual planktivore, *Xenomelanairis venezuelae*. Oecologia 64: 280-283.

Verreth, J., E.H. Eding, G.R.M. Rao, F. Huskens, and H. Segner. 1991. Feeding strategies and nutritional physiology of early life of clariid catfishes. Pages 12-13 *in* P. Lavens, P. Sorgeloos, E. Jaspers, and F. Ollevier, eds. Larvi '91, Special Publication No. 15, European Aquaculture Society, Gent, Belgium.

Vinyard, G.L. 1980. Differential prey vulnerability and predator selectivity: Effects of evasive on bluegill (*Lepomis macrochirus*) and pumpkin seed (*L. gibbosus*) predation. Canadian Journal of Fisheries and Aquatic Sciences 37:2294-2299.

Vinyard, G.L., and W.J. O'Brien. 1976. Effects of light and turbidity on the reactive distance of the bluegill sunfish (*Lepomis macrochirus*). Journal of the Fisheries Research Board of Canada 33:2845-2849.

Walton, W.E., N.G. Hairston, Jr., and J.K. Wetterer. 1992. Growth-related constraints on diet selection by sunfish. Ecology 73:429-437.

Walton, W.E., S.S. Easter, Jr., C. Malinoski, and N.G. Hairston, Jr. 1994. Size-related change in the visual resolution of sunfish (*Lepomis* spp.). Canadian Journal of Fisheries and Aquatic Sciences 51:2017-2026.

Wang, Y.L., R.K. Buddington, and S.I. Doroshov. 1987. Influence of temperature on yolk utilization by the white sturgeon, *Acipenser transmontanus*. Journal of Fish Biology 30:263-271.

Wanzenböck, J. 1992. Ontogeny of prey attack behavior in larvae and juveniles of three European cyprinids. Environmental Biology of Fishes 33:23-32.

Wanzenböck, J., and F. Schiemer. 1989. Prey detection in cyprinids during early development. Canadian Journal of Fisheries and Aquatic Sciences 46:995-1001.

Ware, D.M. 1972. Predation by rainbow trout (*Salmo gairdneri*): The influence of hunger, prey density and prey size. Journal of the Fisheries Research Board of Canada 29:1193-1201.

Ware, D.M. 1975. Relation between egg size, growth, and natural mortality of larval fish. Journal of the Fisheries Research Board of Canada 32:2503-2512.

Watanabe, T., C. Kitajima, and S. Fujita. 1983. Nutritional values of live organisms used in Japan for mass propagation of fish: A review. Aquaculture 34:115-143.

Werner, E.E., 1974. The fish size, prey size, handling time relation in several sunfishes and some implications. Journal of the Fisheries Research Board of Canada 31: 1531-1536.

Werner E.E., and D.J. Hall. 1974. Optimal foraging and the size selection of prey by the bluegill sunfish (*Lepomis macrochirus*). Ecology 55:1042-1052.

Werner, E.E., and G.G. Mittelbach. 1981. Optimal foraging: Field tests of diet choice and habitat switching. American Zoologist 21:813-829.

Werner, E.E., J.F. Gilliam, D.J. Hall, and G.G. Mittelbach. 1983. An experimental test of the effects of predation risk on habitat use in fish. Ecology 64:1540-1548.

Winemuller, K.O., and K.A. Rose. 1992. Patterns of life history diversification in North American fishes: Implications for population regulation. Canadian Journal of Fisheries and Aquatic Sciences 49:2196-2218.

Winkler, H., and C.P. Orellana. 1992. Functional responses of five cyprinid species to planktonic prey. Environmental Biology of Fishes 33:53-62.

Wootton, R.J. 1984. Tactics and strategies in fish reproduction. Pages 1-12 *in* G.W. Potts, and R.J. Wootton, eds. Fish Reproduction: Strategies and Tactics. Academic Press, London, England.

Wright, D.L., and W.J. O'Brien. 1984. The development and field test of a tactical model of the planktivorous feeding of white crappie (*Pomoxis annularis*). Ecological Monographs 54:65-98.

Yin, M.C., and J.H.S. Blaxter. 1987. Feeding ability and survival during starvation of marine fish larvae reared in the laboratory. Journal of Experimental Marine Biology and Ecology 105:73-83.

Yufera, M., E. Pascual, and A. Polo. 1991. The importance of prey sequencing during early growth of the gilthead seabream *Sparus auratus* L. Pages 116-118 *in* P. Lavens, P. Sorgeloos, E. Jaspers, and F. Ollevier, eds. Larvi '91, Special Publication No. 15, European Aquaculture Society, Gent, Belgium.

Yufera, M., E. Pascual, A. Polo, and M.C. Sarasquete. 1993. Effect of starvation on the feeding ability of gilthead seabream (*Sparus aurata* L.) larvae at first feeding. Journal of Experimental Marine Biology and Ecology 169:259-272.

Yufera, M., E. Pascual, and C. Fernandez-Diaz. 1999. A highly efficient microencapsulated food for rearing early larvae of marine fish. Aquaculture 177:249-256.

Zaret, T.M. 1980. Predation and Freshwater Communities. Yale University Press, New Haven, Connecticut.

Breeding Programs for Sustainable Aquaculture

I. Olesen
T. Gjedrem
H. B. Bentsen
B. Gjerde
M. Rye

SUMMARY. Definition of breeding goals for sustainable fish production is considered, with emphasis on non-market (e.g., ethical) as well as market values. The need for long-term biologically, ecologically, and sociologically sound breeding goals is emphasized, because animal breeding determined only by short-term market forces has lead to unwanted side effects. Farmed fish is at an early stage of domestication and breeding, but rapid selection responses for growth have already been documented for several species. Reports of selection responses for fish and shellfish in both temperate and tropical environments are reviewed. Growth-rate responses of 4-20% have been obtained per generation. Broad breeding goals, including health and functional traits, in addition to production traits, are required. More basic knowledge of, e.g., animal welfare and behavioral disorders of fish is also needed.

Less than 1% of the aquaculture fish material in 1993 originated from selection programs. For most species under improvement, only one or very few programs are running, and the effective population sizes are often limited. Such populations may however easily gain sufficient advantage above non-improved populations to capture much of the market.

I. Olesen, T. Gjedrem, H. B. Bentsen, B. Gjerde and M. Rye, AKVAFORSK, P.O. Box 5010, N-1432 Ås, Norway.

[Haworth co-indexing entry note]: "Breeding Programs for Sustainable Aquaculture." Olesen, I. et al. Co-published simultaneously in *Journal of Applied Aquaculture* (Food Products Press, an imprint of The Haworth Press, Inc.) Vol. 13, No. 3/4, 2003, pp. 179-204; and: *Sustainable Aquaculture: Global Perspectives* (ed: B. B. Jana, and Carl D. Webster) Food Products Press, an imprint of The Haworth Press, Inc., 2003, pp. 179-204. Single or multiple copies of this article are available for a fee from The Haworth Document Delivery Service [1-800-HAWORTH, 9:00 a.m. - 5:00 p.m. (EST). E-mail address: getinfo@haworthpressinc.com].

This will also discourage further genetic introductions into the breeding nucleus. Long-term inbreeding and loss of genetic variability because of genetic drift may then affect performance and further genetic progress. A sufficiently large and genetically diverse breeding population with appropriate family structure is therefore fundamental when establishing and running a selection program.

Important prerequisites for breeding programs for sustainable production are appropriate governmental policies and awareness of our way of thinking about aquaculture, nature and society. A more communal worldview informed by a subjective epistemology (how we learn about/analyze nature) and a holistic ontology (belief about what/how nature is) is also required.

KEYWORDS. Breeding programs, sustainable aquaculture, domestication, selection

INTRODUCTION

The world-wide demand for fish protein is expected to increase, because the harvest of wild fish populations has already reached (and in some cases exceeded) the carrying capacity. The need for aquaculture production in 2025 has been estimated to become 63 million tons (New 1991), whereas the production in 1997 was 27 million tons (FAO 1998). Unless the fish supply is increased through successful aquaculture programs, fish protein will become a scarce and costly commodity. High fish prices will result in increased incentives for over-fishing of wild stocks and reduced food security for a large number of poor consumers that depend on fish as a source of animal protein.

The preservation of wild fish populations may consequently depend on an increased development of fish farming operations. Today, aquaculture contributes less than 30% of the total fish supply, and the fish grown are basically undomesticated and poorly suited for farming (Gjedrem 1997). This is in striking contrast to the agricultural production based on a variety of highly bred and cultivated plants and livestock strains and breeds.

The objective of this paper is to review genetic improvement programs for sustainable fish production. Definition of breeding goals and design of breeding programs for long-term selection response and gains obtained in both temperate and tropical environments are considered.

SUSTAINABLE AQUACULTURE AND FISH PRODUCTION

Sustainability is based on a holistic philosophy–a set of values and principles–but may also involve a specific set of practices. When discussing breeding programs for sustainable fish production, it is necessary to regard and describe the fish as an integrated part of sustainable production systems. Also, we must make clear what we see as sustainable systems.

Thompson and Nardone (1999) considered two different methodological approaches to sustainable livestock production: resource sufficiency and functional integrity. Olesen et al. (2000) concluded that in our approach to breeding programs for sustainable production, functional integrity should be preferred, because both ecological and social elements are taken into account explicitly. Also, this means that we recognize that the threat to sustainable animal production is not only scarcity of environmental or natural resources, but also the increasing fragility of human-ecosystem interactions. The Holmenkollen Guidelines for Sustainable Aquaculture (1999) is an example of a functional integrity approach to sustainable aquaculture, because both ecological and social elements and limits are considered. These guidelines adopt the principles of: (1) sustainable development endorsed in the Rio-Declaration of 1992, comprising the inter-relation of natural and technological aspects on the one hand with socio-economic and value-based considerations on the other; (2) precautionary–in the light of uncertain or inconclusive scientific knowledge, strategies which effectively reduce the possibility of future harm to the environment are called for; (3) human equity.

In a review, Torp Donner and Juga (1997) screened different criteria used to describe sustainable livestock production and discussed animal breeding methods that could enhance it. Environmental and economical aspects were also considered in addition to biodiversity and ethical aspects. The new idea in the term "sustainability" is that environmental, genetic diversity, ethical, and social aspects should be accounted for, in addition to short- and long-term economic values (Olesen et al. 2000). Vavra (1996) suggested that sustainable systems exist in the overlap of what the current generation wants for itself and future generations and what is biologically and physically possible in the long run. This reminds us that we have to make some choices or priorities among the many conflicting goals and interests included in the term sustainability, e.g., between the interests of present and future generations.

Francis and Callaway (1993) summarized common elements connected to the phrase "sustainable agriculture," at least among those who are seeking long-term and equitable solutions to the challenge of food production: resource efficiency, profitability, productivity, environmental soundness, and social viability. A stronger time and space definition can be added, such that future generations have at least as many options as we have to explore alternative solutions to the food challenge. A more critical look is needed at the globalization of food systems and the energy involved in specialization and distant transport of high volume food or feed stuffs.

FISH BREEDING PROGRAMS

Prerequisites for Breeding Programs

In order to control fertilization, hatching, and first feeding, it is essential that the entire reproduction cycle for a species is controlled in captivity. Furthermore, an inexpensive marking or tagging system is advantageous and a necessity for a breeding program based on testing and selection within and between families. In order to select the best individuals for breeding, we need records of the traits in the breeding objective or traits correlated to the breeding goal. The records should be obtained for many animals at a reasonable cost during a limited time interval. Knowledge and methodology to optimize selection by predicting and using breeding values is also essential. A description of methods to predict breeding values is given by Cameron (1997).

A breeding objective needs to be defined for a breeding program, and the traits included must show genetic variation. For a selection program, additive genetic variation is required, whereas non-additive genetic variation is needed for utilizing heterosis via crossbreeding.

The breeding objectives should be concerned with the individual producer's objectives, because the producer's primary reason to buy a certain stock at a certain price will be based upon his assessment of how animals will contribute to the efficiency of his farm (Harris 1970). For example, in temperate zones breeding objectives for intensive milk production have been developed for producers or groups of producers to maximize profit rather than for taxpayer-financed national programs, where emphasis on, e.g., environmental and animal welfare issues may be desired (Pearson 1986). The socio-economic (market) attitude of the decision-maker influences the perspective to be considered and is, therefore, crucial in defining the breeding goal. For example, the individual agricultural producer deals with a competitive market, with no individual setting prices (Stonier and Hague 1964), and individual producers must be expected to act according to these (external) prices, even if the prices are misleading from a broader, national perspective (Horring 1948). A common interest for society as a whole may not be a sufficient incentive for the individual farmer when he makes decisions about a breeding program (Brascamp et al. 1998). In order to promote breeding goals and breeding programs based on a holistic, long-term perspective, additional governmental policies could be required. Subsidies and legislation, such as the environmental legislation in Western Europe may contribute to such incentives for the individual farmer (Steverink et al. 1994). For example, the environmental legislation in The Netherlands strongly reduced specialized beef production, because profit per unit of environmental pollution (e.g., N surplus) is lower in beef than in dairy production.

This also illustrates how important it is for breeding operations to communicate and cooperate with farmers as well as the consumers and governmental representatives to ensure a breeding program that contributes to a sustainable production. Sriskandarajah and Bawden (University of Western Sydney, Hawkesbury Ricgmond, Australia, pers. comm.) considered a model based on a communal, cultural, or people-centered worldview, informed by a subjective epistemology (how we learn about/analyze nature) and a holistic ontology (belief about what/how nature is). At one extreme, we may believe that we exist in a world of objective reality only, and that, for example, the choice of breeding goal can and should be based on objective calculations and figures only. The opposite extreme, a subjective view, is that we can only be interpreters, and that our values and emotions affect how we come to know. This makes a single objective truth an illusion, and we might accept that a farmer adopts a breeding strategy only according to his personal preferences. The opposite of a holistic ontology is reductionism and a highly fragmented view of nature, where we may believe that when we know enough about the component parts (e.g., plant, animals, land, water, and air) we can put them all together and understand the whole. The authors mentioned above argued that we should move towards the subjective perspective, the cultural/social perspective of our worldview, to accommodate people, their biases, values, attitudes, and aspirations. This is necessary to cooperate as a society, as farming communities, and to begin protecting the environment, which also provides the basis for further food production. This awareness should influence our thinking when we define fish breeding goals and design breeding programs, and it is necessary in order to accommodate sustainable production systems.

Breeding Goals

Olesen et al. (2000) discussed animal breeding goals for sustainable production systems. They stressed the need for whole-system analysis when considering sustainability in animal selection and breeding. We need animals that can contribute to optimize the system. Additionally, evaluation of various criteria (e.g., economy, pollution and resource depletion rate) at higher level than farm level is necessary.

Most animal breeding programs have focused on cumulative short-term genetic changes of production traits, because breeding optimization has to a very large extent been based on market economic values. Sustainable genetic improvement by animal breeding is a long-term and complex process. Therefore, we also need to focus on long-term biological, ecological, and sociological solutions. Farmed fish is at an early stage of domestication and selection, but rapid selection responses for growth rate are already documented for several species (Gjedrem 1997; Eknath et al. 1998; Gjerde and Korsvoll 1999). Precautions should be taken to avoid the same unwanted side effects, such as increased frequencies of diseases (e.g., mastitis in dairy cattle) and leg problems, that we have often seen in breeding programs for agricultural livestock. Careful monitoring of possible correlated responses is needed. More basic knowledge of, e.g., animal welfare and behavioral needs of fish may also be required, as we know very little about its perception of pain and satisfaction, and what it prefers.

In order to achieve a functional integrity approach of animal breeding, Olesen et al. (2000) suggested a procedure where:

- The ethical aspects and priorities should be made clear.
- The system should be defined with respect to limits and structure, resource efficiency, environment, economics and social effects.
- Indicators should be defined that measure or characterize the above ethical priorities and critical effects of the production system.
- Performance traits and characters that are important or critical to meet these criteria or objectives should be identified and weighed.

The system must be optimized according to ethical priorities, and the objective of animal genetic improvement is to fit animals' traits according to this. So far, weighing of traits has mainly been dependent on economic values and frequency of expression or mean of the trait (e.g., mean growth rate and survival rate). The methodology to weigh the traits with respect to resource efficiency and economy is well developed.

Changes in the quality and/or quantity of animal traits have value in that they could change the benefits associated with human activities or change the costs of those activities. These changes in benefits and costs have an impact on human welfare either through established markets or through non-market activities, such as ecosystem services (Costanza et al. 1997). This implies that traits affecting product value (e.g., feed efficiency and product quality resulting from prices and the market's supply and demand changes) should be taken into account. However, important non-market values should also be included (e.g., ethical values of improved animal welfare) (Olesen et al. 2000). There may also be other criteria that could be considered such as slower depletion of fossil energy and reduced degradation of the atmosphere, that are not easily traceable using traditional economics (Costanza et al. 1997).

Olesen et al. (2000) suggested to split the trait values in the aggregate genotype into non-market values (NV) and pure market economic values (MeV). This gives the following breeding goal or aggregate genotype (H, considering two traits Y_1 and Y_2):

$$H = [NV_1 \times Y_1 + MeV_1 \times Y_1] + [NV_2 \times Y_2 + MeV_2 \times Y_2]$$

Correspondingly, a genetic gain of non-market value will be obtained in addition to a market economic genetic gain. The value of the non-market gain is $NV_1 \times \Delta G_1 + NV_2 \times \Delta G_2$ and likewise for the market economic gain ($MeV_1 \times \Delta G_1 + MeV_2 \times \Delta G_2$). The total genetic gain is a sum of the non-market genetic gain and market economic genetic gain.

As shown, a trait may have both non-market and market value. Reduced disease frequency increases value through improved animal welfare and increases market economic value through reduced economic costs of treatments and reduced yield. When fish production relies on long-distance transport of feeds, feed efficiency should be emphasized through both market and non-market values. However, a trait may have only non-market value or only market economic value. The resulting non-market and market genetic gains provide an opportunity to evaluate the breeding programs in a holistic perspective, where both social, cultural (including subjective values), ecological, and economic objectives and effects can be taken into account. Also, there is already developed methodologies for quantifying values of, e.g., environmental services and animal welfare, where subjective opinions may be strong and vary considerably among individuals and cultures (Braden and Kolstad 1991; Freeman 1993; Smith 1993; Bennett 1996). Olesen et al. (1999) considered these methods, and concluded that they can be applied for estimating non-market values of traits included in animal breeding goals. However, further development and adjustment of these methods to animal breeding is needed, as little experiences have been made and have been documented on this so far.

Genetic Diversity and Breeding Designs

Genetic variation is essential for selection response, and a sufficiently large and genetically diverse breeding population is, therefore, fundamental when establishing and running a breeding program. A breeding design with appropriate family structure is critical to maintain a large effective population size and obtain a long-term selection response with low rates of inbreeding. For mass selection, Bentsen and Olesen (2000) concluded from a simulation study that a minimum of 50 families (pairs of parents) are required to prevent inbreeding and obtain a long-term response in a mass selection program. Gjerde et al. (1996) presented optimum designs for fish breeding programs with constrained inbreeding and mass selection. Various breeding designs for between-family, within-family, and combined selection (between- and within-family) are presented and evaluated by Bentsen and Gjerde (1994). Combined selection designs may improve the accuracy of selection substantially, particularly for traits with low heritability. However, the probability of selecting large numbers of sibs from a few families will be higher than in, e.g., a mass selection or within-family selection program. Hence, the number of broodstock selected per family needs to be restricted to avoid a high rate of inbreeding and reduced genetic variation.

For most livestock species, several populations (or breeds) with similar performance abilities have evolved in parallel. In spite of this, the trend of genetic erosion is strong in livestock due to aggressive marketing of highly productive breeds in intensive systems and loss of less productive breeds (Hammond 1994). A similar or more severe situation may be possible for many farmed fish species in the future due to the low number of breeds and large reproductive capacity.

Less than 1% of the fish stocked for aquaculture in 1993 originated from selection programs (Gjedrem 1997). This situation has not improved much today. For most species considered for selection programs, only a few programs are active. Furthermore, the effective population sizes are often limited and in some cases too low, because the high reproductive capacity allows use of a low number of broodstock. Such populations may still gain sufficient short-term advantage above non-improved populations and capture much of the market share. In turn, this discourages further genetic introductions into the breeding nucleus. Long-term inbreeding and loss of genetic variability because of genetic drift may then affect performance and further the long-term genetic progress. In such populations, strategies for continuous (re)introduction of genetic variability from outside the breeding nucleus without adverse performance consequences are, therefore, required. Furthermore, initiation of additional breeding programs is expected for different environments for the most important farmed species and this may improve the situation.

Tave (1993 and 1995) have also discussed design and development of breeding programs in aquaculture.

Selection Limits

One should distinguish between two situations where a selection limit is reached.

No genetic variance remains: In a closed population undergoing selection, inbreeding will always accumulate. This may result in fixation of genes and a reduction of genetic variance for the traits selected for. Fitness in such a population may also be reduced as inbreeding increases. Natural selection against these less fit and inbred individuals may, therefore, counteract the artificial selection and thus reduce the genetic gain obtained per generation. However, if the population under selection is large and the rate of inbreeding is kept low by using a sufficiently large number of breeders per generation, selection can continue for many generations without severe loss of genetic gain. This is particularly true when there is a large number of genes contributing to the genetic variation in the trait undergoing selection.

Genetic variance is present but the population fails to respond to selection: The following are some possible reasons for lack of response to selection in the presence of genetic variance: Certain biological or physical limits exist. Egg number per day in laying hens can be regarded as having a biological limit; the speed of the egg-shell synthesis is a physical bottleneck. Loss of fitness in the best performing individuals because of antagonistic genetic correlations between fitness traits and the traits undergoing selection, e.g., unfavorable correlations between production and reproduction traits or between production and behavioral traits.

Many examples show that selection and breeding for high production efficiency in farm animals has lead to unwanted side effects (Rauw et al. 1998). This has been explained by the resource allocation theory (Beilhartz et al. 1993), which states that fitness is a product of many components. The metabolic resources used by one fitness component, as well as all other traits (e.g., growth, activity, and coping with a stressful situation), sum to no more than the animal can obtain from its environment. Resources consumed by one process are not available for other processes. Intensive production and management result in more resources available by providing more feed, improved feed quality, or by reducing the amount of resources needed for traits that are important under less intensive management. Under such conditions, rapid genetic progress in both fitness and production traits can be achieved without any negative side effects until a level is reached where the environment constrains further response because of restricted resources. The theory predicts that at this point, any further rise in a given resource demanding production trait will result in a negative response in other traits because of reallocation of resources from these traits to the production trait. Artificial selection may be suspended by natural selection when the natural selection favors other traits and animals than the artificial selection.

Recessive alleles that are unfavorable may be brought to low frequencies by selection. Selection is then less efficient and the response will decline although the alleles may remain unfixed and contribute to some additive genetic variance. For traits affected by many loci, the number of generations will be higher before all loci are brought to low frequencies or are fixed and selection limits are reached.

Overdominance at some loci affecting the character under selection will influence response. Overdominant genes will respond until maximum performance is reached when they are at intermediate frequencies; further selection responses cannot be obtained. The variance they produce is non-additive, and this cannot be exploited by selection only. Selection may also favor heterozygotes through the joint action of artificial and natural selection.

Some long-term selection experiments in laboratory and livestock animals have shown responses continuing over many generations without any sign of approaching a limit. However, many experiments have shown responses that end at a limit. Falconer and Mackay (1996) state that selection limits are often reached after 20-30 generations and that reproduction is then often reduced in selected lines. Eisen (1980) showed that selection limits in mice were reached after approximately 30 generations, when the response was 2.4 standard deviations for growth rate and 1.9 standard deviations for litter size. In most selection experiments, the response seems higher for production traits than for reproduction traits, and reproduction appears to be reduced in lines selected for production traits (Bakken et al. 1998).

In several long-term selection experiments the selection response decreased over generations and finally ceased. Falconer and MacKay (1996) summarized the results from four experiments with two-way selection, two with *Drosophila melanogaster* (Robertson 1955; Clayton and Robertson 1957) and two with mice (MacArthur 1949; Falconer 1955). In these experiments, response to selection ceased after 30 generations. The number of genes that could account for the observed response varied from 35 to 99.

Enfield (1979) selected for increased pupa weight in tribolium, *Tribolium castaneum,* for 120 generations without marked decrease in genetic gain (Figure 1). Both the additive genetic and the phenotypic variance increased during the course of the experiment, but the heritability showed only a very slight decline. The total response to selection represented a change of 28 genetic or 17 phenotypic standard deviations, based on the original estimates of these parameters. Number of genes affecting pupa weight was estimated to be a minimum of 170. The experiment shows that it is possible to change a population dramatically by selection, when the breeding population is large, increase of inbreeding is kept low and a considerable number of genes regulate the trait under selection.

Selection programs in farm animal species are usually directed towards several traits at the same time, and selection is seldom as intense as what can be achieved in the laboratory. It seems to be a consensus that there are, as yet, no signs for selection limits due to depleted genetic variation in commercial cattle (Kennedy 1984), pig (Webb 1991), and poultry (Albers 1998) populations. Knap and Luiting (1999) stated that obstruction of genetic change in production traits may be caused by: undesirable changes in fitness traits acting as constraining factors; or by environmental conditions becoming limiting for the expression of the

FIGURE 1. Selection response for pupa weight of tribolium (Enfield 1979). S_1 and S_2 are selected lines for pupa weight, C_1, C_2, R_1, and R_2 are control populations.

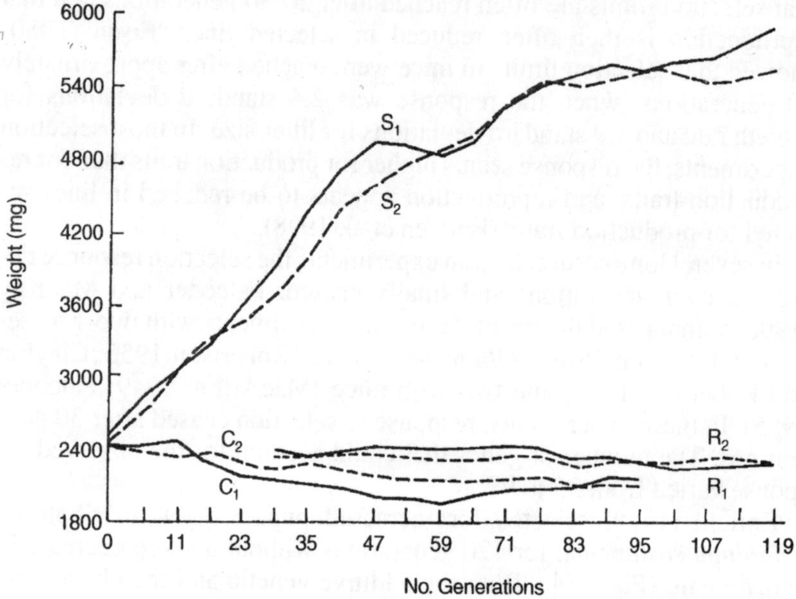

trait (and hindering the identification of the best breeders). The undesirable changes in fitness traits cause reduced selection intensity (animals not surviving or reproducing at off-test or while in production) or reduced selection accuracy (animals prevented from expressing their full genetic potential). However, further genetic progress may be achieved by including fitness traits in the selection objective, or by adjusting the environmental conditions to meet the increased requirements of the animals.

Genetic Gains

In 1925, Embody and Hyford (1925) selected surviving brook trout, *Salvelinus fontinalis*, from a population with endemic furunculosis and increased the survival rate from 2% in the initial population to 69% after three generations of selection. Ehlinger (1977) obtained reduced mortality rates due to furunculosis in brown trout, *Salmo trutta*, and brook trout after selection. Kirpichnikov et al. (1993) reported moderate responses from a selection program started in 1965 against dropsy disease in three stocks of common carp, *Cyprinus carpio*. Schaperclause (1962) reported that a selection experiment in common carp reduced average mortality to 11.5% compared to 57% in non-selected fish.

Lewis (1944) reported selection responses in early spawning, egg number and yearling weight in rainbow trout, *Oncorhynchus mykiss*. Selection in rainbow trout over 23 years (6-7 generations) is reported to have had large effect on growth rate, egg production and early spawning (Donaldson and Olson 1955). In these two studies, the reported response to selection was, however, confounded with environmental changes due to the lack of a control population. Kincaid et al. (1977) selected for increased body weight at 147 days post-fertilization in rainbow trout. The genetic gain during three generations of selection was 0.98 g or 5% per year.

Moav and Wohlfarth (1973, 1976) summarized the results from several mass selection experiments for growth rate in common carp in Israel, and inferred that selection for lower growth rate gave a response, while selection for higher growth rate did not yield a positive response. Kinghorn (1983) reviewed the results from the Israeli selection experiments, and concluded that it seems that the report of no response to selection for higher growth rate is not conclusive. In selection experiments with channel catfish, *Ictalurus punctatus*, growth rate has been increased with 12-20% after one generation of selection (Bondari 1983; Dunham 1986). In rohu carp, *Labeo rohita*, combined selection (between- and within-family) yielded 34% higher growth rate in two generations of selection (Mahapatra et al. 2000).

Some resulting responses in growth rate from fish selection experiments and breeding programs with large numbers of animals involved are presented in Table 1. Gjerde and Korsvoll (1999) reported a realized accumulated selection differential in Atlantic salmon, *Salmo salar*, of 83.9% after 6 generations or 14% per generation for growth rate and a reduction of 12.5 percent-units in frequency of early sexual maturity.

In Nile tilapia, *Oreochromis niloticus*, a selection project (Genetic Improvement of Farmed Tilapia or GIFT) was carried out in the Philippines to improve growth rate. The response obtained after the first 5 generations of selection was 12-17% per generation (Eknath et al. 1998). Production trials and socio-economic surveys in five Asian countries revealed that the cost of production per unit of fish produced was 20-30% lower for the GIFT Nile tilapia than for other Nile tilapia strains in current use. Price elasticity data indicated that the gain obtained by using GIFT Nile tilapia, will benefit mostly the vast group of

TABLE 1. Response to selection for growth rate.

Species	Mean weight	No. of generations	Gain per generation (%)	Reference
Coho salmon, *Oncorhynchus kisutch*	250 g	4	10.1	Hershberger et al. (1990)
Rainbow trout, *Oncorhynchus mykiss*	4.0 kg	2	13.0	Gjerde (1986)
Atlantic salmon, *Salmo salar*	4.5 kg	1	14.4	Gjerde (1986)
Atlantic salmon	5.7 kg	6	14 *	Gjerde and Korsvoll (1999)
Channel catfish, *Ictalurus punctatus*	-	1	12-18	Dunham (1986)
Channel catfish	67 g	1	20	Bondari (1983)
Nile tilapia, *Oreochromis niloticus*	ca. 80 g	5	12-17	Eknath et al. (1998)
Rohu carp, *Labeo rohita*	400 g	1	13-15	Gjerde (pers. comm.)
Whiteleg shrimps, *Penaeus vannamei*	18 g	1	4.4	Fjalestad et al. (1997)
Golden shiner, *Notemigonus crysoleucas*	-	1	5.3	Tave (1994)

*Estimated from realized selection differentials.

poor consumers. In a bi-directional selection experiment with GIFT Nile tilapia, the frequency of early sexually maturing females differed with 24.5 percent-units after one generation of selection (Longalong et al. 1999).

High selection responses for growth rate have also been reported in oysters (Haley et al. 1975; Newkirk 1980; Nell et al. 1999). In whiteleg shrimp, *Penaeus vannamei*, the estimated response after one generation of selection was 4.4% for growth rate and 12.4% for survival after challenge testing against the Taura Virus Syndrome (Fjalestad et al. 1997). Tave (1994) report a response to selection of 9.1% (unadjusted) and 5.3% (adjusted) for body length per generation for golden shiner, *Notemigonus crysoleucas*.

In general, the response to selection for growth rate in fish is very good and much higher than obtained for growth rate in farm animals. The main reasons for these differences are: a higher genetic variance in fish and shellfish (CV = 20-35%) compared to farm animals (CV = 7-10%) (Gjedrem 1998); the high fecundity of aquatic organisms allows for a higher selection intensity than in farm animals; and the domestication and selection in fish is still at an early stage, where problems with reallocation of limited resources to increase, e.g., growth rate has not yet occurred.

Thodesen et al. (1999) reported a favorable correlated response in feed conversion efficiency when selecting for growth rate in Atlantic salmon. Wild salmon showed 20% higher intake of energy and protein per kg gained body weight, 19% lower retention of protein, and 30% lower retention of energy compared to the 5th generation salmon selected for increased growth rate. This shows that fish selected for growth rate have a better utilization of feed resources compared to non-selected animals.

Effects of Selection on Domestication

Domestication can be defined as "that process by which a population of animals become adapted to humans and to the captive environment by some combination of genetic changes occurring over generations and environmentally induced developmental events re-occurring during each generation" (Price 1984). Many behavioral, physiological, and morphological components of the phenotype can be changed through selection. In fish and shellfish, directional selection is usually applied to improve important traits like growth rate, disease resistance, age at maturity, and meat quality. Response to selection is also expected in traits that are genetically correlated to these production traits. In addition, natural selection will contribute to adaptation of the animals to the farm environment. According to Ruzzante (1994), behavioral traits are among the first traits to be affected by the domestication process. Although the behavior repertoire has been maintained through several thousand years of domestication, the threshold at which many behavior patterns occur have been changed in domesticated animals (Wood-Gush 1983).

Purdom (1974) argued that selection for fast growth rate may increase competition among fish as demonstrated when food availability was limited (Moyle 1969; Swain and Riddell 1991). However, selection for fast growth rate in medaka, *Oryzias latipes*, showed that when food is available in excess, agonistic behavior decreased (Ruzzante and Doyle 1991). Under such conditions, the dominant and more aggressive fish tended to waste their energy on unnecessary attempts to monopolize the food supply by chasing competitors and thus allow less aggressive fish to grow. This result is in agreement with Doyle and Talbot (1986) who concluded that selection for rapid growth gives a correlated response in tameness rather than aggression. The decline in aggression resulting from domestication is consistent with the experiences from farm animals.

Breeding Programs Implemented

Today there are only a few large-scale breeding programs in fish and shellfish which aim at producing improved stocks for the industry in the respective countries. In 1975, a breeding program for Atlantic salmon and rainbow trout was started in Norway (Gjøen and Bentsen 1997). Today, it supplies about 70% of the industry with improved eyed eggs. In Canada, a similar breeding program for Atlantic salmon is running at Atlantic Salmon Federation, St. Andrews (Friars 1993). In 1993, the Philippines National Tilapia Breeding Program (PNTBP) was started with broodstock from the GIFT project. It is now organized by the GIFT Foundation (Rye and Eknath, 1999). In Israel (Wohlfarth 1983) and Hungary (Bakos 1979), crossbreeding programs with common carp are established. A national breeding program for shrimp was initiated in Colombia in 1997 (Suárez et al. 1999). It is designed as a combined between-and within-family selection scheme to improve growth rate and resistance to TSV. Breeding programs run by private companies have also been established in several countries, e.g., Atlantic salmon in Norway, Scotland, and Chile.

Utilization of Genetic Gain

Genetic changes in traits resulting from selective breeding programs can be utilized to:

- Increase yield and production income.
- Maintain yield while reducing time and costs of production; e.g., costs of labor, feed, and drugs. This may, in turn, lead to lower product prices for the consumer.
- Improve the resource efficiency of production factors, e.g., improved feed efficiency and reduced use of land needed for ponds.
- Improve the ethical standards of the production; e.g., by improving animal welfare through reduced frequency of diseases.
- Reduce use of drugs and chemicals to combat diseases and parasites, and minimize the development of drug resistant organisms.
- Improve the product quality and increase the market economic and (or) non-market value of the product. Market economic value can be increased through, e.g., a higher price obtained in the market for a leaner fillet. Non-market economic value can be the value of a more ethical and environmental friendly production.

The first three points are, in most cases, largely determined by market economic forces. For example, selection for growth rate will shorten the time required reaching market size, as illustrated in Figure 2. In this example, a population of Nile tilapia reaching market size at 1 kg in 8

FIGURE 2. Growth curves of Nile tilapia not selected (Alt. I) and selected for 6 (Alt. II) and 12 generations (Alt. III) for growth rate.

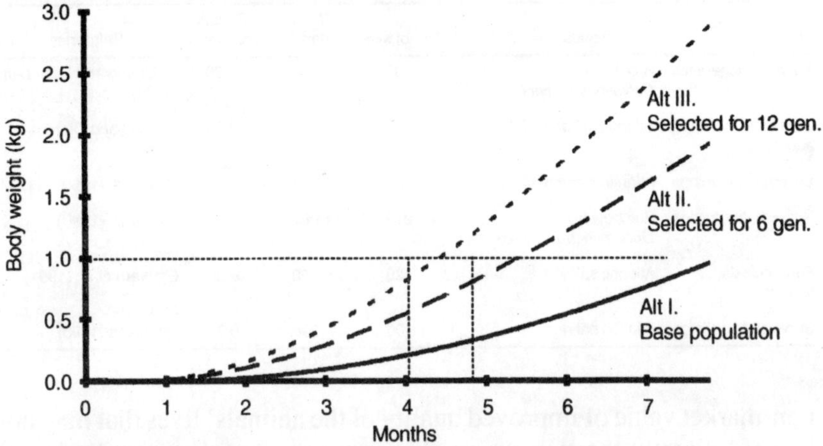

months is considered. Selection for increased growth rate gave a genetic gain of 13% per generation, which doubled growth rate in 6 generations. This means that it will take 8 months to produce a fish of 2 kg and 5.5 months to produce a fish of 1 kg. If selection continues for another 6 generations with the same response, growth rate would be three times greater than growth rate in the base population, and it would take only 4.3 months to produce a fish of 1 kg.

Examples of the benefit of increasing growth rate in fish through selective breeding are: increased turnover rate resulting in reduced production cost; reduced maintenance requirement and higher retention of energy and protein, which reduces feeding costs; reduced mortality due to shortened production time; a positive genetic correlation between growth rate and diseases resistance, as reported in Atlantic salmon, rainbow trout and brook trout (Table 2), will also lead to increased survival rate, and reduced costs and drug consumption.

In addition to economic profit, the non-market value of improving resource efficiency (e.g., feed efficiency) may be important as it may contribute to less transport, less pollution, and less consumption of non-renewable fossil energy. Both market and non-market economic values can be obtained for benefits of genetic changes for points 3-6. For example, improved ethical acceptance of production among the consumers may give the producers goodwill among the public, which can be paid back in a market economy. In addition, it may have a

TABLE 2. Genetic correlations between growth rate and survival traits.

Trait	Species	No. of sires	No. of dams	Genetic correlation	Reference
Survival, fingerlings	Brook trout, *Salvelinus fontinalis*	32	32	0.30	Robinson and Luempert (1984)
Coldwater vibriosis, adults	Atlantic salmon, *Salmo salar*	53	329	0.18	Standal and Gjerde (1987)
Survival, fingerlings	Atlantic salmon	187	1,404	0.37	Rye et al. (1990)
Survival, fingerlings	Rainbow trout, *Oncorhynchus mykiss*	213	1,062	0. 23	Rye et al. (1990)
Furunculosis, fingerlings	Atlantic salmon	25	50	0.30	Gjedrem et al. (1991)
Survival, fingerlings	Atlantic salmon	100	298	0.31	Jonasson (1993)

non-market value of improved quality of the animals' lives that may not be paid directly through the market economy, but that may be found beneficial. Breeding for healthier, well-functioning and robust fish may also be important insurance and may have a long-term strategic value.

A better utilization of genetically improved fish can be expected in more optimal environmental conditions. This will often imply more intensive production systems. It also means that the efficiency of high-input production systems can be further improved by stocking fish from selective breeding programs.

Dissemination of Genetic Gain

Bentsen (1990) considered commercial applications of fish breeding. The large reproductive capacity of farmed fish species facilitates a centralized breeding structure. In most cases, one breeding station may produce enough fry or eggs to supply the entire fish farming industry on a national level, either directly or through a network of multiplier stations. For safety reasons, the breeding nucleus may be split in two or more locations, but a controlled exchange of breeders should be practiced to expand the effective population size and increase the selection intensity.

A centralized breeding system will only be important for the commercial fish farming if the improved genetic material is appreciated and utilized by fish farmers. In most species, farms may easily be restocked by reproducing any available broodstock. To ensure a close communication between the breeding nucleus and fish farmers, some integration should be established. Participation of a fish farmers' organization is particularly important when defining the breeding goal and strategies. The ultimate integration is, of course, a farmer-owned breeding system.

INTERACTION WITH WILD FISH

There has been much concern and controversy about the possible negative effects of escaped farmed salmon on wild populations (e.g., Bentsen 1991; Hindar et al. 1991; Hutchinson 1997; Anon. 1999). Genetic interactions between escaped farmed salmon and wild populations has been observed (Crozier 1993; Clifford et al. 1998). Genetic and environmental differences between wild and farmed salmon include many fitness-related traits such as survival, body shape, growth, competitive ability, risk sensitivity, migratory behavior, and reproductive performance (Fleming et al. 1994; 1996; Cross 1998). Even if it is shown that escapees compete poorly with wild salmon in fitness traits, the potential for genetic and ecological interaction is significant. The estimated numbers of escaped salmon in Norway has however been reduced from 1.6-2 million (2-3% of farmed stock) in the late 1980s to 0.65 million (0.5% of farmed stock) in 1997 (Anon. 1999). This positive development is likely to continue, since it benefits both the industry and the environment.

Most aquaculture systems are likely to result in some level of farmed stock escaping, and genetic interactions with wild stocks may occur. Maintaining high levels of genetic variability in the farmed stocks will provide a base for natural selection to counteract loss of fitness in the affected wild stocks. The use of broodstock from a variety of local strains when establishing a base population for aquaculture breeding programs will reduce the risk of introducing new alleles into local strains by escapees from fish farms and at the same time secure a wide, long-term genetic variability in the farmed stock. Also, establishing and maintaining a sufficient number of families will be critical to obtain a sufficiently large effective population size in order to avoid inbreeding and loss of genetic variation. Selecting towards a broad breeding goal by including more than a few productivity traits will contribute to a higher genetic diversity and reduced risk of fixation of alleles.

The use of transgenic fish in aquaculture is likely to increase the risk of introduction of alleles that do not exist in wild populations, and escapees may cause a more dramatic ecological change. In a selection program, the genetic changes are smaller, slower and more predictable than what may be obtained in genetically modified organisms (GMO). Because the variation utilized in a selection program is a result of natural segregation of gametes and because the selection is carried out on fish that are developing normal and biologically balanced traits, this minimizes the risk for introducing genotypes with deleterious effects on the ecosystem or other unforeseen negative side effects. Furthermore, skepticism to and reactions against GMO products among the consumers can be taken into account by using traditional selection methods.

Use of sterile triploid fish in fish farming has been suggested in order to avoid genetic interaction with wild fish. However, the costs of such techniques may be high due to reduced growth rate of the sterile fish and possible consumers' resistance against such techniques due to ethical or precautionary reasons. Experiments have shown that sterile triploid females of some species continue to grow long after they have reached the normal size and age at sexual maturation, and when the life of normal fish is ended. The ecological effects of such escaped fish are unknown.

Reductions in the frequency of alleles with favorable effects on fitness traits in wild populations caused by escaped farmed fish will be counteracted by natural selection balancing the effects of migration from farmed fish populations. Tufto (Tufto, pers. comm.) modeled the effects of migration on population size and evolution of a quantitative trait. He concluded that the genetic difference between the wild and the farmed stocks determined the effects of escapees from farmed fish populations on quantitative traits of wild fish populations. Hence, a moderate genetic improvement in the farmed populations will reduce the chances for negative effects on the wild populations. However, this will not be very attractive for the fish farmers. A rapid and large cumulative genetic change may be more favorable for the wild stocks if this reduces the fitness of the escaped farmed fish sufficiently to prevent the escapees and their progeny to survive, mate, and contribute to the next generations of wild stocks.

ACKNOWLEDGMENTS

Financial support provided for this work by the Norwegian Research Council (Grant no. 122859/122) is acknowledged and appreciated.

REFERENCES

Albers, G.A.A. 1998. Future trends in poultry breeding. Pages 16-20 *in* Proceedings of the 10th WPSA European Poultry Conference, Vol. 1, Jerusalem, Israel.

Anon. 1999. Til laks åt alle. Norges Offentlige Utredninger 1999:9.

Bakken, M., O. Vangen, and W. Rauw. 1998. Biological limits to selection and animal welfare. Pages 381-389 *in* Proceedings of the 6th World Congress of Genetics Applied to Livestock Production, Vol. 27, University of New England, Armidale, Australia.

Bakos, J. 1979. Crossbreeding Hungarian races of common carp to develop more productive hybrids. Pages 633-635 *in* T.V.R. Pillay and W.A. Dill, eds. Advances in Aquaculture. Fishing News Books Ltd, Farnham, Surrey, England.

Beilhartz, R.G., B.G. Luxford, and J.L. Wilkinson. 1993. Quantitative genetics and evolution: Is our understanding of genetics sufficient to explain evolution? Journal of Animal Breeding and Genetics 110:161-170.

Bentsen, H.B. 1990. Application of breeding and selection theory on farmed fish. Pages 149-158 *in* Proceedings of the 4th World Congress of Genetics Applied to Livestock Production, Volume 16, Edinburgh, Scotland.

Bentsen, H.B. 1991. Quantitative genetics and management of wild populations. Aquaculture 98:263-266.

Bentsen, H.B., and B. Gjerde. 1994. Design of fish breeding programs. Pages 353-359 *in* Proceedings of the 5th World Congress of Genetics Applied to Livestock Production, Vol. 19, University of Guelph, Guelph, Canada.

Bentsen, H.B., and I. Olesen. 2000. Designing aquaculture mass selection programs to avoid high inbreeding rates. Proceedings of the 7th World Congress of Genetics in Aquaculture, Townsville, Australia.

Bondari, K. 1983. Response to bidirectional selection for body weight in channel catfish. Aquaculture 33:73-81.

Braden, J.B., and C.D. Kolstad. 1991. Measuring the demand for environmental quality. Elsevier Science Publishers, Amsterdam, The Netherlands.

Brascamp, E.W., A.F. Groen, I.J.M. De Boer, and H. Udo. 1998. The effect of environmental factors on breeding goals. Pages 129-136 *in* Proceedings of the 6th World Congress on Genetics Applied to Livestock Production, Vol. 27, University of New England, Armidale, Australia.

Cameron, N.D. 1997. Selection Indices and Prediction of Genetic merit in Animal Breeding. CAB International. Oxon, England.

Clayton, G.A., and A. Robertsson. 1957. An experimental check on quantitative genetical theory. II. The long-term effects of selection. Journal of Genetics 55: 152-170.

Clifford, S.L., P. McGinnity, and A. Ferguson. 1998. Genetic changes in Atlantic salmon (*Salmo salar*) populations of northwest Irish rivers resulting from escapes of adult farm salmon. Canadian Journal of Fisheries and Aquatic Sciences 55: 358-363.

Costanza, R., R. D'Arge, R. De Groot, S. Farber, M. Grasso, B. Hannon, K. Limburg, S. Naeem, R.V. O'Neill, J. Paruelo, R.G. Raskin, P. Sutton, and M. van den Belt.

1997. The value of the world's ecosystem services and natural capital. Nature 387:253-260.

Cross, T.F. 1998. Genetic implications of translocation and stocking of fish and aquatic invertebrates. Pages 27-29 *in* Abstracts, Conference on Genetics in the Aquaculture Industry: pages 27-29, Perth, Australia.

Crozier, W. 1993. Evidence of genetic interaction between escaped farmed salmon and wild Atlantic salmon (*Salmo salar* L.) in a northern Irish river. Aquaculture 113:19-29.

Donaldson, L.R., and P.R. Olson. 1955. Development of rainbow trout broodstock by selective breeding. Transactions of the American Fisheries Society 85:93-101.

Doyle, R.W., and A.J. Talbot. 1986. Artificial selection on growth and correlated selection on competitive behavior in fish. Canadian Journal of Fisheries and Aquatic Sciences 43:1059-1064.

Dunham, R.A. 1986. Selection and crossbreeding responses for cultured fish. Pages 391-400 *in* Proceedings of the 3rd World Congress in Genetic Applied to Livestock Production, Vol. X, Lincoln, Nebraska.

Ehlinger, N.F. 1977. Selective breeding of trout for resistance to furunculosis. New York Fish and Game Journal 24:25-36.

Eisen, E.G. 1980. Conclusions from long-term selection experiments with mice. Zeitschrift für Tierzucht and Züchtungsbiologie 97:305-319.

Eknath, A.E., M.M. Dey, M. Rye, B. Gjerde, T.A. Abella, R. Sevilleja, M.M. Tayamen, R.A. Reyes, and H.B. Bentsen. 1998. Selective breeding of Nile tilapia for Asia. Pages 89-96 *in* Proceedings of the 6th World Congress on Genetics Applied to Livestock Production, Vol. 27, University of New England, Armidale, Australia.

Embody, G.C., and C.D. Hyford. 1925. The advantage of rearing brook trout fingerlings from selected breeders. Transactions of the American Fisheries Society 55:135-138.

Enfield, F.D. 1979. Long term effects of selection; the limits to response. Pages 69-86 *in* Proceedings, Symposium on Selection Experiments in Laboratory and Domestic Animals. Commonwealth Agricultural Bureau.

Falconer, D.S. 1955. Patterns of response in selection experiments with mice. Cold Spring Harbor Symposium on Quantitative Biology 20:178-196.

Falconer, D.S., and T.F.C. Mackay. 1996. Introduction to Quantitative Genetics. Longman, Essex, England.

FAO. 1998. Aquaculture Production Statistics 1987-1996. FAO Fisheries Circular No. 815, Rev 10. FAO, Rome, Italy.

Fjalestad, K.T., T. Gjedrem, W.H. Carr, and J.N. Sweeney. 1997. Final report: The shrimp breeding program. Selective breeding of *Penaeus vannamei*. AKVAFORSK, Report No. 17/97, Ås, Norway.

Fleming, I.A., B. Jonsson, and M.R. Gross. 1994. Phenotypic divergence of sea-ranched, farmed, and wild salmon. Canadian Journal of Fisheries and Aquatic Sciences 51:2808-2824.

Fleming, I., B. Jonsson, M. Gross, and A. Lamberg. 1996. An experimental study of the reproductive behavior and success of farmed and wild Atlantic salmon (*Salmo salar*). Journal of Applied Ecology 33:893-905.

Francis, C. A., and M.B. Callaway. 1993. Crop improvement for future farming systems. Pages 1-18 *in* B. Callaway and C.A. Francis, ed. Crop Improvement for Sustainable Agriculture. University Nebraska Press, Lincoln, Nebraska.

Freeman, A.M. 1993. The Measurement of Environmental and Resource Values: Theory and Methods. Resources for the Future, Washington, DC.

Friars, G.W. 1993. Breeding Atlantic salmon: A primer. Atlantic Salmon Federation, St. Andrews, New Brunswick, Canada.

Gjedrem, T. 1997. Selective breeding to improve aquaculture production. World Aquaculture 22(1):33-45.

Gjedrem, T. 1998. Developments in fish breeding and genetics. Acta Agriculurae Scandinavica Section A, Animal Science 28:19-26.

Gjedrem, T., R. Salte, and H.M. Gjøen. 1991. Genetic variation in susceptibility of Atlantic salmon to furunculosis. Aquaculture 97:1-6.

Gjerde, B. 1986. Growth and reproduction in fish and shellfish. Aquaculture 57:37-55.

Gjerde, B., and S.A. Korsvoll. 1999. Realized selection differentials for growth rate and early sexual maturity in Atlantic salmon. Pages 73-74 *in* Abstracts, Aquaculture Europe 99. European Aquaculture Society, Special publication no. 27, Trondheim, Norway.

Gjerde, B., H.M. Gjøen, and B. Villanueva. 1996. Optimum designs for fish breeding programmes with constrained inbreeding. Mass selection for a normally distributed trait. Livestock Production Science 47:59-72.

Gjøen, H.M., and H.B. Bentsen. 1997. Past, present and future of genetic improvement in salmon aquaculture. ICES Journal of Marine Science 54:1009-1014.

Haley, L.E., G.F. Newkirk, D.W. Waugh, and R.W. Doyle. 1975. A report on the quantitative genetics of growth and survivalship of the American oyster, *Crassostera virginica* under laboratory conditions. Pages 221-228 *in* Proceedings of the 10th European Symposium on Marine Biology, Vol. 1, Ostend, Belgium.

Hammond, K. 1994. Conservation of Domestic Animal Diversity: Global Overview. Pages 423-437 *in* Proceedings of the 5th World Congress of Genetics Applied to Livestock Production, Vol. 21, University of Guelph, Guelph, Canada.

Harris, D.L. 1970. Breeding for efficiency in livestock production: defining the economic objectives. Journal of Animal Science 30:860-865.

Hershberger, W.K., J.M. Meyers, W.C. McAuley, and A.M. Saxton. 1990. Genetic changes in growth of coho salmon *(Oncorhynchus kisutch)* in marine netpens, produced by ten years of selection. Aquaculture 85:187-197.

Hindar, K., N. Ryman, and F. Utter. 1991. Genetic effects of cultured fish on natural fish populations. Canadian Journal of Fisheries and Aquatic Sciences 48:945-957.

Holmenkollen Guidelines. 1999. Holmenkollen Guidelines. Pages 343-346 *in* Svennevig, M. New, and H. Reinertsen, eds. Sustainable Aquaculture. Proceedings of the 2nd International Symposium on Sustainable Aquaculture, Oslo, Norway. A.A. Balkema, Rotterdam, Holland/Brookfield, The Netherlands.

Horring, J. 1948. Methode van kostprijsberekening. Ten Kate, Emmen., Holland.

Hutchinson, P. 1997. ICES/NASCO Symposium on Interaction Between Salmon Culture and Wild Stocks of Atlantic Salmon: The Scientific and Management Issues. ICES Journal of Marine Science 54(6):963-1225.

Jonasson, J. 1993. Selection experiments in salmon ranching: I. Genetic and environmental sources of variation in survival and growth in freshwater. Aquaculture 109:225-236.

Kennedy, B.W. 1984. Selection limits: Have they been reached in dairy cattle? Canadian Journal of Animal Science 64:207-215.

Kinghorn, B.P. 1983. A review of quantitative genetics in fish breeding. Aquaculture 31:283-304.

Kirpichnikov, V.S., Y.I. Ilyassov, L.A Shart, A.A. Vikhman, M.V. Ganchenko, L.A. Ostashevsky, V.M. Smirnov, G.F. Tikhonov, and V.V. Tjurin. 1993. Selection of krasnodar common carp (*Cyprinus carpio* L.) for resistance to dropsy: principal results and prospects. Aquaculture 111:7-20.

Kincaid, H.L., W.R. Bridges, and B. Von Limbach. 1977. Three generations of selection forgrowth rate in fall spawning rainbow trout. Transactions of the American Fisheries Society 106:621-629.

Knap, P.W., and P. Luiting. 1999. Selection limits and fitness constraints in pigs. Page 53 in Book of Abstracts of the 50th Annual Meeting of the European Association of Animal Production, Zürich, Switzerland. Book of Abstracts No. 5. Wageningen Pers, Wageningen, The Netherlands.

Lewis, R.C. 1944. Selective breeding of rainbow trout at Hot Creek Hatchery. California Fish and Game 30: 95-97.

Longalong, F.M., A.E. Eknath, and H.B. Bentsen. 1999. Response to bi-directional selection for frequency of early maturing females in Nile tilapia (*Oreochromis niloticus*). Aquaculture 178:13-25.

MacArthur, J.W. 1949. Selection for small and large body size in the house mouse. Genetics 34:194-209.

Mahapatra, K.D., Meher, P.K., Saha, J.N., Gjerde, B., Reddy, P.V.G.K., Jana, R.K., Sahoo, M. and M. Rye. 2000. Selection response of rohu, *Labeo rohita*, for two generations of selective breeding. The Fift Indian Fisheries Forum. 17-20 January, 2000, Abstracts.

Moav, R., and G.W. Wolfarth. 1973. Carp breeding in Israel. Pages 295-318 *in* R. Moav, ed. Agricultural Genetics. Selected Topics. J. Wiley, New York, New York.

Moav, R., and G.W. Wolfarth. 1976. Two way selection for growth rate in the common carp (*Cyprinus carpo* L.). Genetics 82:83-101.

Moyle, P.B. 1969. Comparative behavior of young brook trout of domestic and wild origin. Progressive Fish-Culturist 31:51-59.

Nell, J.A., I.R. Smith, and A.K. Sheridan. 1999. Third generation evaluation of Sydney rock oyster *Saccostrea commercialis* (Iredale and Roughley) breeding lines. Aquaculture 170:195-203.

New, M. 1991. Turn of the millennium aquaculture. World Aquaculture 22(3):28-49.

Newkirk, G.F. 1980. Review of the genetics and the potential for selective breeding of commercially important bivalves. Aquaculture 19:209-228.

Olesen, I., B. Gjerde, and A.F. Groen. 1999. Methodology for deriving non-market trait values in animal breeding goals for sustainable production systems. Proceedings International Workshop on EU Concerted Action on Genetic Improvement of Functional Traits in Cattle (GIFT); Breeding Goals and Selection Schemes, Wageningen, The Netherlands. Interbull Bulletin No. 23 1999:13-21.

Olesen, I., A.F. Groen, and B. Gjerde. 2000. Definition of animal breeding goals for sustainable production systems. Journal of Animal Science 78:570-582.

Pearson, R.E. 1986. Economic evaluation of breeding objectives in dairy cattle. Intensive specialized milk production in temperate zones. Pages 11-17 *in* Proceedings of the 3rd World Congress on Genetics Applied to Livestock Production, Vol. 9, Lincoln, Nebraska.

Price, E.O. 1984. Behavioral aspects of animal domestication. Quarterly Review of Biology 59:1.

Purdom, C.E. 1974. Variation in fish. Pages 347-355 *in* F.R. Harden Jones, ed. Sea Fisheries Research, Elek Science, London, England.

Rauw, W.M., E. Kanis, E.N. Noordhuizen-Stassen, and F.J. Grommers. 1998. Undesirable side effects of selection for high production efficiency in farm animals. A review. Livestock Production Science 56:15-33.

Robertson, A. 1955. Selection response and the properties of genetic variation. Cold Spring Harbor Symposium on Quantitative Biology 20:166-177.

Robinson, O.W., and L.G. Luempert. 1984. Genetic variation in weight and survival of brook trout (*Salvelinius fontinalis*). Aquaculture 38:155-170.

Ruzzante, D.E. 1994. Domestication effects on aggressive and schooling behavior in fish. Aquaculture 120:1-24.

Ruzzante, D.E., and R.W. Doyle. 1991. Rapid behavioral changes in medaka, *Oryzias latipes*, during selection for competitive and noncompetitive growth. Evolution 45:1936-1946.

Rye, M., K.M. Lillevik, and B. Gjerde. 1990. Survival in early life of Atlantic salmon and rainbow trout: estimates of heritabilities and genetic correlations. Aquaculture 89:209-216.

Rye, M., and A.E. Eknath. 1999. Genetic improvement of tilapia through selective breeding-Experience from Asia. European Aquaculture Society, Special Publication. No. 27; June 1999: 207-208.

Schaperclause, W. 1962. Trate de Pisciculture en Etang. Vigot Freres, Paris, France.

Smith, V.K. 1993. Nonmarket valuation of environmental Resources: An Interpretive Appraisal. Land Economics 69:1-26.

Standal, M., and B. Gjerde. 1987. Genetic variation in survival of Atlantic salmon during the sea rearing period. Aquaculture 66:197-207.

Steverink, M.H.A., A.F. Groen, and P.B.M. Berentsen. 1994. The influence of environmental policies for dairy farms on dairy cattle breeding goals. Livestock Production Science 40:251-261.

Stonier, A.W., and D.C. Hague. 1964. A Textbook of Economic Theory. Longmans and Green, London, England.

Suarez, J.A., T. Gitterle, M.R. Angarita, and M. Rye. 1999. Genetic parameters for harvest weight and pond survival in *Litopenaeus vannamei*. Pages 232-233 *in* Abstracts, Aquaculture Europe 99. European Aquaculture Society, Special Publication No. 27, Trondheim, Norway.

Swain, D.P., and B.E. Riddell. 1991. Domestication and agonistic behavior in coho salmon: reply to Ruzzante. Canadian Journal of Fisheries and Aquatic Sciences 48:520-522.

Tave, D. 1993. Genetics for fish hatchery managers. An AVI Book Published by Van Nostrand Reinhold, New York, New York.

Tave, D. 1994. Response to selection and realized heritability for length in golden shiner, *Notemigonus crysoleucas*. Journal of Applied Aquaculture 4(4):55-63.

Tave, D. 1955. Selective breeding programmes for medium-sized fish farms. FAO Fish Tech. Paper, 352 pp.

Thodesen, J., B. Grisdale-Helland, S. Helland, and B. Gjerde. 1999. Feed intake, growth and feed utilization of offspring from wild and selected Atlantic salmon. Aquaculture 180:237-246.

Thompson, P.B., and A. Nardone. 1999. Sustainable livestock production: Methodological and ethical challenges. Livestock Production Science 61:111-119.

Torp Donner, H., and J. Juga. 1997. Sustainability–a challenge to animal breeding. Journal of Agriculture and Food Science in Finland 6:229-239.

Vavra, M. 1996. Sustainability of animal production systems: An ecological perspective. Journal of Animal Science 74:1418-1423.

Webb, A.J. 1991. Near the limit for genetic progress? Pig International, August 1991:11-14.

Wohlfarth, G.W. 1983. Genetics of fish: Application to warm water fishes. Aquaculture 33:373-381.

Wood-Gush, D.G.M. 1983. Elements of Ethology. Chapman and Hall, London, England.

Recent Advances in Hormonal Induction of Sex-Reversal in Fish

T. J. Pandian
S. Kirankumar

SUMMARY. A neuroendocrine peptide can induce sex-reversal in protogynous labrid teleosts, similar to androgenic and antiestrogenic steroids. Culture water containing metabolites of 17α-methyltestosterone (MT), when recirculated, are more potent to induce sex-reversal than dietary supplemented MT; human estrogens present in sewage water, or other pollutants derail the normal course of sex differentiation. Steroid administration by immersion is 200-1,000 times cheaper than dietary supplementation; discrete immersion for 2-8 hours per day for 1-3 weeks after hatching induces maximal sex-reversal in salmonids like coho salmon, *Oncorhynchus kisutch*, while others like chinook salmon, *O. tshawytscha*, require dietary supplementation. Studies on uptake, disposition in the organs, and elimination of unlabeled or labeled steroids have indicated that the androgen is absorbed within one hour after intake, and selectively retained it in gall bladder, pyloric caeca and liver at higher concentrations for a longer time; however, 90-95% of the steroid is eliminated within 1-2 weeks. Estimated residual steroid of < 5 ng/g fish is too low to be a hazard to human health. Steroid delivered through water can be absorbed and retained by egg yolk of salmonids and is gradually released into the developing embryos. Temperature and autosomal gene(s) alter the process of sex differentiation in hormonally sex-reversing fish. Suboptimal treatment of steroids results in the production of hermaphro-

T. J. Pandian and S. Kirankumar, School of Biological Sciences, Madurai Kamaraj University, Madurai-625 021, India.

[Haworth co-indexing entry note]: "Recent Advances in Hormonal Induction of Sex-Reversal in Fish." Pandian, T. J., and S. Kirankumar. Co-published simultaneously in *Journal of Applied Aquaculture* (Food Products Press, an imprint of The Haworth Press, Inc.) Vol. 13, No. 3/4, 2003, pp. 205-230; and: *Sustainable Aquaculture: Global Perspectives* (ed: B. B. Jana, and Carl D. Webster) Food Products Press, an imprint of The Haworth Press, Inc., 2003, pp. 205-230. Single or multiple copies of this article are available for a fee from The Haworth Document Delivery Service [1-800-HAWORTH, 9:00 a.m. - 5:00 p.m. (EST). E-mail address: getinfo@haworthpressinc.com].

ditic and intersex individuals. Hormonally sex-reversed fish also suffers from low survival and functional deficiences. The Indian scenario on the use of hormones for monosex aquaculture is described.

KEYWORDS. Hormonal sex-reversal, steroid residues, temperature and autosomal genes

INTRODUCTION

The process of sex differentiation in teleosts is diverse and labile (Francis 1992) rendering hormonal induction of sex-reversal possible. The induction involves administration of an optimum dose of a sex steroid, such as 17α-methyltestosterone (MT), during the labile period which reverses the phenotypic expression of a genetic female into a male but a genetic male remains a male (Pandian and Varadaraj 1988).

The ease with which various hormones can be administered through diet or immersion has led to a large number of publications; these have been comprehensively summarized by Hunter and Donaldson (1983) and Yamazaki (1983). Subsequently, Pandian and Sheela (1995) have critically reviewed available treatment protocols for 47 species (15 families) of gonochores (34 species; 9 families) and hermaphrodites, using one of the 31 (16 androgens and 15 estrogens) steroids.

Protocols to produce 90-100% sex-reversed fish have been described for European sea bass, *Dicentrachus labrax* (Blazquez et al. 1995; Chatain et al. 1999); pejerrey, *Odontesthes bonariensis* (Strussman et al. 1996); dusky grouper, *Epinephelus marginatus* (Glamuzina et al. 1998); black crappie, *Pomoxis nigromaculatus* (Al-ablani and Phelps 1997); mud loach, *Misgurnus mizolepis* (Kim et al. 1997); silver barb, *Puntius gonionotus* (Pongthana et al. 1999); and pikeperch, *Stizostedion lucioperca* (Demska-Zakes and Zakes 1999). This review intends to highlight the recent findings on the following: (1) use of new steriods such as trenbolone acetate, TBA (Galvez et al. 1995) and non-steroids such as fadrozole (Piferrer et al. 1994a), aromatase gene (D'Cotta et al. 1999), and neuropeptides (Kramer and Imbriano 1997) to successfully induce sex-reversal; (2) administration of diet, which includes animal and/or plant ingredients containing high concentration of steroids

(Pelissero and Sumpter 1992) and/or absorption of "hormone disrupters" (Kaiser 1996) and other steroid-analogues like the pesticides (Chlordecone, DDT) (Gimeno et al. 1996) from habitats modifying sex differentiation and its ratio; (3) discrete immersion of the steroid-sensitive embryonic stage to induce the desired sex-reversal, especially in those species which attain sexual maturity at, or after, the age of one year; (4) uptake and elimination of steroids from the treated fish (Piferrer and Donaldson 1994); (5) overriding influence of temperature and/or autosomal gene(s) on hormonal induction of sex-reversal (Varadaraj et al. 1994; Baroiller et al. 1995, 1999); (6) occurrence of hermaphroditism (Liu and Yao 1995) and intersex (Gomelsky et al. 1994); and (7) on the poor reproductive performance of hormonally sex-reversed fish (Piferrer and Donaldson 1992, 1994).

New Chemicals for Sex Reversal

The search for less expensive and safer chemicals, that can ensure the desired sex-reversal in treated fish has become a necessity for following reasons: the use of steroids like diethylstilbestrol can be carcinogenic and has been legally prohibited; the use of steroids like mesterolone (Alam and Monwar 1998) which induce sex-reversal and anabolic growth, may prove to be more cost-effective; and many currently used steroids are not capable of inducing complete reversal in all treated individuals. It is not unusual for authors to claim successful sex-reversal in fish at the optimal/maximal treatment conditions, when not all the treated individuals have undergone the expected sex-reversal. For instance, George and Pandian (1998) could achieve a maximum production of 59% males of black molly, *Poecilia sphenops,* even at the highest dose of 400 mg/kg diet containing natural (androstenedione) or synthetic (MT) steroid. Likewise, they could achieve 100% feminization, but only 82% masculinization, in the convict cichlid, *Cichlasoma nigrofasciatum*, at the highest dose (200 mg/kg diet) of estadiol-17β (E_2) and MT, respectively (George and Pandian 1996). It is not clear whether 100% masculinization can be achieved in these species with a suitable androgen, which not only promotes androgenic activity but also inhibits estrogenic activity (Hines and Watts 1995). Trenbolone acetate (TBA), a synthetic anabolic androgenic steroid, which is 10-15 times more potent than testosterone, induces both androgenic and anti-estrogenic activity (Neumann 1976) and its administration may result in 100% masculinization in species, in which not all the treated individuals are shown to have undergone complete sex-reversal. Indeed, Galvez et al. (1995) have generated 98-100% males by administering 150 mg TBA/kg diet for 60 days to sexually undifferentiated fry of channel catfish, *Ictalurus punctatus*, when earlier attempts with MT or 17α-ethynyltestosterone or dihydrotestosterone have always led to paradoxical feminization (Goudie at al. 1983; Davis et al. 1990, 1992).

Occurrence of feminization after treatment with androgen either at high doses or for extended period, is known as paradoxical feminization. A reason for the observed paradoxical feminization is that androgens, like MT, can be aromatized into estrogens, reducing the number of males produced (Piferrer and Donaldson 1991). For instance, 17α-methyldihydrotestosterone (MDHT) is a non-aromatizable form of MT; hence, MDHT induced a high percentage (91%) of masculinization in chinook salmon, *Oncorhynchus tshawytscha* (Piferrer et al. 1993) and Nile tilapia, *Oreochromis niloticus* (Gale et al. 1999). Aromatase, a turn-key enzyme, located at the end of gonadal steroidogenic pathway, catalyses the conversion of androgens into estrogens. Piferrer et al. (1994a) demonstrated that aromatase acts as a regulatory switch in the process of sex differentiation. Administration of an aromatase inhibitor (AI) increased the number of individuals masculinized; when AI is administered alone in increasing dosages or in combination with MT, it increased the number of individuals masculinized to 98%. Use of TBA, MDHT, or androgen in combination with AI (e.g., fadrozole) (Afonso et al. 1999) may provide a solution to the paradoxical feminization phenomenon. More recently, D'Cotta et al. (1999) obtained a Nile tilapia aromatase cDNA of 1,300 bp by screening an ovary cDNA library. High levels of aromatase gene expression were revealed in all-female populations, when compared to males. In future transgenic tilapia with aromatase gene may serve as an important mechanism for sex control.

Besides AI, a series of peptides and gonadotropins, secreted in the neuroendocrine axis have been administered to induce sex-reversal. In the protogynous labrid bluehead wrasse, *Thalassoma bifasciatum*, Koulish and Kramer (1989a,1989b) induced sex-reversal by three weekly injections of human chorionic gonadotropin (hCG), which resulted in 80% gonad reversal after six weeks of treatment (Yeung et al. 1993). Subsequently, 92% gonad reversal was also achieved after six weeks of administering sGnRH-A in combination with domperidone (Kramer et al. 1993). Thus, the neuroendocrine control of sex-reversal could be traced to the level of the hypothalamus (Figure 1). Apparently, an increase in GnRH promoted a rise in GtH, which is a pre-requisite for sex-reversal

FIGURE 1. Proposed pathway in the neuroendocrine-gonadal axis that is likely to be in operation in sex-reversing protogynous hermaphrodite fish. The symbol ☿ indicates intersex.

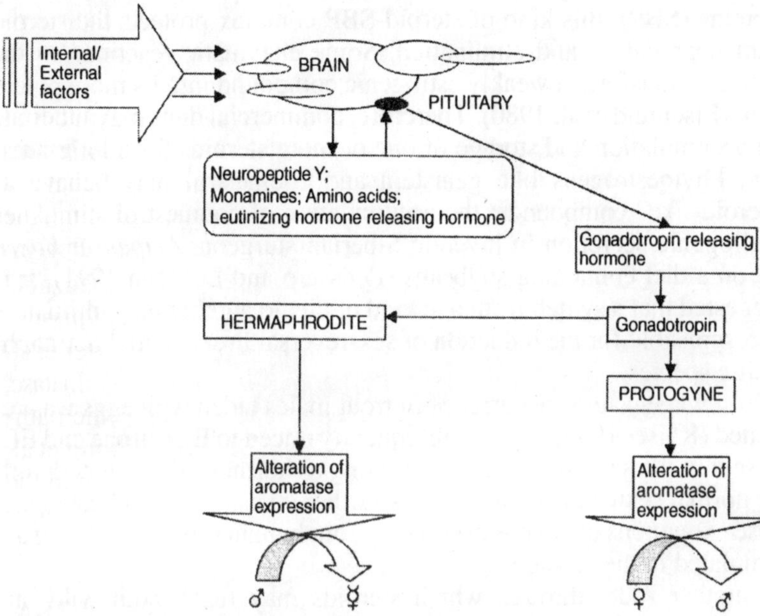

in the bluehead wrasse. Since sex-reversal in this species is triggered by social cues monitored by visual and/or olfactory senses, one or more regulatory factors such as neuroendocrine peptide (NPY) may regulate the release of gonadotropin by interacting with various components along the neuroendocrine axis from receptor to effector. True to this expectation, Kramer and Imbriano (1997) observed signs of sex-reversal in 82% individuals of the bluehead wrasse, which received three weekly injections of 0.02 μg NPY/g fish for eight weeks.

Other Sources of Steroids

Fish may also acquire steroids and their analogues from other sources, namely diet and/or surrounding water. Fish meal and diet can contain both estrogens and androgens; the levels of these steroids vary from diet to diet depending on their ingredients. Steroids can resist the low pH encountered in fish stomachs (Western and Jennings 1970). They are small molecules, stable, and capable of crossing the intestinal wall easily; their solubility in lipids also allows them to pass across cell membranes by simple diffusion (Labaune 1984). On entering the blood stream, steroids may bind themselves specifically with steroid binding proteins (SBP); this kind of steroid-SBP complex protects the steroid from degradation and elimination. Some enzymatic reactions in the liver can transform a weakly estrogenic compound into its more potent form (Eisenfeld et al. 1980). Therefore, commercial diets may facilitate the accumulation and storage of one or more steroids for a long duration. Phytoestrogens like genistein and coumestrol may behave as "steroid-like" compounds; the accumulation of coumestrol stimulates vitellogenic secretion in juvenile Siberian sturgeon, *Acipenser baeri*, fed on a diet containing soybeans (Pelissero and LeMenn 1991). It is suggested that a switch from one kind of diet to another may ultimately be responsible for the induction of sex-reversal in sequential hermaphroditic species.

In sewage waters, occurrence of trout males laden with eggs was recorded (Kaiser 1996). It was subsequently traced to E_2, estrone and EE_2 present in the sewage, which were acting as hormone disrupters derailing normal sexual differentiation. A further investigation indicated that these exogenous estrogens originated from human urine, which had accumulated in the sewage.

Another route through which steroids may reach both wild and farmed fish, is by absorption of steroids and their metabolites from the surrounding water. About 99% of the administered hormone is metabolized and released in < 24 hours; hence, large-scale field applications of steroid may lead to a synergic effect on sex-reversal in fishfarms and natural aquatic systems (Johnstone et al. 1983). Gomelsky et al. (1994) demonstrated that MT metabolites present in culture water were more potent than the dietary MT in masculinizing the meiotic and mitotic gynogens of common carp, *Cyprinus carpio*. They recorded a higher percentage of sex-reversal in gynogens exposed to recirculated water which contained MT metabolites than those which were administered MT through their diet; the gonadosomatic index of the former was also higher than that of the latter. Clearly, MT metabolites administered through recirculating water are more potent in not only inducing sex-reversal but also more fecund females. These observations have also been confirmed by Abucay and Mair (1997).

Many pesticides and other industrial pollutants structurally resemble steroids. These steroid analogues act as hormone disrupters and cause partial or complete sex-reversal in fish. This has been reviewed by Anon (1995). The following example is chosen to attract the attention of fish and pollution biologists to this serious problem. Gimeno et al. (1996) exposed 50-day-old undifferentiated all-male (XY) common carp, *Cyprinus carpio*, fry to a nominal concentration of 1 mg 4-tert-pentylphenol (TPP)/L for a period of 90 days under intermittent flow through conditions; the period of exposure was synchronized to the steriod-sensitive age of common carp. A histological analysis indicated the appearance of female phenotypic characteristics (oviduct) in the gonads of the genetic male. Investigations on the modifying effects of xenosteroids on sex differentiation are required to know whether these chemicals can induce complete sex-reversal.

Immersion Technique

Dietary administration is the most common method of administering steroids. However, Pandian and Sheela (1995) have listed the following as the limitations to steroid administration through diet: more than 90% steroid suffers degradation in the digestive tract; uniformity of its distribution in diet may vary; and size hierarchy may lead to differential diet uptake and hence hormone intake. Nakamura (1975) produced 100% male Mozambique tilapia, *Oreochromis mossambicus*, using dietary (50 mg/kg for 19 days) or immersion (10 μg/L) modes of MT administration. Likewise, Simpson (1976) produced 100% male rainbow trout, *Oncorhynchus mykiss*, using immersion (250 μg/L) or dietary (3 mg/kg diet for 90 days) modes of MT administration. Approximate calculations of the basic data provided by these authors indicate that the steroid required is 200-1,000 times less, when immersion technique was chosen. Clearly, immersion is cheaper, less time-consuming, and produces far less hormone as a pollutant into the culture water. However, most authors have chosen dietary supplementation of steroids as the preferred method of sex-reversal. This has led to the use of very high doses (up to 120 mg/kg diet; Jalabert et al. 1975) for longer durations (504 days; Fagerlund and McBride 1975).

However, continuous immersion of eyed eggs and/or alevins in steroid-treated water for a long period (67 days) (Nakamura 1984) has not led to the production of 100% females or males in many cases. Hence, efforts have been made to identify the most sensitive period of steroid action during the ontogenic development (Kavumpurath and Pandian 1993a), which will make sex-reversal more successful and less expensive. In most fish, ovarian differentiation usually precedes testicular differentiation; for instance, ovarian and testicular differentiation commences on the 20th and 42nd day after hatching, respectively in the African catfish, *Clarias gariepinus* (Van den Hurk et al. 1989), and 19th and 90th day after hatching, respectively, in channel catfish (Patino et al. 1996). Likewise, alevin stage of salmonids (from 1 day before hatching to 1-2 weeks after hatching) has been recognized as the most sensitive to steroid action (Table 1). Treatment after the sensitive labile period may require higher dosages to counteract normal sex differentiation and may not be effective. Apparently, steroid administration by diet or immersion after the sensitive period is the reason for the inability to sex reverse salmonids (Hunter and Donaldson 1983).

A combination of immersion and dietary supplementation was shown to enhance the production of 100% females and produce higher survival in lumpfish, *Cyclopterus lumpus* (Martin-Robichaud et al. 1994). Since immersion has been most frequently adopted to induce hormonal sex-reversal of salmonids, available information on this aspect indicates three categories of protocols (Table 1): one developed for coho salmon, *Oncorhynchus kisutch*, requires the discrete immersion in the desired

TABLE 1. Examples for protocols for three categories of immersion technique adopted to induce hormonal sex-reversal in salmonids.

Species	Protocol
Discrete immersion alone	
O. kisutch	400 μg E_2/L for 2 hour immersion 2nd and 9th day after-hatching yielded 97% females (Hunter et al. 1986).
O. kisutch	10 mg MT/L for 2 hour immersion per day on 3rd, 10th, 17th, and 24th day after hatching produced 92% steriles (Piferrer et al. 1994c).
O. tshawytscha	400 μg E_2/L for 8 hour or 400 μg 17α EE_2/L for 2 hour immersion yielded 82% 2n females (Piferrer and Donaldson 1992).
Discrete immersion + diet	
O. tshawytscha	400 μg MT/L for 2 hour immersion on 4th and 11th days after hatching + dietary administration of 9 μg MT/kg for 3 week following hatching yielded highest percentage of males; cryopreserved sperm from sex-reversed males fertilized normal ova (Hunter et al. 1983).
O. mykiss	400 μg MT/L for 2 hour immersion per day for 2 week or 400 μg OHA/L for 2 hour per day + 3 mg OHA/kg diet feeding produced 100% males; males did not suffer the lack of or incomplete sperm ducts, as those produced after dietary MT application alone by Bye and Lincoln (1986) (Feist et al. 1995).
Continuous immersion	
O. masou	0.5 μg E_2/L from 5th to 22nd day after hatching; low percentage of males and intersex observed (Nakamura 1984).
O. keta	1 μg E_2/L for 67 days after hatching; females with ovaries having the efferent duct observed (Nakamura 1984).

steroid for 2 or 8 hours per day on specific days (2nd and 9th day) within 2 weeks after hatching (Hunter et al. 1986; Piferrer and Donaldson 1991); the second includes immersion and dietary administration for a maximum of 3 weeks after hatching (rainbow trout, Fiest et al. 1995; chinook salmon, *O. tshawytscha*, Hunter et al. 1983); the third method involves the continuous immersion protocol developed for masu salmon, *O. masou*, and chum salmon, *O. keta* (Nakamura 1984); this protocol has not been as successful as the other two (Table 1). Like salmonids, commercially-important cyprinids too attain sexual maturity at the age of 1+ or 2 years; administration of high doses of steroids through diet or subcutaneous implantation of steroid containing silastic capsules in cyprinids, especially the grass carp, *Ctenopharyngodon idella*, has not been successful. Investigations on induction of sex-reversal in cyprinids using immersion techniques could improve success rates.

Residual Steroid Analysis

Natural and synthetic steroids have long been used in the livestock industry to enhance feed conversion. Dietary administration of anabolic steroids (a few µg/kg diet) in aquaculture was commenced in the 1970s (Fagerlund and McBride 1975). Subsequently, hormones have been applied on a large scale to induce sex-reversal. The safety of this technology is of concern, as consumption of the steroid may lead to health hazards, including liver disfunction and hepatic adenocarcinoma in man (Murad and Hayes 1985). However, few contributions are available on steroid uptake, and elimination in the hormone-treated fish (Table 2), compared to the large number of publications and reviews on hormonal control and sex-reversal in fish (Pandian and Sheela 1995). A reason for this paucity can be traced to the fact that most authors have uniformly concluded that the steroid residues present in the treated juvenile and adult fish are too low to be hazardous to human health.

The species subjected to residual analysis include two salmonids, three cichlids, and one cyprinid, and the hormones tested are also limited to testosterone (T), 17α-methyltestosterone (MT) and mibolerene (MB) (Table 2). Most investigators have limited their study to the pattern of uptake, organ-wise disposition, and elimination of the administered labeled or unlabeled steroid. However, Goudie et al. (1986a) used a mixture of ^{3}H- and ^{14}C-labeled MT in a known ratio in the diet to evaluate potential differential metobolism of the steroid nucleus and the

TABLE 2. Estimation of uptake, organ-wise disposition, and elimination of steroids administered to induce sex reversal in fish. (T–Testosterone; MB–Mibolerone; MT–17α-Methyltestosterone)

Species and age/size	Steroid and treatment protocol	Observations
Oncorhynchus kistuch 30 g	^{3}H T; 3-day isotope feeding followed by 10-day analysis	> 1% residue on the 4th day of depuration (Fagerlund and McBride 1978).
O. kisutch 30 g	^{3}H MT; 3-day isotope feeding followed by 10-day observation	Identification and quantification of isotope in 16 organs within 2 hours of feeding; 86% residue in gall bladder-pyloric caeca-intestine complex; 4.5% in muscles; on the 10th day 0.1 ng MT residue/g fish (Fagerlund and Dye 1979).
O. mykiss alevins	fed 30 mg MT/kg feed for 21 days; dosed with ^{3}H MT for 1 day; analysis for 400 hour following the dosing	> 99% ^{3}H MT eliminated within 4 days (Johnstone et al. 1983).
O. mykiss	^{14}C MT; dosed for a day	95% of ingested isotope absorbed within 24 hours; 67% excreted via gills, 22% via feces and 0.5% via urine. Bile accumulated 200-2,000 times more isotope than other organs (Cravedi et al. 1989).
Oreochromis mossambicus 8 mg fry	MT fed 37 days after hatching; dosed with ^{3}H MT for 1 day; analysis for 400 hour following the dosing	100% elimination from tilapia cercass within 50 hours (Johnstone et al. 1983).
O. aureus fry	^{3}H MT and ^{14}C MT; dosed at 30 µg MT/g feed for 21 days followed by 21 day analysis	Identification and quantification of the isotope in 12 organs within 1 hour of feeding; 90% activity in viscera; 5 ng isotope residue/g fish on the 20th day (Goudie et al. 1986a).
O. niloticus; 10-day-old fry	^{3}H MT; 30 days MT feeding followed by 1-day closing and analysis until 10th day	Identified and quantified 97% residues as polar metabolites especially in bile; 100% elimination on 10 day but parent from of ^{3}H MT detectable even on the 10th day. (Curtis et al. 1991)
Cyprinus carpio 3-day-old fry	15 mg MB/kg diet fed for 30 days; bioassay of residue by feeding treated fish to castrated rats	No detectable sexual activity in the castrated rats (Das et al. 1990).
C. carpio 10-day-old fry	50 mg MT/kg diet fed for 45 days. T assayed in liquid scintellation counter	No detectable residual activity (Rao et al. 1990).
C. carpio 60 g	^{3}H T; 10 mg ^{3}H T fed for 12 day followed by 12-day analysis	Identification and quantification of ^{3}H T in 14 organs within 1 hour of feeding after isotope feeding 91.5% in lower kidney and spleen, 3.8 ^{3}H T residue/g fish on the 20th day (Lone and Matty 1981).

17α-methyl group. As tritium atoms at the commonly labeled positions of the steroid nucleus (carbon 1, 2, 6, 7) are labile and give no indication of the fate of the 17α-methyl group, they labeled the steroid nucleus with ^{3}H and the 17α-methyl group with ^{14}C. From the relatively constant ^{3}H:^{14}C ratio in the carcass and similarity to the dietary ratio,

Goudie et al. (1986a,1986b) concluded that MT or its metabolite was absorbed into the muscle and remained intact.

Three techniques have been used to evaluate residual levels of steroids in fish: direct identification and quantification of the administered steroid by the high performance liquid chromotography (HPLC); indirect estimation of radio-activity, assumed to reflect the mass of steroid; and bioassay by monitoring the revival of sexual activity in castrated rats fed on the carcass of steroid-treated fish. Rothbard et al. (1990) recognized that the high performance liquid chromatography (HPLC) technique could not detect residual steroid, when it was < 50 ng/g fish; likewise the bioassay technique of Das et al. (1990) is a useful tool in studying steroid potency but cannot quantify the residual level.

The uptake of labeled MT was very fast; within 1-2 hour after administration, the isotope could be quantified in several organs (Fagerlund and Dye 1979; Johnstone et al. 1983; Piferrer and Donaldson 1994). The gall bladder, pyloric caeca, and posterior intestine accumulated 80-95% of the residue (Fagerlund and Dye 1979; Lone and Matty 1981). However, about 80-95% of the isotope was eliminated within a day after the withdrawal of steroid-treated diet (Johnstone et al. 1983). Consequently, the residual steroid levels were at 0.1 ng MT/g coho salmon on the 10th day (Fagerlund and Dye 1979), 3.85 ng T/g common carp on the 12th day (Lone and Matty 1981), 5 ng MT/g blue tilapia, *Oreochromis aureus*, on the 21st day (Goudie et al. 1986a) after the withdrawal of steroid administration (Table 2). A tilapia egg has as much as 3 ng testosterone and an equal amount of estradiol; hence, a tilapia having 1 g ovarian tissue with 200 eggs contains more than 600 ng of each these steroids in its gonads (Rothbard et al. 1987). Therefore, the estimated residual steroids of less than 5 ng/g fish is too low to cause any concern or hazard to humans, especially as the level is likely to decrease as the fish grow. However, the fact that 90-99% of the administered hormone is eliminated within a day of administration, must be taken more seriously, especially in the light of findings by Gomelsky et al. (1994) that metabolites of MT in the culture water are more potent to induce sex-reversal than dietary MT.

Elimination of steroid was rapid initially and gradually stabilized, as time elapsed. Trends obtained for the elimination were comparable, but the rate at which the elimination decreased was dependent on species, steroid potency, organ, and treatment protocol. In Mozambique tilapia and rainbow trout, the isotope level decreased to 9% within 50 hours of dosing. A comparison of the residual levels of T observed for different organs of common carp suggests that gall bladder selectively retained the administered steroid at higher concentration for a relatively longer duration. A comparison of the residual MT retention in coho salmon (Fagerlund and Dye 1979) and that of T in common carp (Lone and Matty 1981) also indicates that MT was retained at higher concentration in gall bladder for a relatively longer period. Piferrer and Donaldson (1994) traced the levels of steroids in alevins of coho salmon subjected to discrete immersion during different stages of the alevin. After the second immersion, the cumulative uptake stood 2 times higher than the first one.

Being the first and perhaps the only publication on the use of culture water as the route of administration of steroid, especially estradiol 17β (E_2), the contribution by Piferrer and Donaldson (1994) merits special consideration. Administering unlabeled and radio-labeled steroids at 400 µg/L through culture water, the uptake and elimination of E_2 or testosterone (T) were estimated in eyed eggs, newly hatched alevins, and first feeding fry of coho salmon. The uptake and elimination were stage-dependent, and were fastest in fry and slowest in eggs. A double immersion (each lasting for 2 hour) in E_2 at 400 µg/L, administered 2 days apart, maximized exposure during the alevin stage and induced the production of 100% females. This observation confirms that the most sensitive period to the action of exogenous steroids is at the alevin stage. A 10-day depuration for alevins was required to eliminate the administered steroid. However, the maximum amount of E_2 (2.5 µg) and T (1.8 µg) accumulated by eggs after 96 hour incubation, was much higher than the respective amount of E_2 and T from maternal origin. As yolk can retain liposoluble steroid, it can provide the developing embryos with extended supply of these steroids well after treatment is completed.

Temperature and Autosomal Genes

There are reports on the occurrence of unexpected sex in fish, which were subjected to hormonal induction for sex-reversal at the optimal dose/maximal dose precisely during the labile period (Kim et al. 1993). A possible solution to minimize such occurrence is the selection to eliminate autosomal sex-modifying genes (Komen and Richter 1990). Melard (1995) intended to generate a broodstock to produce all-male progenies in the female-heterogametic blue tilapia by crossing hormonally sex-reversed females with normal males (ZZ) but obtained only 68-100% F_1 male progenies. He assumed that the sex-reversed females were characterized by one of the following genotypes: normal female (ZW), sex-reversed female (ZZΔ) and sex-reversed female with the autosomal sex gene (Z^fZ^fΔ). Z^fZ^fΔ females were identified by progeny testing of such females siring 100% F_1 males; these Z^fZ^fΔ females were crossed with normal males (presumed to carry Z^fZ^f genotype) to generate F_2 progenies, which were feminized. When these F_2 sex-reversed females were crossed with normal males, 97.5% F_3 male progenies were obtained (Figure 2). In this protocol, had Melard (1995) induced gynogenesis instead of inducing hormonal feminization of F_2 progenies, the yield of Z^fZ^fΔ females could have been doubled.

Besides the effect of autosomal sex-modifying gene(s) (Komen and Richter 1990), the sex differentiation process in the hormonally-treated populations is subjected to one or other external factors, which tend to modify the expected sex ratio. In recent years, temperature has been identified as a major factor responsible for modifying the sex-reversing potency of steroid. Varadaraj et al. (1994) showed that temperature can alter the expected sex ratio in Mozambique tilapia which were treated at the optimum MT dose during the labile period (Pandian and Varadaraj 1988); the MT treatment led to the production of more females than males at high temperature (36°C). This observation was confirmed by Mair et al. (1990). Baroiller et al. (1995) also observed the overriding effect of temperature on the sex ratio of Nile tilapia treated with 30 μg MT/g diet during the labile period at different temperatures; in general, treatment at higher temperatures significantly increased the proportion of males. More recently, Nomura et al. (1998) recorded the preponderance of males among gynogenetic cyprinid loach, *Misgurnus anguillicaudatus*, exposed to higher temperatures (25, 30°C); at 20°C, sex ratio of loach was 1:1 for diploids but 1 female:0 male for diploid gynogens; however, when gynogens were raised at 25°C or 30°C, both males and intersexes appeared at high frequency. This shows that temperature alters not only the sex-reversing potency of a steroid but also the genotypic expression of XX gynogenetic females.

Table 3 lists selected examples of altered sex ratios of induced gynogenetic and androgenetic fish. Occurrence of males and intersexes in gynogenetic populations of male-heterogametic species is a puzzling feature and may be attributed to one or more of the following: (1) paternal genetic contamination (Volckaert et al. 1994); (2) assumed presence of minor sex-genes (Komen and Richter 1990); and (3) environmental factors. A temporary rise from 30 to 39°C for 3 hour on the 9th day of

FIGURE 2. Melard's (1995) protocol for generation of highest percentage of males in the female heterogametic blue tilapia, *Oreochromis aureus*. Hormonal feminization to generate all female F_1 progenies assumed to carry ZW, ZZΔ, Z^fZ^fΔ females. Z^fZ^fΔ♀'s were identified by progeny testing that produced 100% males; crossed with normal ZZ♂'s and their progenies were feminized to produce F_2 Z^fZ^fΔ♀'s; these F_2 sex-reverse females were crossed with normal males to generate F_3 male progenies.

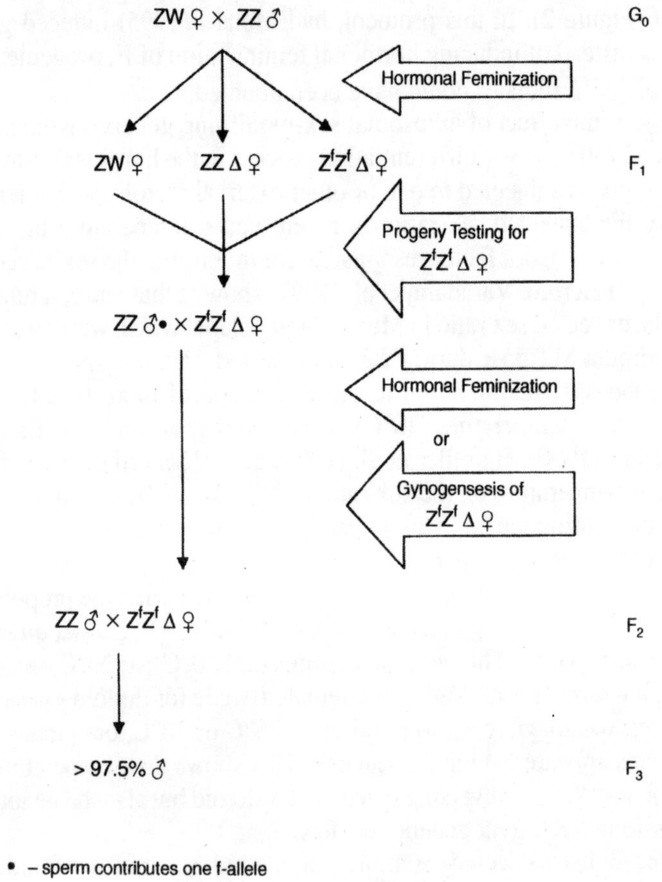

development resulted in the production of all-male progenies in Chinese catfish, *Clarias lazera* (Liu and Yao 1995).

Temperature seems to alter not only the process of sex differentiation of normal XY and XX genotypes (including gynogens) but also that of YY supermales. Varadaraj and Pandian (1989) and Scott et al. (1989)

TABLE 3. Unexpected sex ratios observed in meiotic gynogens belonging to male heterogametic fish species and for F_1 progenies sired by crossing different stains of female and androgenetic males of *Oncorhynchus mykiss*.

Species	Sex Ratio			References
	♂	♀	☿	
Gynogens				
Brachydanio rerio	0.5	9.5	-	Streisinger et al. (1981)
Oncorhynchus mykiss		9.6	0.4	Chourrout and Quillet (1982)
Paralicthys olivaceus	0.3	9.7	-	Tabata (1991)
Misgurnus anguillicaudatus	0.5	9.5	-	Suzuki et al. (1985)
Cyprinus carpio	0.6	9.4	-	Komen and Richter (1990)
Oreochromis niloticus	0.6	9.4	-	Belecke and Schwark (1995)
Oncorhynchus mykiss androgenates from different habitats				
Spokane pool ♀ × Hot Creek 57 ♂ A	0	10	-	Scheerer et al. (1991)
Spokane pool ♀ × Hot Creek 60 ♂ A	1	9	-	Scheerer et al. (1991)
Spokane pool ♀ × Spokane 1 ♂ A	2.5	7.5	-	Scheerer et al. (1991)

reported the generation of YY supermales in Mozambique tilapia and Nile tilapia, respectivley. These supermales were capable of siring all-male progenies when crossed with normal females. However, observations indicated 2-12% were females in Mozambique tilapia (Pandian 1993). More recent studies on the progenies sired by supermales have established the occurrence of a high percentage of females, in Nile tilapia belonging to different strains (Abucay et al. 1999). These studies on the sex-determining mechanism have concentrated on intraspecific sex ratios, particularly from single pair matings of normal and hormone-treated fish (Abucay et al. 1999; Tuan et al. 1999). Abucay et al. (1999) showed that the process of sex differentiation in progenies sired by supermales and normal females is more labile than that of the progenies sired by normal males and normal females. Using Nile tilapia supermales belonging to four different strains, they found strain-dependent sensitivity to temperature; higher temperature led to the production of higher percentage of males in a few strains but produced more females in others, indicating that higher temperature can influence sex ratio not only in the direction of male, as has been reported by Baroiller et al. (1995) for Nile tilapia, but also to female, as has been observed by Varadaraj et al. (1994) in Mozambique tilapia. Unexpected sex ratios observed among progenies resulting from the crosses between normal female (XX) and normal male (XY), normal female (XX) and sex-reversed male (XXΔ), and normal female (XX) and supermale (YY) indicates that the modifying genetic factor is autosomal gene(s), or contaminated broodstock.

This kind of labile sex differentiation process of genetic females and males can be stabilized by hormonal treatment. For instance, the sex differentiation process in female olive flounder, *Paralichthys olivaceus*, is so unstable that spontaneous sex-reversal to male occurs frequently. Yamamoto (1999) showed that hormonal treatment of the olive flounder from 41st to the 70th day after hatching by immersing in water containing 1 µg E_2/L prevented such sex-reversal. Incidentally, the stabilizing property of the steroid on sex differentiation can be replaced by appropriate thermal exposure. In some species exogenous factors like temperature or hormone administration appear to alter the expected sex ratio by stimulating autosomal gene(s).

Occurrence of Hermaphroditism and Intersex

Sex-reversal is known to cause undesirable side effects on sexuality and alterations in structure and function of reproductive organs. An extreme case is reported on the occurrence of self-fertilizing hermaphrodites in the Chinese catfish (Liu and Yao 1995), when fed at the suboptimum dose of 30 µg MT/g diet for a period of 60 days; the treatment resulted in the production of 40% hermaphrodites, and self-fertilization; the observed 96% fertilization and 88% hatchability showed that these hermaphrodites were functional. In this catfish, a pair of functional ovotestis was recognized, but in walleye, *Stizostedion vitreum*, subjected to MT-treatment, an intersex displayed one apparently normal gonad of each sex (Malison et al. 1998).

In rainbow trout, MT-treatment has led to complete masculinization but some masculinized females had a pair of ovary-like testis (Kim et al. 1993). Bye and Lincoln (1986) masculinized genetic female rainbow trout with 100% success, but the males lacked sperm ducts.

In simultaneously functional intersex walleye, mature oocytes intermingled with mature sperm were discernible (Malison et al. 1998). The proportion of females with affected ovaries, as a consequence of estrogen treatment, was 35 and 40% in E_2- and EE_2-treated chinook salmon (Piferrer and Donaldson 1992). When gynogenetic hatchlings of common carp were treated with MT, only 20% developed into males, but 3, 17, and 62% became females, steriles, and intersex, respectively (Komen et al. 1993). Jensen et al. (1978) showed that an increasing number of gynogenetic grass carp suffered abnormal gonad development due to the combination of genetic and hormonal effects. In coho salmon, treatment of triploids with estrogen resulted in 82% sex-reversed triploid females; the remaining 18% were partially sex-reversed individuals. Gonads from these individuals were intermediate in size of those for triploid ovaries and triploid testes (Piferrer et al. 1994b).

Poor Reproductive Performance

When a sex steroid is administered suboptimally, the fish is known to grow faster than the untreated controls (Pandian and Sheela 1995). Apart from reversing sex, the treated fish is known to suffer degeneration of renal tubules and corpuscles (Simone 1990). Hormonally sex-reversed fish suffered from functional deficiencies. For instance, the levels of sex steroids (11 ng/mL) in sex-reversed males, 3-4 months after androgen treatment, were lower than those in untreated mature Nile tilapia (38 ng/ml) and tilapia, *Tilapia hornorum* (42 ng/mL; Rothbard et al. 1983).

Survival of hormonally sex-reversed fish is adversely affected, especially in those treated with higher doses; for instance, survival of MT-treated mud loach, *M. mizolepis*, decreased from 95% to 85, 75, and 55% in groups treated with 200 µg/L for 1, 2, and 3 weeks, respectively (Nam et al. 1998). Data in Table 4 also confirm that irrespective of the genotype and irrespective of natural or synthetic estrogen used, treated oviparous or viviparous fish produced significantly fewer eggs or young. MT-treatment also lowered sperm production. George and Pandian (1998) treated black molly with synthetic MT or natural androstenedione at different doses and observed that sperm count and frequency of courtship activities decreased with increasing dosage. Estradiol-treated black molly produced far fewer (32) young than the untreated controls (80), as evidenced by low percentage of fertilization success (George and Pandian 1998).

INDIAN SCENARIO

In India, a well known commercial fish farm cultured masculinized red tilapia through MT-supplemented diet. Culturing this hormonally

TABLE 4. Reproductive capacity of normal and endocrine sex-reversed fish (data adapted from Kavumpurath and Pandian, 1993b).

Species	Steroid	Genotype	Age at puberty (day)	Young produced (No)	Eggs produced (No)	Milt volume (mL)
Betta splendens	Control	XX	178		518	
	17β-Estradiol	XX	198		485	
		XY	194		487	
	Diethylstilbestrol	XY	219		405	
		XY	241		425	
Poecilia sphenops	Control	ZW	231	17		
	17β-Estradiol	ZW	188	9		
		ZZ	190	7		
	Diethylstilbestrol	ZW	191	10		
		ZZ	+	+		
Oncorhynchus tshawytsha	Control	XY	720			0.38
	17α-Methyl-testosterone	XX,XY	1080			0.25

+ = did not survive

sex-reversed fish helped the farm to monopolize this fast growing exotic strain (Pandian 1988). Through a series of experiments undertaken both in laboratory and field, Rao and his collaborators (Rao and Rao 1983; Ali and Rao 1989) produced > 90% sterile common carp. Fully developed ovaries of the female carp contributed 28% of its body weight. On being hormonally sterilized, the fish was shown to channalize the equivalent ovarian energy into the formation of muscle. Since residual level of MT was negligible in these steriles (Das et al. 1990), many farmers in the Karnataka state have chosen hormone-induced sterilization in the common carp.

By combining endocrine sex-reversal and/or gynogenetic technique, supermales have been produced in oviparous Mossambique tilapia (Varadaraj and Pandian 1989) and viviparous guppy (Kavumpurath and Pandian 1992). Pandian and his collaborators have also developed appropriate protocols for producing all-male or all-female progenies in oviparous male heterogametic Siamese fighting fish, *Betta splendens* (Kavumpurath and Pandian 1994) and viviparous female heterogametic black molly (George and Pandian 1995).

Scientists at the Central Marine Fisheries research institurte, Cochin have developed protocol for masculinizing the protogynous grouper, *Epinephelus tauvina*.

ACKNOWLEDGMENTS

The authors gratefully acknowledge the financial support from Indian Council of Agricultural Research, New Delhi and technical assistance by R. Koteeswaran, B. Guhan, and A. Tamilselvi.

REFERENCES

Abucay, J.S., and G.C. Mair. 1997. Hormonal sex reversal of tilapias: implications of hormone treatment application in closed water systems. Aquaculture Research 28:841-845.

Abucay, J.S., G.C. Mair, D.O.F. Skibinski, and J.A. Beardmore. 1999. Environmental sex determination: the effect of temperature and salinity on sex ratio in *Oreochromis niloticus* L. Aquaculture 173:219-234.

Afonso, L.O.B., G.K. Iwama, J. Smith, and E.M. Donaldson. 1999. Effects of the aromatase inhibitor fadrozole on plasma sex steroid secretion and ovulation rate in female coho salmon, *Oncorhynchus kisutch*, close to final maturation. General and Comparative Endocrinology 113:221-229.

Al-ablani, S.A., and R.P. Phelps. 1997. Sex reversal in black crappie *Pomoxis nigromaculatus*: effect of oral administration of 17α-methyltestosterone on two age classes. Aquaculture 158:155-165.

Alam, M.S., and M.K. Monwar. 1998. Effects of mesterolone on growth and sex-ratio of the nile tilapia (*Oreochromis niloticus* Linnaeus). Indian Journal of Fisheries 45:293-299.

Ali, P.K.M.M., and G.P.S. Rao. 1989. Growth improvement in carp, *Cyprinus carpio* (Linnaeus) sterilized with 17α-methyltestosterone. Aquaculture 76:157-167.

Anon. 1995. Estrogens in the environment *in* Environmental Health Perspectives. Vol.103, Supplement 7. National Institute of Environmental Health Science, National Institutes of Health, Bethesda, Maryland.

Baroiller, J.F., D. Chourrout, A. Fostier, and B. Jalabert. 1995. Temperature and sex chromosomes govern sex ratios of the mouth-brooding cichlid fish *Oreochromis niloticus*. Journal of Experimental Zoology 273:216-223.

Baroiller, J.F., Y. Guiguen, and A. Fostier. 1999. Endocrine and environmental aspects of sex differentiation in fish. Cellular and Molecular Life Sciences 55:910-931.

Belecke, A.M., and G.H. Schwark. 1995. Sex determination in tilapia (*Oreochromis niloticus*) sex ratio in homozygous gynogenetic progeny and their offspring. Aquaculture 137:57-65.

Blazquez, M., F. Piferrer, S. Zanuy, M. Carrillo, and E.M. Donaldson. 1995. Development of sex control techniques for European seabass (*Dicentrachus labrax* L.) aquaculture: effects of dietary 17α-methyltestosterone prior to sex differentiation. Aquaculture 135:329-342.

Bye, V.J., and R.F. Lincoln. 1986. Commercial methods for the control of sexual maturation in rainbow trout (*Salmo gairdneri* R.). Aquaculture 57:299-309.

Chatain, B., E. Saillant, and S. Peruzzi. 1999. Production of monosex male populations of European seabass, *Dicentrachus labrax* L. by use of the synthetic androgen 17α-methyltestosterone. Aquaculture 178:225-234.

Chourrout, D., and E. Quillet. 1982. Induced gynogenesis in the rainbow trout: sex and survival of progenies. Production of all-triploid populations. Theoretical and Applied Genetics 63:201-205.

Cravedi, J.P., G. Delons, and D. Rao. 1989. Disposition and elimination routes of 17α-methyltestosterone in rainbow trout (*Salmo gairdneri*). Canadian Journal of Fisheries and Aquatic Sciences 46:159-165.

Curtis, L.R., F.T. Diren, M.D. Hurley, W.K. Seim, and R.A. Tubbe. 1991. Disposition and elimination of 17α-methyltestosterone in Nile tilapia (*Oreochromis niloticus*). Aquaculture 99:193-201.

Das, S.K., H.P.C. Shetty, K. Narayana, M.C. Nandeesha, and G.P.S. Rao. 1990. Androgenic bioassay for the residual effect of administered mibolerone in common carp, *Cyprinus carpio* (Linn). Pages. 85-86 *in* Proceedings of Second Indian Fisheries Forum, Mangalore, India.

Davis, K.B., B.A. Simco, C.A. Goudie, N.C. Parker, W. Cauldwell, and R. Snellgrove. 1990. Hormonal sex manipulation and evidence for female homogamety in channel catfish. General and Comparative Endocrinology 78:218-223.

Davis, K.B., C.A. Goudie, B.A. Simco, T. Tiersch, and G.J. Carmichael. 1992. Influence of dihydrotestosterone on sex determination in channel catfish and blue catfish: period of developmental sensitivity. General and Comparative Endocrinology 78:218-223.

D'Cotta, H., Y. Guiguen, M. Govoroun, O. McMeel, and J.F. Baroiller. 1999. Aromatase gene expression in temperature induced gonadal sex differentiation of tilapia *Oreochromis niloticus*. Pages. 197 *in* Proceedings of Sixth International Symposium on Reproductive Physiology of Fish, Bergen, Norway.

Demska-Zakes, K., and Z. Zakes. 1999. The effect of 11β-hydroxyandrostenedione on pikeperch *Stizostedion lucioperca* (L.). Aquaculture Research 30:731-735.

Eisenfeld, A.J., R.F. Aten, and R.B. Dickson. 1980. Estrogen receptors in the mammalian liver. Pages. 69-96 *In* J.A. Mchachlan, ed. Estrogens in the Environment, Elsevier, Amsterdam, The Netherlands.

Fagerlund, U.H.M., and H.M. Dye. 1979. Depletion of radioactivity from yearling coho salmon (*Oncorhynchus kisutch*) after extended ingestion of anabolically effective doses of 17α-methyltestosterone-1,2-^{3}H. Aquaculture 18:303-315.

Fagerlund, U.H.M., and J.R. McBride. 1975. Growth increments and some flesh and gonad characteristics of juvenile coho salmon receiving diets supplemented with 17α-methyltestosterone. Journal of Fish Biology 7:305-314.

Fagerlund, U.H.M., and J.R. McBride. 1978. Distribution and disappearance of radioactivity in blood and tissues of coho salmon (*Oncorhynchus kisutch*) after oral administration of ^{3}H-testosterone. Journal of the Fisheries Research Board of Canada 35:893-900.

Feist, G., C.G. Yeoh, M.S. Fitzpatrick, and C.B. Schreck. 1995. The production of functional sex reversed rainbow trout with 17α-methyltestosterone and 11β-hydroxyandrostenedione. Aquaculture 131:145-152.

Francis, R.C. 1992. Sexual lability in teleosts: developmental factors. Quarterly Review of Biology 67:1-18.

Gale, W.L., M.S. Fitzpatrick, M. Lucero, W.M. Contreras-Sanchez, and C.B. Schreck. 1999. Masculinization of Nile tilapia (*Oreochromis niloticus*) by immersion in androgens. Aquaculture 178:349-357.

Galvez, J.I., P.M. Mazik, R.P. Phelps, and D.R. Mulvaney. 1995. Masculinization of channel catfish *Ictalurus punctatus* by oral administration of trenbolone acetate. Journal of the World Aquaculture Society 26:378-383.

George, T., and T.J. Pandian. 1995. Production of ZZ females in the female heterogametic black molly *Poecilia sphenops* by endocrine sex reversal and progeny testing. Aquaculture 136:81-90.

George, T., and T.J. Pandian. 1996. Hormonal induction of sex reversal and progeny testing in the zebra cichlid *Cichlasoma nigrofasciatum*. Journal of Experimental Zoology 275:374-382.

George, T., and T.J. Pandian. 1998. Dietary administration of androgens induces sterility in the female-heterogametic black molly, *Poecilia sphenops* (Cuvier & Valenciennes, 1846). Aquaculture Research 29:167-175.

Gimeno, S., A. Gerritsen, T. Bowmer, and H. Komen. 1996. Feminization of male carp. Nature 384:221-222.

Glamuzina, B., N. Glavic, B. Skaramuca, and V. Kozul. 1998. Induced sex reversal of dusky grouper, *Epinephelus marginatus* (Lowe). Aquaculture Research 29:563-567.

Gomelsky, B., N.B. Cherfas, Y. Peretz, N. Ben-Dom, and G. Hulata. 1994. Hormonal sex inversion in the common carp (*Cyprinus carpio* L.). Aquaculture 126:265-270.

Goudie, C.A., B.D. Redner, B.A. Simco, and K.B. Davis. 1983. Feminization of channel catfish by oral administration of steroid sex hormones. Transactions of the American Fisheries Society 112:670-672.

Goudie, C.A., W.L. Shelton, and N.C. Parker. 1986a. Tissue distribution and elimination of radiolabeled methyltestosterone fed to adult blue tilapia. Aquaculture 58:227-240.

Goudie, C.A., W.L. Shelton, and N.C. Parker. 1986b. Tissue distribution and elimination of radiolabeled methyltestosterone fed to sexually undifferentiated blue tilapia. Aquaculture 58:215-226.

Hines, G.A., and S.A. Watts. 1995. Nonsteroidal chemical sex manipulation of tilapia. Journal of the World Aquaculture Society 26:98-102.

Hunter, G.A., and E.M. Donaldson. 1983. Hormonal sex control and its application to fish culture. Pages. 223-291 *in* W.S. Hoar, D.J. Randall, and E.M. Donaldson, eds. Fish Physiology, Vol. IX, Academic Press, New York, New York.

Hunter, G.A., E.M. Donaldson, J. Stoss, and I. Baker. 1983. Production of monosex female groups of chinook salmon (*Oncorhynchus tshawytscha*) by the fertilization of normal ova with sperm from sex-reversed females. Aquaculture 33:355-364.

Hunter, G.A., I.I. Solar, I.J. Bakerr, and E.M. Donaldson. 1986. Feminization of coho salmon (*Oncorhynchus kisutch*) and chinook salmon (*O. tshawytscha*) by immersion of alevins in a solution of estradiol-17β. Aquaculture 53:295-302.

Jalabert, B., R. Billard, and B. Chevassus. 1975. Preliminary experiments on sex control in trout: Production of sterile fishes in simultaneous self-fertilization hermaphrodites. Annales of Biology: Animalia, Biochimie and Biophysique 15:19-28.

Jensen, G.L., W.L. Shelton, and L.O. Wilken. 1978. Use of methyltestosterone silastic implants to control sex in grass carp. Pages. 200-219 *in* R.O. Smitherman, W.L. Shelton, and J.H. Grover, eds. Culture of Exotic Fishes, Fish Culture Section, American Fisheries Society, Auburn University, Alabama.

Johnstone, R., D.J. Macintosh, and R.S. Wright. 1983. Elimination of orally administrated 17α-methyltestosterone by *Oreochromis mossambicus* (tilapia) and *Salmo gairdneri* (rainbow trout) juveniles. Aquaculture 35:249-259.

Kaiser, J. 1996. Scientists angle for answers. Science 274:1837-1838.

Kavumpurath, S., and T.J. Pandian. 1992. Production of YY male in the guppy *Poecilia reticulata* by endocrine sex reversal and progeny testing. Asian Fisheries Science 5:265-276.

Kavumpurath, S., and T.J. Pandian. 1993a. Determination of labile period and critical dose for sex reversal by oral administration of estrogens in *Betta splendens*. Indian Journal of Experimental Biology 31:16-20.

Kavumpurath, S., and T.J. Pandian. 1993b. Production of YY female guppy (*Poecilia reticulata*) by endocrine sex reversal and selective breeding. Aquaculture 116:183-189.

Kavumpurath, S. and T.J. Pandian. 1994. Masculinization of fighting fish, *Betta splendens* Regan using synthetic or natural androgens. Aquaculture and Fisheries Management 25:373-381.

Kim, D.S., C.H. Jeong, and I. Kim. 1993. Induction of all-female triploid in rainbow trout (*Oncorhynchus mykiss*). Korean Journal of Genetics 15: 213-218.

Kim, D.S., Y.K. Nam, and J.Y. Jo. 1997. Effect of Oestradiol-17β immersion treatments on sex reversal of mud loach, *Misgurnus mizolepis* (Gunther). Aquaculture Research 28:941-946.

Komen, J., and C.J.J. Richter 1990. Sex control in carp (*Cyprinus carpio* L). Recent Advances in Aquaculture 4:78-86.

Komen, J., E.H. Eding, A.B.J. Bongers, and C.J.J. Richter. 1993. Gynogenesis in common carp (*Cyprinus carpio*) 4. Growth, phenotypic variation and gonad differentiation in normal and methyltestosterone-treated homozygous clones and F_1 hybrids. Aquaculture 111:271-280.

Koulish, S., and C.R. Kramer. 1989a. HCG induces sex reversal in *Thalassoma bifasciatum*, a protogynous fish. Journal of Cell Biology 107:483.

Koulish, S., and C.R. Kramer. 1989b. Human chorionic gonadotropin (hCG) induces gonad reversal in a protogynous fish, the bluehead wrasse, *Thalassoma bifasciatum* (Teleostei, Labridae). Journal of Experimental Zoology 252:156-168.

Kramer, C.R., and M.A. Imbriano. 1997. Neuropeptide Y (NPY) induces gonad reversal in the protogynous blue head wrasse, *Thalassoma bifasciatum*. Journal of Experimental Zoology 279:133-144.

Kramer, C.R., M.T. Caddell, and L. Bubenheimer-Livolsi. 1993. sGn RH-A (D-Arg^6, Pro^9, NEt-)LHRH) in combination with domperidone induces gonad reversal in a protogynous fish, the bluehead wrasse, *Thalassoma bifasciatum*. Journal of Fish Biology 42:185-195.

Labaune, J.P. 1984. Pharmacocinetique Pricipes Fondamentaux. Masson, Paris, France.

Liu, S., and Z. Yao. 1995. Self-fertilization of hermaphrodites of the teleost *Clarias lazera* after oral administration of 17α-methyltestosterone and their offspring. Journal of Experimental Zoology 273:527-532.

Lone, K.P., and A.J. Matty. 1981. Uptake and disappearance of radioactivity in blood and tissues of carp (*Cyprinus carpio*) after feeding ^{3}H-testosterone. Aquaculture 24:315-326.

Mair, G.C., J.A. Beardmore, and D.O.F. Skibinski. 1990. Experimental evidence for environmental sex determination in *Oreochromis* species. Pages. 555-558 *in* R. Hirano, and I.Hanyu, eds. Proceedings of the 2nd Asian Fisheries Forum, Asian Fisheries Society, Manila, Philippines.

Malison, J.A., J.A. Held, L.S. Procarione, and M.A.R. Garcia-Abiado. 1998. Production of monosex female populations of walleye from intersex broodstock. Progressive Fish-Culturist 60:20-24.

Martin-Robichaud, D.J., R.H. Peterson, T.J. Benfey, and L.W. Crim. 1994. Direct feminization of lumpfish (*Cyclopterus lumpus* L.) using 17β-oestradiol enriched Artemia as food. Aquaculture 123:137-151.

Melard, C. 1995. Production of a high percentage of male offspring with 17α-ethynylestradiol sex reversed *Oreochromis aureus*. I. Estrogen sex-reversal and production of F_2 pseudofemales. Aquaculture 130:25-34.

Murad, F., and R.C. Hayes, Jr. 1985. Androgens. Pages. 81-172 *in* A.G. Gilman, L.S. Goodman, T.W. Rall, and F. Murad, eds. Goodman and Gilman's, The Pharmocological Basis of Therapeutics, 7th ed. MacMillan Publishers, New York, New York.

Nakamura, M. 1975. Dosage-dependent changes in the effect of oral administration of methyltestosterone on gonadal sex differentiation in *Tilapia mossambica*. Bulletin of the Faculty of Fisheries, Hakkaido University 26:99-108.

Nakamura, M. 1984. Effects of estradiol-17β on gonadal sex differentiation in two species of salmonids, the masu salmon, *Oncorhynchus masou* and the chum salmon, *O. keta*. Aquaculture 43:83-90.

Nam, Y.K., C.H. Noh, and D.S. Kim. 1998. Effects of 17α-methyltestosterone immersion treatments on sex reversal of mud loach, *Misgurnus mizolepis*. Fisheries Science 64:914-917.

Neumann, F. 1976. Pharmacological and endocrinological studies on anabolic agents. Pages. 253-263 *in* F. Coulston and F. Corte, eds. Anabolic Agents in Animal Production, Proceedings of FAO/WHO Veterinary Symposium, Supplement. Gerog Thieme, Stuttgart, Germany.

Nomura, T., K. Arai, T. Hayashi, and R. Suzuki. 1998. Effect of temperature on sex ratios of normal and gynogenetic diploid loach. Fisheries Science 64:753-758.

Pandian, T.J. 1988. Scope for commercial culture of tilapia. Journal of Indian Fisheries 18:109-119.

Pandian, T.J. 1993. Endocrine and chromosome manipulation techniques for the production of all-male and all-female populations in food and ornamental fishes. Proceedings of Indian National Science Academy 59B:549-566.

Pandian, T.J., and S.G. Sheela. 1995. Hormonal induction of sex reversal in fish. Aquaculture 138:1-22.

Pandian, T.J., and K. Varadaraj. 1988. Techniques for producing all-male and all-triploid *Oreochromis mossambicus*. Pages. 243-349 *in* R.S.V. Pullin, T. Bhukaswan, K. Tonguthai, and J.L. Maclean, eds. The Second International Symposium on Tilapia in Aquaculture. ICLARM Conference Proceedings, 15, Department of Fisheries,

Bangkok, Thailand and International Center for Living Aquatic Resources Management, Manila, Philippines.

Patino, R., K.B. Davis, J.E. Schoore, C. Uguz, C.A. Strussmann, N.C. Parker, B.A. Simco, and C.A. Goudie. 1996. Sex differentiation of channel catfish gonads: Normal development and effects of temperature. Journal of Experimental Zoology 276:209-218.

Pelissero, C., and F. Le Menn. 1991. Evolution of sex steroid levels in male and first time maturing female of the Siberian sturgeon *Acipenser baeri*. Pages. 87-97 *in* P. Williot, ed. Acipenser. Cemagref Publishers.

Pelissero, C., and J.P. Sumpter. 1992. Steroids and "steroid-like" substances in fish diets. Aquaculture 107:283-301.

Piferrer, F., and E.M. Donaldson. 1991. Dosage-dependent changes in the effect of aromatizable and nonaromatizable androgens on the resulting phenotype of coho salmon (*Oncorhynchus kisutch)*. Fish Physiology and Biochemistry 9:145-150.

Piferrer, F., and E.M. Donaldson. 1992. The comparative effectiveness of the natural and a synthetic estrogen for the direct feminization of chinook salmon (*Oncorhynchus tshawytscha*). Aquaculture 106:183-193.

Piferrer, F., and E.M. Donaldson. 1994. Uptake and clearance of exogenous estradiol-17β and testosterone during the early development of coho salmon (*Oncorhynchus kisutch*) including eggs, alevins and fry. Fish Physiology and Biochemistry 13: 219-232.

Piferrer, F., I.I. Baker, and E.M. Donaldson. 1993. Effects of natural, synthetic, aromatizable and nonaromatizable androgens in inducing male sex differentiation in genotypic female chinook salmon (*Oncorhynchus tshawytscha)*. General and Comparative Endocrinology 91:59-65.

Piferrer, F., S. Zanuy, M. Carrillo, I.I. Solar, R.H. Devlin, and E.M. Donaldson. 1994a. Brief treatment with an aromatase inhibitor during sex differentiation causes chromosomally female salmon to develop as normal, functional males. Journal of Experimental Zoology 270:255-262.

Piferrer, F., T.J. Benfey, and E.M. Donaldson. 1994b. Production of female triploid coho salmon (*Oncorhynchus kisutch*) by pressure shock and direct estrogen treatment. Aquatic Living Resources 7:127-131.

Piferrer, F., M. Carrillo, S. Zanuy, I.I. Solar, and E.M. Donaldson. 1994c. Induction of sterility in coho salmon, *Oncorhynchuss kisutch*, by androgen immersion before first feeding. Aquaculture 119:409-423.

Pongthana, N., D.J. Penman, P. Baoprasertkul, M.G. Hussain, M.S. Islam, S.F. Powell, and B.J. McAndrew. 1999. Monosex female production in the silver barb (*Puntius gonionotus* Bleeker). Aquaculture 173:247-256.

Rao, H.N.S., and G.P.S. Rao. 1983. Hormonal manipulation of sex in the common carp, *Cyprinus carpio* var. *communis* (L.). Aquaculture 35:83-88.

Rao, G.P.S., P.V.K. Sharma, and G.Y. Keshavappa. 1990. Elimination of testosterone in the fry of common carp, *Cyprinus carpio* (Linn.). Pages. 87-89 *in* M.M. Joseph, ed. Second Indian Fisheries Forum Proceedings, Mangalore, India.

Rothbard, S., E. Solnik, S. Shabbath, R. Amado, and I. Grabie. 1983. The technology of mass production of hormonally sex-inversed all-male tilapias. Pages. 425-434 *in*

L. Fishelson, and Z. Yaron, compilers. International Symposium on Tilapia in Aquaculture. Tel Aviv University, Tel Aviv, Israel.

Rothbard, S., B. Moav, and Z. Yaron. 1987. Changes in steroid concentrations during sexual androgenesis in tilapia. Aquaculture 61:59-74.

Rothbard, S., Y. Zohar, N. Zmora, B. Levavi-Sivan, B. Moav, and Z. Yaron. 1990. Clearance of 17α-ethynyltestosterone from muscle of sex reversed tilapia hybrids treated for growth enhancement with two doses of the androgen. Aquaculture 89:365-376.

Scheerer, P.D., G.H. Thorgaard, and F.W. Allendorf. 1991. Genetic analysis of androgenetic rainbow trout. Journal of Experimental Zoology 260:382-390.

Scott, A.G., D.J. Penman, J.A. Beardmore, and D.O.F. Skibinshi. 1989. The 'YY' supermale in *Oreochromis niloticus* and its potential for aquaculture. Aquaculture 28:237-251.

Simone, D.A. 1990. The effects of the synthetic steroid 17α-methyltestosterone on the growth and organ morphology of the channel catfish (*Ictalurus punctatus*). Aquaculture 84:81-93.

Simpson, T.H. 1975-1976. Endocrine aspects of salmonid culture. Proceedings of the Royal Society, Edinburgh, Series B 75:241-252.

Streisinger, G., C. Walker, N. Dower, D. Knauber, and F. Singer. 1981. Production of clones of homozygous diploid zebrafish (*Brachydanio rerio*). Nature 291:293-296.

Strussmann, C.A., F. Takashima, and K. Toda. 1996. Sex differentiation and hormonal feminization in pjerrey *Odontesthes bonariensis*. Aquaculture 139:31-45.

Suzuki, R., T. Oshiro, and T. Nakanishi. 1985. Survival, growth and fertility of gynogenetic diploids induced in the cyprinid loach, *Misgurnus anguillicaudatus*. Aquaculture 48:45-55.

Tabata, K. 1991. Application of chromosomal manipulation in aquaculture of hirame, *Paralichthys olivaceus*. Bulletin of the Kyogo Perfecture on Fisheries Experimental Station 28:1-134.

Tuan, P.A., G.C. Mair, D.C. Little, and J.A. Beardmore. 1999. Sex determination and the feasibility of genetically male tilapia production in the Thai-chitralada strain of *Oreochromis niloticus* (L). Aquaculture 173:257-269.

Van Den Hurk, R., C.J.J. Richter, and J. Janssen-Dommerholt. 1989. Effects of 17α-methyltestosterone and 11β-hydroxyandrostenedione on gonad differentiation in the African catfish, *Clarias gariepinus*. Aquaculture 83:179-191.

Varadaraj, K., and T.J. Pandian. 1989. First report on production of supermale tilapia by integrating endocrine sex reversal with gynogenetic technique. Current Science 58: 434-441.

Varadaraj, K., S.S. Kumari, and T.J. Pandian. 1994. Comparison of conditions for hormonal sex reversal of Mozambique tilapias. Progressive Fish-Culturist 56:81-90.

Volckaert, F.A.M., P.H.A. Galbusera, B.A.S. Hellemant, C. van den Heute, D. Vanstaen, and F. Ollevier. 1994. Gynogenesis in the African catfish (*Clarias gariepinus*). I: Induction of meiogynogenesis with thermal and pressure shocks. Aquaculture 128:221-233.

Western, J.R.H., and J.B. Jennigs. 1970. Histochemical demonstration of hydrochloric acid in gastric tubules of teleosts using *in vivo* Prussian blue technique. Comparative Biochemistry and Physiology 35:879-884.

Yamamoto, E. 1999. Studies on sex manipulation and production of cloned populations in hirame, *Paralichthys olivaceus* (Temminck et Schlegal). Aquaculture 173:235-246.

Yamazaki, F. 1983. Sex control and manipulation in fish. Aquaculture 33:329-354.

Yeung, W.S.B., H. Chen, and S.T.H. Chan. 1993. Effects of LH and LHRH analog on gonad development and *in vitro* steriodogenesis in the protogynous *Monopterus albus*. General and Comparative Endocrinology 89:323-332.

Disease Problems Affecting Fish in Tropical Environments

Iddya Karunasagar
Indrani Karunasagar
Subhendu Kumar Otta

SUMMARY. Aquaculture produce in Asian countries contributes a major share to total global fish production. However, disease due to biotic and abiotic factors results in considerable losses to the industry. The most frequently encountered bacterial agents associated with fish diseases in tropical environments are *Vibrio* sp. in marine and brackishwater systems and motile aeromonads in freshwater systems. Virulence mechanisms of these bacterial species have been widely studied. Though a few viral agents such as rhabdovirus, reo-like virus, infectious pancreatic necrosis virus, picornavirus, and irido-like virus have been recorded in different parts of Asia, such reports are scanty. Among fungal pathogens, *Aphanomyces invaderis* is the most virulent and has been reported to be associated with epizootic ulcerative syndrome. Other fungal agents such as *Achlya* sp. and *Saprolegnia* sp. have been recorded in various disease conditions. In addition to various conventional microbiological and serological methods, nucleic acid-based methods are being widely adopted for diagnosis of different fish diseases. Few successful vaccines have been developed to protect against disease conditions, and some of

---

Iddya Karunasagar, Indrani Karunasagar, and Subhendu Kumar Otta, Department of Fishery Microbiology, College of Fisheries, Mangalore-575 002, India.

Address correspondence to: Iddya Karunasagar, Department of Fishery Microbiology, University of Agricultural Sciences, College of Fisheries, Mangalore-575 002, India.

[Haworth co-indexing entry note]: "Disease Problems Affecting Fish in Tropical Environments." Karunasagar, Iddya, Indrani Karunasagar, and Subhendu Kumar Otta. Co-published simultaneously in *Journal of Applied Aquaculture* (Food Products Press, an imprint of The Haworth Press, Inc.) Vol. 13, No. 3/4, 2003, pp. 231-249; and: *Sustainable Aquaculture: Global Perspectives* (ed: B. B. Jana, and Carl D. Webster) Food Products Press, an imprint of The Haworth Press, Inc., 2003, pp. 231-249. Single or multiple copies of this article are available for a fee from The Haworth Document Delivery Service [1-800-HAWORTH, 9:00 a.m. - 5:00 p.m. (EST). E-mail address: getinfo@haworthpressinc.com].

the better known ones are against vibriosis and furunculosis. Many other vaccines are still in the experimental stages. Early diagnosis of disease and development of successful vaccines are important for the future development of aquaculture.

KEYWORDS. Aquaculture, tropics, bacteria, fungi, virus, diagnosis, immunoprophylaxis

INTRODUCTION

Many tropical countries in Asia have a long tradition of aquaculture, and according to FAO statistics, over 80% of fish produced by aquaculture comes from Asia, where the production was 31.07 million metric tons valued US$38.855 billion (FAO 1996). Nine of the 14 top aquaculture producers in the world are from Asia. Cyprinids, tilapia, and catfishes are among the major freshwater fishes cultured and contribute as much as 72% of the total aquaculture production in Asia (Csavas 1994). The share of cultured diadromous fish in Asia, which include salmonids, eels, milkfish, and hilsa, is 42.9%. Major marine fish that are cultured include red fish, basses, congers, jacks, mullets, and sauries. However, culture of such marine fish is not on a big scale, and hence their contribution is low. One of the major problems limiting fish production in Asia is loss due to diseases. Aquatic animal disease and environmental related problems may cause annual losses of more than US$3 billion annually to aquaculture production in Asian countries (FAO 1996). Fish in freshwater, brackishwater, and marine environments are susceptible to a number of bacterial, viral, and parasitic diseases. In this review, the major disease problems of tropical fish are presented.

BACTERIAL DISEASES

Major Bacterial Groups Involved

Mass mortalities of fish due to bacterial diseases have been reported from a number of countries in Asia. Among the various bacterial agents associated with fish disease in tropical environments, motile aeromonads in fresh water and *Vibrio* spp. in brackish and marine water are most frequently encountered. Gopalakrishnan (1961) reported mass mortalities in India is major carps, catla, *Catla catla*; rohu, *Labeo rohita;* and mrigala, *Cirrhenius mrigala;* due to motile aeromond infection and noted that catla is most susceptible, followed by mrigal and rohu. Diseases due to motile aeromonads may manifest in a number of ways. Snieszko and Axelrod (1971) classified disease signs caused by *Aeromonas hydrophila* into four categories: (1) acute, rapidly fatal septicaemia with a few gross symptoms–an acute form with dropsy; (2) blisters, abscesses, and scale protrusion; (3) a chronic ulcerous form with furuncles and abscesses; and (4) a latent form with no signs. In tropical fish, motile aeromonad infections manifesting in all these forms have been reported (Gopalakrishnan 1961; Karunasagar et al. 1986, 1989). Motile aeromonads are commonly found in freshwater ecosystems, including fish farms, and it is not yet established whether strains associated with fish disease are different from other environmental strains. The taxonomy of aeromonads is receiving considerable attention, due to the increased recognition of the members of this genera as human pathogens (Carnahan 1993). Currently ten different species and 14 DNA hybridization groups or biovars are recognized (Joseph and Carnahan 1994), but in moribund tropical fish, most commonly isolated species are *A. hydrophila* and *A. sobria.* Though *A. caviae* are also isolated from aquaculture systems, this species has low virulence in experimental studies and may represent a secondary flora (Karunasagar et al. 1995).

Motile aeromondads are also the most commonly associated bacteria in cases of epizootic ulcerative syndrome (Llobrera and Gacutan 1987; Pal and Pradhan 1990; Torres et al. 1990; Lio-Po et al. 1991; Karunasagar and Karunasagar 1994; Karunasagar et al. 1995). However, the precise role of motile aeromonads in this syndrome is not understood. Strains with varying virulence are commonly found associated with the syndrome (Karunasagar et al. 1995), but motile aeromonads could be isolated from internal organs of fish showing reddish lesions of skin before ulceration (Karunasagar and Karunasagar 1994), and therefore, it is possible that these bacteria have a role in causing ulcerations. During recent years, some workers have suggested that epizootic ulcerative syndrome (EUS) is a fungus-associated syndrome with an invasive fungus, *Aphanomyces invaderis,* the predominant species (Roberts et al. 1992, 1993; Willoughby et al. 1995). However, it is not yet established that the fungus is the primary causative agent. In fact, the isolation rate of fungus in cases of EUS is rather low, and the presence of the fungus is inferred from histological sections (Roberts et al. 1993). It has been summarized that the fungus is killed by host tissue reactions (Chinabut and Roberts 1999). On the other hand motile aeromonads isolated from EUS showed dermonecrotic properties, and produced a number of proteolytic enzymes associated with virulence (McGarey et al. 1991; Karunasagar et al. 1995; Leano et al. 1996); therefore, it is possible that motile aeromonads have a role in causing ulcers, though they may not be the primary causative agent of EUS.

Among pathogens of brackishwater and marine fish, *Vibrio* spp. are the most important agents causing infections in a number of cultured fish species (Table 1). However, the species involved have not been properly characterized. In temperate waters, *V. anguillarum, V. ordalii,* and *V. salmonicida* are important pathogens (Austin and Austin, 1999). Though the former two species have been reported from marine fishes of Japan (Muroga et al. 1986), it has been shown that the susceptibility of the Asian marine fish to *V. anguillarum* is much lower compared to that of salmonids and ayu, *Plecoglossus altivelis*, and that this bacterium could be categorized as a facultative pathogen (Muroga 1997). Vibriosis is generally observed in stressed fish, and the species observed are mainly *V. parahaemolyticus* in grouper, *Epinephalus salmoides* and *V. parahaemolyticus* and *V. alginolyticus* in seabass, *Lates calcarifer* (Cheong et al. 1982). The disease may occur in the form of dermatitis or enteritis in a variety of cultured marine finfish, regardless of size (Leong 1992). The clinical signs of dermatitis are hemorrhages, lesions, and progressive development of deep ulcerative myositis, while internally, congestion, hemorrhage of liver, and enlargement and liquifaction of kidney might occur (Leong 1992). Enteritis is the less common form of vibriosis manifesting as hemorrhage of the intestine and destruction of the tunica mucosa (Leong 1992).

Other bacterial pathogens such as *Flavobacterium* spp., *Flexibacter* spp., *Pseudomonas* spp., *Citrobacter freundii*, *Mycobacterium* spp., *Edwardsiella tarda*, and *Staphylococcus aureus* have been reported to cause diseases in tropical fishes (Table 1). These are generally involved in sporadic cases.

Bacterial Pathogenicity and Virulence Factors

Most of the bacterial pathogens described above are commonly found in aquatic systems, and whenever these bacteria are isolated from moribund fish, the question remains as to whether they are primary

TABLE 1. Important bacterial diseases reported from tropical fish.

Causative agent	Fish affected	Reference
Aeromonas hydrophila	Catla, rohu, mrigal	Gopalakrishnan (1961)
		Karunasagar et al. (1986,1989)
	Catfish	Djajadiredja et al. (1982)
	Crucian carp	Miyazaki and Kage (1985)
	Ornamental fish	Hettiarchi and Cheong (1994)
	Murrels, barbs pearlspot	Karunasagar and Karunasagar (1994)
Aeromonas salmonicida	Common carp, tilapia	Reddy et al. (1994)
	Common carp	Bootsma et al. (1977)
Citrobacter fruendii	Common carp	Karunasagar et al. (1992)
Edwardsiella tarda	Eels	Miyazaki and Egusa (1976)
Flexibacter columnaris	Eels	Lia-Po et al. (1982)
Flexibacter maritimus	Seabass	Chong and Chao (1986)
	Red sea bream	Wakabayashi et al. (1984)
Flavobacterium spp.	Salmonids	Wakabayashi et al (1980)
Mycobacterium spp.	Snakehead	Adams et al. (1996)
Pseudomonas spp.	Eels	Lia-Po et al. (1982)
Staphylococcus aureus	Silver carp	Shah and Tyagi (1986)
Vibrio spp.	Seabass	Kasornchandra and Boonyaratpalin (1984)
		Chong and Chao (1986)
	Grouper	Chua and Teng (1987)
		Chong and Chao (1986)
	Milkfish	Muroga et al. (1986)
	Snapper	Leong (1992)
	Eels	Miyazaki et al. (1977)
Yersinia ruckeri	Rainbow trout	Sharma et al. (1995)

pathogens or opportunistic invaders of stressed fish. The virulence mechanisms of some of the fish pathogens have been intensively studied, though the mechanism by which these pathogens cross the epithelial barrier and overcome the natural defences of the fish is not yet clearly understood.

The virulence factors of motile aeromonads have been investigated by a number of workers, and this interest is perhaps driven by the fact that these organisms are emerging human pathogens (Carnahan 1993). Some of the suspected virulence-associated characteristics are serotype (Mittal et al. 1980), surface characters (Dooley and Trust 1988), and ability to produce various extra-cellular products (Thune et al. 1993).

Mittal et al. (1980) noted that strains of *A. hydrophila* virulent for fish shared a common O antigenic group, did not settle after boiling, and were resistant to the bactericidal action of mammalian serum. Studies done on tropical isolates suggested that pelleting after boiling and resistance to serum may be associated with virulent strains (Karunasagar et al. 1988). Pai (1992) studied a number of factors such as hemolytic activity, surface hydrophobicity, pelleting after boiling, lack of agglutination in acriflavine, congo red reaction, suicidal phenomenon, elastase production, and resistance to the bactericidal action of serum of motile aeromonads associated with tropical fish ponds and did not find direct correlation between many of these factors and virulence to fish. Mittal et al. (1980), Lallier et al. (1981), and Corral et al. (1990) noted that virulent strains of *A. hydrophila* were resistant to complement-mediated lysis of fresh serum, and this was attributed to either lipopolysaccharide or surface protein components of gram negative bacteria (Munn et al. 1982; Buchanan and Pearce 1997). Pai (1992) noted that resistance to bactericidal action of serum was associated with tropical strains having a low LD50, supporting the view that this activity constitutes an important virulence factor of motile aeromonads. It has also been suggested that the virulent strains show higher elastase and gelatinase activities (Hsu et al. 1981; Wakabayashi et al. 1981; Pai 1992).

During recent years, cell culture systems are being used to study how pathogens interact with host cells, and in combination with molecular biological studies of gene inactivation and complementation, such studies have greatly contributed to the understanding of the pathogenesis of a number of diseases, particularly salmonellosis, listeriosis, infections caused by pathogenic *Escherichia coli*, etc. (Henderson et al. 1999). Interaction of fish pathogenic *A. hydrophila* with fish cells has been studied by some investigators. Krovacek et al. (1987) noted that *A. hydrophila* and *V. anguillarum* adhere to fish cells and to glass slides coated with fish mucus. Leung et al. (1995, 1996) suggested that *A. hydrophila* is capable of invading and multiplying inside tilapia phagocytes and carp epithelial tissue cultured cells. They further noted that virulent strains induced cytopathic changes in cells during invasion, and the internalization was accompanied by the rearrangement of micro-filaments (Low et al. 1998). However, studies with more virulent and avirulent strains are required to assess whether the cell culture assay could be used to differentiate virulent from avirulent strains of *A. hydrophila*.

Compared to motile aeromonads, fish pathogenic *Vibrio* sp. have been little investigated, with the possible exception of *V. anguillarum*. Crosa (1980) noted that highly virulent strains of *V. anguillarum* carry a

65 kilobase plasmid named pJM1, which encodes an efficient iron uptake system that allows the organism to grow under iron restricted conditions. On losing the plasmid, the organism loses both the ability to grow under such conditions and also the virulence (Crosa and Hodges 1981). In some virulent strains of this organism, the genes encoding the high affinity iron transport system are located in the chromosome (Toranzo et al. 1983). Milton et al. (1996) noted that a 40.1 kDa flagellin A protein encoded by *fla A* gene is essential for the virulence of *V. anguillarum*, since the virulence could be reduced 500-fold when flagella was lost by transposon mutagenesis. Lee (1995) and Balebona et al. (1995) suggested that extracellular products such as proteases are involved in the pathogenicity of *V. alginolyticus* to cultured grouper and seabream. *In vitro*, *V. alginolyticus* has been shown to adhere to skin mucus and skin cells of seabream; however, the infection experiments using bath immersion suggested that *V. alginolyticus* could be considered a pathogen for seabream, only when the mucus layer is removed and the skin damaged (Balebona et al. 1998).

Detailed studies on the pathogenicity of *Flexibacter* group are still lacking, though challenge experiments have been conducted with a few strains (Wakabayashi et al. 1980, 1984). It has been suggested that the virulence of this group is associated with extracellular protein (ECP), subcellular components, hemolysin, and lipopolysaccharides (Austin and Austin 1999).

VIRAL DISEASES

Though a number of viral fish disease problems have been reported from both freshwater as well as marine habitats in North America, Canada, Australia and Japan, reports of such incidences in warmer parts of Asia are very scanty (Table 2). This could also reflect the lack of facilities required to investigate viral diseases in this region. In the freshwater systems, a few viral agents, particularly, rhabdoviruses have been isolated from EUS-affected fish (Frerichs et al. 1986, 1989; Kasornchandra et al. 1991; Roberts et al. 1994). However, their precise role in the syndrome is not yet established. Inability to isolate viruses from many epizootics could be due to various reasons. It is possible that the viruses are present early during the syndrome (Kanchanakhan 1997). Other viral agents isolated include reo-like viruses (Roberts et al. 1994). In freshwater systems, there have been two reports of detection of infectious

TABLE 2. Viral agents reported from tropical fish.

Pathogen	Host	Reference
Rhabdovirus	Snakehead, swamp eel	Frerichs et al. (1986, 1989)
	Blackfish, three spot gourami	Kasornchandra et al. (1991)
	Giant gourami	Roberts et al. (1994)
		Lilley and Frerichs (1994)
		Kanchanakhan (1997)
Reo-like virus	Snakehead	Roberts et al. (1994)
Snakehead fish virus	Snakehead	Saitanu et al. (1986)
Infectious pancreatic	Giant snakehead,	Wattanavijarn et al. (1988)
Necrosis virus (IPNV)	Snakehead, eye-spot barb	Subramaniam et al. (1993)
Sand goby virus	Sand goby	Hedrick et al. (1986)
Picorna-like virus	Grouper	Boonyaratpalin et al. (1996)
Iridovirus	Grouper	Kanchanakhan (1996)
	Seabass	Limsuwan et al. (1983)
		Chao (1984)
		Leong (1992)

pancreatic necrosis virus (Wattanavijarn et al. 1988; Subramanian et al. 1993).

In marine fishes two major viral disease problems have been recorded. A picorna-like virus has been confirmed to be the causative agent of a paralytic syndrome affecting grouper culture in Thailand (Kanchanakhan 1996). The disease is a major problem at fry and fingerling stage. The affected fish exhibit dark coloration, anorexia, loss of equilibrium with corkscrew-like swimming motion, and high mortalities (Danayadol et al. 1995). Another virus has been isolated from diseased grouper exhibiting lethargy, dark coloration of the tail and fins, and loss of balance. An iridovirus-like agent has been confirmed as the etiological agent (Kanchanakhan 1996). Lymphocystis caused by an iridovirus has been found to occur in seabass in Thailand (Limsuwan et al. 1983), Singapore (Chao 1984), and Malaysia (Leong 1992).

FUNGAL DISEASES

Though fungal agents such as *Achlya* sp. and *Saprolegnia* sp. have been reported in diseased fishes, these are generally regarded as secondary pathogens. Interestingly, Limsuwan and Chinabut (1983) described severe chronic granulomatous mycoses in Thailand from which these fungi were isolated, but later, they were considered secondary pathogens in EUS (Tonguthai 1985). Before EUS was a serious problem in Asia, Srivastava (1979) reported aphanomycosis due to *Aphanomyces pisci* in mrigal and suggested that this disease could be a new threat to fish populations, but it was only after the report of Roberts et al. (1993) that the association between an invasive fungus, *Ap. invaderis* and EUS has been increasingly recognized and debated on its role as the causative agent. They isolated a strain of *Aphanomyces* from the muscle of EUS-affected fish, and this fungus was shown to be slow-growing and thermolabile when compared to other saprophytic *Aphanomyces*. When placed below the dermis of striped snakehead fish, *Channa striatus*, the mycelium of slow-growing *Aphanomyces* succeeded in migrating into the tissues causing severe myonecrosis with chronic epithelial reaction typical of EUS (Roberts et al. 1993). This fungus was named *Ap. invaderis* by Willoughby et al. (1995). Interestingly, *Ap. invaderis* is not detected on the surface of EUS-affected fish. Lilley and Roberts (1997) compared the infectivity of several *Ap. invaderis* isolates with that of other *Ap.*, *Achlya*, and *Saprolegnia* found on the surface of EUS-affected fish. Only *Ap. invaderis* produced invasive lesion typical of EUS. However, they point out that *Ap. invaderis* cannot be considered the primary pathogen unless the infective zoospore stage can be shown to breach the skin unaided. Therefore, many investigators still consider EUS to be a syndrome of complex aetiology. Under experimental conditions, this fungus could infect juvenile snakehead fish, only when it was injected or inserted into the body of the fish. Several workers have shown that it is unable to infect the fish when challenged by bathing in the fungus (Kanchanakhan 1997). It has been suggested, therefore, that the fungus may need special environmental conditions and/or other fish pathogens to induce an infection (Roberts et al. 1993; Chinabut et al. 1995). The ecology of *Ap. invaderis* is still a mystery, since no isolates have been reported from natural waters. Moreover, the fungus has not been shown to sporulate in fish tissue, and therefore the question remains as to where the fungus sporulates and how it spreads from fish to fish to cause an epizootic.

An epizootic infection due to *A. debaryana* in the *Mastacembelus armantus,* has been reported from an artificial reservoir in India (Khulbe et al. 1994). In the same reservoir, when water temperatures were moderate, mycotic infections were observed to be severe, with fungi like *A. flagellata* and *S. parasitica* being more virulent (Khulbe et al. 1995).

METHODS FOR DIAGNOSIS OF MICROBIAL DISEASES

So far, diagnosis of microbial diseases of tropical fish has been based on classical methods such as clinical signs, histopathology, and isolation and identification of pathogens by conventional methods. However, during recent years, application of nucleic-acid based and antibody-based diagnostic methods are being increasingly applied in medical and veterinary fields and have been widely used for a number of fish and shellfish diseases (Karunasagar and Karunasagar 1999). Even in tropical countries such as Taiwan, India, Thailand, and Malaysia, polymerase chain reaction (PCR)-based detection methods are being routinely used for whitespot syndrome virus affecting shrimp (Lo et al. 1996; Kasornchandra et al. 1998; Otta et al. 1999). However, such methods are not yet used for diseases of finfishes. There may be several reasons for this. Shrimp is a high-value product, and the farmers have a high financial stake. They are therefore willing to spend on expensive diagnostic methods, but the economics of a fish farm is on a different scale. Nevertheless, these methods will be extremely useful, particularly in cases where traditional methods are also expensive and time consuming, e.g., in viral diseases.

DNA-based methods, such as hybridization with an oligonucleotide probe or PCR/reverse transcriptase PCR, has been reported to be useful for detection of fish viral pathogens such as Infectious Pancreatic Necrosis Virus (Rimstaad et al. 1990; Rodriguez et al. 1995), fish rhabdovirus (Estepa et al. 1995), fish iridovirus (Tamai et al. 1997), and red seabream iridovirus (Oshima et al. 1996), and for bacteria that are difficult to culture or slow growing, such as mycobacteria (Talaat et al. 1997). Therefore, adopting these methods for viruses or bacteria affecting tropical fishes should not be difficult.

Similarly, antibody-based methods such as fluorescent antibody test (FAT) and enzyme-linked immunosorbant assay (ELISA) have been reported for detection of a number of fish pathogens including bacteria, viruses, and parasites (Karunasagar and Karunasagar 1999). However, these are mostly experimental studies, and commercial kits for diagnosis of fish diseases are still not available for routine use. There are also problems in application of antibody-based methods for bacterial pathogens, such as *A. hydrophila*, which are serologically heterogenous. The serological structure of *Vibrio* spp., such as *V. alginolyticus*, affecting tropical fishes has not been studied in great detail. Such studies would be important for development of diagnostic tests suitable for pathogens of tropical fishes.

IMMUNOPROPHYLAXIS OF FISH DISEASES

Prevention of diseases would be the major goal of scientific aquaculture. The possibility of vaccinating fishes against diseases has been investigated for nearly half a century, but successful vaccines have been developed to protect against only a few disease conditions (Karunasagar and Karunasagar 1999). One of the most successful vaccines has been against vibriosis, and this generally contains *V. anguillarum* serotype 1 and 2, but polyvalent vaccines containing other antigens such as *V. salmonicida*, *A. salmonicida* or *V. ordalii* are also available (Toranzo et al. 1997). Whether they are effective against vibriosis of tropical marine fishes is not known. Leong et al. (1998) tested the effect of an imported vaccine (not specified) and a locally produced vibrio vaccine on the survival of grouper, *Epinephelus cocciodes*, cultured in floating net cages in Malaysia. After 12 weeks, the overall survival for the control grouper ranged between 20.5% to 35%, while grouper vaccinated with locally produced vaccine showed a survival range of 8.5% to 43.5% and the group vaccinated with imported vaccine showed a survival range of 30% to 80%. However, high mortalities occurred in the first 4 weeks, regardless of whether groupers were vaccinated or not (Leong et al. 1998). These results suggest that more appropriate vaccines need to be developed for vibriosis affecting tropical fish.

Among diseases of freshwater fish, motile aeromonad septicaemia is very important and there have been a number of attempts to develop vaccines against motile aeromonads. Vaccination against the non-motile aeromonad *A. salmonicida* affecting temperate fish has been possible because of the serological homogeneity in this species. Motile aeromonads affecting tropical fish are, on the other hand, serologically very heterogenous. However, experimental studies have shown that bacterins delivered by injection or immersion can induce antibody production and protection against homologous challenge in a number of fish, including common carp, *Cyprinus carpio* (Baba et al. 1988), catla, rohu, mrigal (Karunasagar et al. 1991, 1997), and tilapia, *Aureochromis mosambicus* (Ruangpan et al. 1986). There was good correlation between the levels of agglutinating antibody and extent of protection against challenge; cross-reacting antibodies induced by monovalent vaccines showed varying degrees of protection against heterologous challenge (Karunasagar et al. 1997). When polyvalent vaccines were used, antibody response against all the component strains was observed; however, the level of antibody was less compared to that in fishes immunized with monovalent vaccines (Karunasagar et al. 1997).

Since inclusion of all serotypes of the organism in polyvalent vaccines would not be possible, it is necessary to understand whether there are any common protective antigens in the fish pathogenic strains of motile aeromonads. Therefore, many basic studies are necessary before a successful vaccine for this organism can be developed. The possibility of using live attenuated vaccine has been proposed by some investigators. Karunasagar et al. (1991) noted that an aerolysin deletion mutant induced significant protection, as evidenced by challenge experiments. Protection against heterologous strain was only partial. Azad et al. (1997) reported that *A. hydrophila* grown as a biofilm on substrates such as chitin and bagasse could be used as an oral vaccine; however, they did not address the question of protection against heterologous strains or attenuation of the vaccine strain. Hernanz Moral et al. (1998) developed an aromatic-dependent mutant of *A. hydrophila* as a candidate for live vaccine and noted that the strain conferred significant protection against the wild-type strain in rainbow trout, *Oncorhynchus mykiss*. However, the extent of protection against heterologous strains was not evaluated.

REFERENCES

Adams, A., K.D. Thompson, H. McEvan, S.C. Chen, and R.H. Richards. 1996. Development of monoclonal antibodies to *Mycobacterium* spp. isolated from Chevron snakehead and Siamese fighting fish. Journal of Aquatic Animal Health 8: 208-215.

Austin, B., and D.A. Austin. 1999. Bacterial Fish Pathogens: Disease of Farmed and Wild Fish. Springer Praxis Publishing, Chichester, United Kingdom.

Azad, I.S., K.M. Shankar, and C.V. Mohan. 1997. Evaluation of an *Aeromonas hydrophila* biofilm for oral vaccination of carp. Pages 181-185 *in* T.W. Flegel, and MacRae, Eds., Diseases in Asian Aquaculture III, Asian Fisheries Society, Manila, Philippines.

Baba, T., J. Imamura, K. Izawa, and K. Ikeda. 1988. Cell mediated protection in carp, *Cyprinus carpio* L., against *Aeromonas hydrophila*. Journal of Fish Diseases 11:171-178.

Balebona, M.C., M.A. Morinigo, and J.J. Borrego. 1995. Role of extracellular products in the pathogenicity of *Vibrio* strains on cultured gilthead seabream (*Sparus aurata*). Microbiologia 11: 439-446.

Balebona, M.C., M.J. Andreu, M.A. Bordas, I. Zorrilla, M.A. Morinigo, and J.J. Borrego. 1998. Pathogenicity of *Vibrio alginolyticus* for cultured gilt-head seabream (*Sparus aurata* L.). Applied and Environmental Microbiology 64: 4269-4275.

Bootsma, R., N. Fijan, and J. Blommaert. 1977. Isolation and identification of the causative agents of carp-erythrodermatitis. Veterinarski Archiv 47: 291-302.

Boonyaratpalin, S., K. Supamattaya, J. Kasornchandra, and R.W. Hoffman. 1996. Picorna-like virus associated with mortality and a spongious encephalopathy in grouper *Epinephelus malabaricus*. Diseases of Aquatic Organisms 26: 75-80.

Buchanan, T.M., and W.A. Pearce. 1997. Pathogenic aspects of outer membrane components of gram negative bacteria. Pages 475-514 *in* M. Inouye, ed. Bacterial Outer Membrane. John Wiley and Sons Inc., New York, New York.

Carnahan, A.M. 1993. *Aeromonas* taxonomy: A sea of change. Medical Microbiology Letters 2: 206-211.

Chao, T.M. 1984. Studies on transmissibility of lymphocystis disease occurring in seabass (*Lates calcarifer* Bloch). Singapore Journal of Pr. Ind. 12: 11-16.

Cheong, L., R. Chou, R. Singh, and C.T. Mee. 1982. Country reports-Singapore. Pages 47-50 *in* Fish Quarantine and Fish Diseases in South East Asia. Proceedings of the Workshop, Jakarta, Indonesia.

Chinabut, S., and R.J. Roberts. 1999. Pathology and histopathology of epizootic ulcerative syndrome. Aquatic Animal Health Research Institute Newsletter, Bangkok, Thailand.

Chinabut, S., R.J. Roberts, G.R. Willoghby, and M.D. Pearson. 1995. Histopathology of snakehead, *Channa striatus* (Bloch), experimentally infected with specific *Aphanomyces* fungus associated with epizootic ulcerative syndrome (EUS) at different temperatures. Journal of Fish Diseases 18: 41-47.

Chong, Y.C., and T.M. Chao. 1986. Common Diseases of Marine Food Fish. Fisheries Handbook No. 2. Primary Production Department, Ministry of National Development, Singapore.

Chua, T.E., and S.K. Teng. 1987. Floating Fish-Pens for Rearing Fishes in Coastal Waters, Reservoirs and Mining Pools in Malaysia. Fisheries Bulletin No. 20, Ministry of Agriculture, Kualalampur, Malaysia.

Corral, F.D., E.B. Shotts, Jr., and J. Brown. 1990. Adherence, haemagglutination and cell surface characters of motile aeromonads virulent for fish. Journal of Fish Diseases 13: 255-268.

Crosa, J.H. 1980. A plasmid associated with virulence in the marine fish pathogen, *Vibrio anguillarum* specifies an iron sequestering system. Nature 284:556-568.

Crosa, J.H., and L.L. Hodges. 1981. Outer membrane proteins induced under conditions of iron limitation in the marine fish pathogen *Vibrio anguillarum*. Infection and Immunity 31:223-227.

Csavas, I. 1994. The Status and Outlook of World Aquaculture with Special Reference to Asia. AQUATECH, 1994, Colombo, Sri Lanka.

Danayadol, Y., S. Direkbusarakom, and K. Supamattaya. 1995. Viral nervous necrosis in brown spotted grouper, *Epinephelus malabaricus*, cultured in Thailand. Pages 227-233 *in* M. Shariff, J.R. Arthur, and R.P. Subasinghe ed. Diseases in Asian Aquaculture II. Asian Fisheries Society, Manila, The Philippines.

Djajadiredja, R., T.H. Panjaitan, A. Rukyani, A. Sarono, D. Satyani, and H. Supriyadi. 1982. Country report-Indonesia. Pages 19-27 *in* Fish Quarantine and Fish Diseases in Southeast Asia. Proceeding of workshop, Jakarta, Indonesia.

Dooley, J.S.G., and T. Trust. 1988. Surface composition of *Aeromonas hydrophila* strains virulent for fish: Identification of a surface array protein. Journal of Bacteriology 170: 499-506.

Estepa, A., C. DeBlas, F. Ponz, and J.M. Coll. 1995. Detection of trout haemorrhagic septicaemia rhabdovirus by capture with monoclonal antibodies and amplification with PCR. Veterinary Research 26: 530-532.

FAO. 1996. FAO Year Book of Fishery Statistics. Vol. 83, FAO, Rome, Italy.

Frerichs, G.N., S.D. Millar, and R.J. Roberts. 1986. Ulcerative rhabdovirus in fish in South-East Asia. Nature 322: 216.

Frerichs, G.N., S.D. Millar, and M. Alexander. 1989. Rhabdovirus infection of ulcerated fish in South-East Asia. Pages 396-410 *in* W. Ahne, and E. Kurstak, eds. Viruses of Lower Vertebrates. Springer-Verlag, Berlin, Germany.

Gopalakrishnan, V. 1961. Observations on infectious dropsy of Indian major carps and its experimental induction. Journal of Scientific and Industrial Research 20: 357-358.

Hedrick, R.P., W.D. Eaton, J.L. Fryer, W. G. Groberg, Jr., and S. Boonyaratpalin. 1986. Characteristics of a birnavirus isolated from cultured sand goby, *Oxyeleotris marmoratus*. Diseases of Aquatic Organisms 1: 219-225.

Henderson, B., M. Wilson, R. McNab, and A.J. Lax. 1999. Cellular Microbiology: Bacteria-Host Interactions in Health and Disease. John Wiley & Sons, Chichester, England.

Hernanz M.C., E.F. Castillo, P.L. Fierro, A.V. Cortes, J.A. Castillo, A.C. Soriano, M.S. Salazar, B.R. Peralta, and G.N. Carrasco. 1998. Molecular characterisation of the *Aeromonas hydrophila aro A* gene and potential use of an auxotrophic *aro A* mutant as a live attenuated vaccine. Infection and Immunity 66: 1813-1821.

Hettiarchi, D.C., and C.H. Cheong. 1994. Some characteristics of *Aeromonas hydrophila* and *Vibrio* species isolated from bacterial diseases outbreak in ornamental fish culture in Sri Lanka. Journal of the National Science Council of Sri Lanka 22: 261-269.

Hsu, T.C., W.D. Waltman, and E.B. Shotts. 1981. Correlation of extracellular enzymatic activity and biochemical characteristics with regard to virulence factors of *Aeromonas hydrophila*. Developments in Biological Standardisation 49: 101-111.

Joseph, S.W., and A. Carnahan. 1994. The isolation, identification and systematics of motile *Aeromonas* sp. Annual Review of Fish Diseases 4: 315-343.

Kanchanakhan, S. 1996. Diseases of cultured grouper. Aquatic Animal Health Research Institute Newsletter, Bangkok, Thailand, 2: 3-4.

Kanchanakhan, S. 1997. Variability in the isolation of viruses from fish affected by epizootic ulcerative syndrome (EUS). Aquatic Animal Health Research Institute Newsletter, Bangkok, Thailand 6: 1-3.

Karunasagar, I., and I. Karunasagar. 1994. Bacteriological studies on epizootic ulcerative syndrome in India. Proceedings ODA Regional Seminar on Epizootic Ulcerative Syndrome. Pages 158-170 *in* R.J. Roberts, and I.H. MacRae, eds. Aquatic Animal Health Research Institute, Bangkok, Thailand.

Karunasagar, I., G. Sugumar, and I. Karunasagar. 1995. Virulence characters of *Aeromonas* spp. isolated from epizootic ulcerative syndrome (EUS). Pages 313-320 *in* M. Shariff, J.R. Arthur, and R. Subasinghe, eds. Diseases in Asian Aquaculture II. Asian Fisheries Society, Manila, Malaysia.

Karunasagar, I., A. Ali, S.K. Otta, and I. Karunasagar. 1997. Immunisation with bacterial antigens: infections with motile aeromonads. Pages 135-141 *in* R. Gudding, A. Lillehaug, P.J. Midtlyng, and F. Brown, eds. Developments in Biological Standardization.

Karunasagar, I., and I. Karunasagar. 1999. Diagnosis, treatment and prevention of microbial diseases of fish and shellfish. Current Science 76: 387-399.

Karunasagar, I., P.K.M.M. Ali, G. Jeyasekaran, and I. Karunasagar. 1986. Ulcerative form of *Aeromonas hydrophila* infection of *Catla catla*. Current Science 55: 1194-1195.

Karunasagar, I., K. Segar, I. Karunasagar, P.K.M.M. Ali, and G. Jeyasekaran. 1988. Virulence of *Aeromonas hydrophila* strains isolated from fish ponds and infected fish. Pages 205-211 *in* S.T. Chang, K.Y. Chan, and N.Y. S. Woo, eds. Recent Advances in Biotechnology and Applied Biology. The Chinese University of Hong Kong, Hong Kong, China.

Karunasagar, I., G.M. Rosalind, I. Karunasagar, and K.G. Rao. 1989. *Aeromonas hydrophila* septicaemia of Indian major carps in some commercial fish farms of West Godavari District, Andhra Pradesh. Current Science 58: 1044-1045.

Karunasagar, I., G. Rosalind, and I. Karunasagar. 1991. Immunological response of Indian major carps to *Aeromonas hydrophila* vaccine. Journal of Fish Diseases 14: 413-417.

Karunasagar, I., I. Karunasagar, and R. Pai. 1992. Systemic *Citrobacter fruendii* infection in common carp, *Cyprinus carpio* fingerlings. Journal of Fish Diseases 15:95-98.

Kasornchandra, J., and S. Boonyaratpalin. 1984. Diseases found in seabass (*Lates calcarifer* Bloch). Thai Fisheries Gazette 37: 426-432.

Kasornchandra, J., H.M. Engelking, C.N. Lannan, J.S. Rohovec, and J.L. Fryer. 1992. Characterisation of the rhabdovirus from snakehead fish *Ophicephalus striatus*. Diseases of Aquatic Organisms 13: 89-94.

Kasornchandra, J., C.N. Lannan, J.R. Rohovec, and J.L. Fryer. 1991.Characterisation of a rhabdovirus isolated from snakehead fish (*Ophicephalus striatus*). Pages 175-182 *in* J.L. Fryer, ed. Proceedings of the Second International Symposium. Viruses of Lower Vertebrates. Oregon State University, Cornvallis, Oregon.

Kasornchandra, J., S. Boonyaratpalin, and T. Itami. 1998. Detection of whitespot syndrome in cultured penaeid shrimp in Asia: Microscopic observation and polymerase chain reaction. Aquaculture 164: 243-251.

Khulbe, R.D., G.S. Bist, and C. Joshi. 1994. Epizootic infection due to *Achlya debaryana* in a catfish. Mycoses 37: 61-63.

Khulbe, R.D., G.S. Bist, and C. Joshi. 1995. Fungal diseases of fish in Nanak Sagar, Nainital, India. Mycopathologia 130: 71-74.

Krovacek, K., A. Faris, W. Ahne, and I. Mansson. 1987. Adhesion of *Aeromonas hydrophila* and *Vibrio anguillarum* to fish cells and mucus coated glass slides. FEMS Microbiology Letters 42: 85-89.

Lallier, R., K.R. Mittal, D. Leblanc, G. Lalonde, and G. Olivier. 1981. Rapid methods for differentiation of virulent and non-virulent *Aeromonas hydrophila* strains. Pages 119-125 *in* S. Karger, ed. International Symposium on Fish biologics, Developments in Biological Standardization.

Leano, E.M., G.D. Lio-Po, and L.A. Dureza.1996. Virulence and production of extracellular proteins (ECP) of *Aeromonas hydrophila* associated with epizootic ulcerative syndrome (EUS) of fresh water fish. University of the Philippines Vistas Journal of National Science 1: 30-38.

Lee, K.K. 1995. Pathogenesis studies on *Vibrio alginolyticus* in the grouper, *Epinephelus malabaricus* Bloch et Schneider. Microbial Pathology 19:39-48.

Leong, T.S. 1992. Diseases of brackishwater and marine fish cultured in some Asian countries. Pages 223-236 *in* M. Shariff, R.P. Subasinghe, and J.R. Arthur, eds. Diseases in Asian Aquaculture I, Fish Health Section, Asian Fisheries Society, Manila, The Philippines.

Leong, T.S., W.C. Yong, and L.H. Hong. 1998. Effect of *Vibrio* vaccine on the survival of grouper *Epinephelus coioides* cultured in floating net cages. Aquatic Animal Health Research Institute Newsletter, Bangkok 7: 1-3.

Leung, K.Y., K.W. Low, T.J. Lam, and Y.M. Sin. 1995. Interaction of fish pathogen *Aeromonas hydrophila* with tilapia, *Oreochromis aureus* (Steindachner), phagocytes. Journal of Fish Diseases 18: 435-447.

Leung, K.Y., T.M. Lim, T.J. Lam, and Y.M. Sin. 1996. Morphological changes in carp epithelial cells infected with *Aeromonas hydrophila*. Journal of Fish Diseases 19: 167-174.

Lilley, J,H., and G.N. Frerichs. 1994. Comparison of rhabdoviruses associated with epizootic ulcerative syndrome (EUS) with respect to their structural proteins, cytopathology and serology. Journal of Fish Diseases 17: 513-522.

Lilley, J.H., and R.J. Roberts. 1997. Pathogenicity and culture studies comparing the *Aphanomyces* involved in epizootic ulcerative syndrome (EUS) with other similar fungi. Journal of Fish Diseases 20: 135-144.

Limsuwan, C., and S. Chinabut. 1983. Histological changes of some freshwater fishes during 1982-83 disease outbreak. Proceedings of the Symposium on Freshwater Fish Epidemic. Chulalongkorn University, Bangkok, Thailand.

Limsuwan, C., S. Chinabut, and Y. Donyadol. 1983. Lymphocystis Disease in Seabass (*Lates calcarifer*). National Inland Fisheries Institute Technical Paper.

Lio-Po, G.D., J.P. Pascual, and J.G. Santos. 1982. Pages 35-43 *in* Fish Quarantine and Fish Diseases in South East Asia, Country reports, Philippines, Workshop, Jakarta, Indonesia.

Lio-Po, G.D., L.J. Albright, and E.V. Alapide-Tendencia. 1991. *Aeromonas hydrophila* in epizootic ulcerative syndrome (EUS) of snake head (*Ophicephalus striatus*) and catfish (*Clarius batrachus*): Quantitative estimation in natural infection and experimental induction of dermonecrotic lesion. Pages 431-436 *in* M. Shariff, R.D. Subasinghe, and J.R. Arthur, eds. Diseases in Asian Aquaculture I. Asian Fisheries Society, Manila, The Philippines.

Llobrera, A.T., and R.Q. Gacutan. 1987. *Aeromonas hydrophila* associated with ulcerative epizootic in Laguna de Bay, Philippines. Aquaculture 67: 237-278.

Lo, C.F., J.H. Leu, C.H. Ho, C.H. Chen, S.E. Peng, Y.T. Chen, C.M. Chou, P.Y. Yeh, C.J. Huang, H.Y. Chou, C.H. Wang, and G.H. Kou. 1996. Detection of baculovirus associated with whitespot syndrome (WSBV) in penaeid shrimps using polymerase chain reaction. Diseases of Aquatic Organisms 25: 133-141.

Low, K.W., S.G. Goh, T.M. Lim, Y.M. Sin, and K.Y. Leung. 1998. Actin rearrangements accompanying *Aeromonas hydrophila* entry into cultured fish cells. Journal of Fish Diseases 21: 55-65.

McGarey, D.J., L. Milanesi, D.P. Foley, B. Reyes Jr., L.C. Frye, and D.V. Lim. 1991. The role of motile aeromonads in the fish disease, ulcerative disease syndrome (UDS). Experimentia 47: 441-444.

Milton, D.L., R. O'Toole, P. Horstedt, and H. Wolf-Watz. 1996. Flagellin A is essential for the virulence of *Vibrio anguillarum*. Journal of Bacteriology 178: 1310-1319.

Munn, C.B., E.E. Ishiguro, W.W. Kay, and T.J. Trust. 1982. Role of surface components in serum resistance of virulent *Aeromonas salmonicida*. Infection and Immunity 36: 1069-1075.

Muroga, K., G. Lio-Po, C. Pitogo, and R. Imada. 1984. *Vibrio* sp. isolated from milkfish (*Chanos chanos*) with opaque eyes. Gyobyo Kenkyu. 19: 81-87.

Mittal, K.R., D. Lalonde, D. Leblanc, G. Olivier, and R. Lallier. 1980. *Aeromonas hydrophila* in rainbow trout: Relation between virulence and surface characteristics. Canadian Journal of Microbiology 26: 1501-1503.

Miyazaki, T., Y. Jo, S.S. Kubota, and S. Egusa. 1977. Histopathological studies on vibriosis of the Japanese eel *Anguilla anguilla*. Part I–National infection. Fish Pathology 12: 163-170.

Miyazaki, T., and S. Egusa. 1976. Histopathological studies on *Edwardsiella tarda* infection of the Japanese eel. Part I–National infection–suppurative interstitial nephritis. Fish Pathology 11: 33-43.

Miyazaki, T., and N. Kage. 1985. A histopathological study on motile aeromonad disease of crucian carp. Fish Pathology 21: 181-185.

Muroga, K. 1997. Recent advances in infectious diseases of marine fish with particular reference to the case in Japan. Pages 21-31 *in* T.W. Flegel, and I.H. MacRae, eds. Diseases in Asian Aquaculture III. Asian Fisheries Society, Manila, The Philippines.

Muroga, K., Y. Jo, and K. Masumura. 1986. *Vibrio ordalii* isolated from diseased ayu (*Plecoglossus altivelis*) and rockfish (*Sebastes schlegeli*). Fish Pathology 21:239-243.

Oshima, S., J. Hata, C. Segawa, N. Hirasawa, and S. Yamashita. 1996. A method for direct DNA amplification of uncharacterised DNA viruses and for development of a viral polymerase chain reaction assay: Application to red seabream iridovirus. Analytical Biochemistry 242: 15-19.

Otta, S.K., G. Shubha, B. Joseph, A. Chakraborty, I. Karunasagar, and I. Karunasagar. 1999. Polymerase chain reaction (PCR) detection of whitespot syndrome virus (WSSV) in cultured and wild crustaceans in India. Diseases of Aquatic Organisms 38: 67-70.

Pai, R. 1992. Pathogenicity of Fish Pond Associated *Aeromonas hydrophila* in the Common Carp. Doctoral dissertation, University of Agricultural Sciences, Bangalore, India.

Pal, J., and K. Pradhan. 1990. Bacterial involvement in ulcerative condition of air-breathing fish from India. Journal of Fish Biology 36:833-839.

Reddy, T.V., K. Ravindranath, P.K. Sreeraman, and M.V.S. Rao. 1994 *Aeromonas salmonicida* associated with mass mortality of *Cyprinus carpio* and *Oreochromis mossambicus* in a freshwater reservoir in Andhra Pradesh, India. Journal of Aquaculture in the Tropics 9: 259-268.

Rimstaad, E., R. Krona, E. Hornes, O. Olsvik, and B. Hyllseth. 1990. Detection of infectious pancreatic necrosis virus (IPNV) RNA by hybridisation with an oligo probe. Vetenary Microbiology 23:211-219.

Roberts, R.J., G.N. Frerichs, and S.D. Millar. 1992. Epizootic ulcerative syndrome–the current position. Pages 413-416 *in* M. Shariff, J.R. Arthur, and R. Subasinghe, eds. Diseases in Asian Aquaculture II. Asian Fisheries Society. Malaysia.

Roberts, R.J., L.G. Willoughby, and S. Chinabut. 1993. Mycotic aspects of epizootic ulcerative syndrome (EUS) of Asian fishes. Journal of Fish Diseases 16:169-183.

Roberts, R.J., G.N. Frerichs, K. Tonguthai, and S. Chinabut. 1994. Epizootic ulcerative syndrome of farmed and wild fishes. Pages 207-239 *in* J.F. Muir, and R.J. Roberts, eds. Recent Advances in Aquaculture. Vol. 5. Blackwell Science Ltd., Oxford, England.

Rodriguez, S., M.P. Vilas, M. Alonso, and S.I. Perez. 1995. Study of viral dual infection in rainbow trout (*Onchorhynchus mykiss*) by sero-neutralisation, western blot and polymerase chain reaction assays. Microbiologia 11: 461-470.

Ruangpan, L., T. Kitao, and T. Yoshida. 1986. Protective efficacy of *Aeromonas hydrophila* vaccine in Nile tilapia. Veterinary Immunology and Immunopathology 12: 345-350.

Saitanu, K., S. Wongsawang, B. Sunyascotcharee, and S. Sahaphong. 1986. Snakehead fish virus isolation and pathogenicity studies. Pages 327-330 *in* J.L. Maclean, L.B. Dizon, and L.V. Hosillos, eds. The First Asian Fisheries Forum. Asian Fisheries Society, Manila, The Philippines.

Shah, K.L., and S.C. Tyagi. 1986. An eye disease in silver carp *Hypophthalmichthys molitrix* held in tropical ponds associated with the bacterium *Staphylococcus aureus*. Aquaculture 55: 1-4.

Sharma, M., R.C. Katoch, K.B. Nagal, D.S. Sambyal, and R.K. Asrani. 1995. Isolation of *Yersinia ruckeri* from rainbow trout suffering from enteric red mouth disease in Himachal Pradesh, India. Journal of Aquaculture in the Tropics 10: 73-77.

Snieszko, S.F., and N.R. Axelrod. 1971. Diseases of Fishes Vol. 2A: Bacterial Diseases of Fishes, TFH Publications, Neptune, New Jersey.

Srivastava, R.C. 1979. Aphanomycosis–a new threat to fish population. Mykosen 22:25-29.

Subramanian, S., M. Chew-lim, S.Y. Chong, J. Howe, G.H. Ngoh, and Y.C. Chan. 1993. Molecular and electron microscopic studies of infectious pancreatic necrosis virus from snakehead. Abstracts of IXth International Congress of Virology, Glasgow, United Kingdom.

Talaat, A.M., R. Reimschussel, and M. Trucksis. 1997. Identification of mycobacteria infecting fish to the species level using polymerase chain reaction and restriction enzyme analysis. Veterinary Microbiology 58: 229-237.

Tamai, T., K. Tsujimura, S. Shirahata, H. Oda, T. Noguchi, R. Kusuda, N. Sato, S. Kimura, Y. Katakura, and H. Murakami. 1997. Development of DNA diagnostic methods for the detection of new fish iridoviral diseases. Cytotechnology 23: 211-220

Thune, R.L., L.A. Stanley, and R.K. Cooper. 1993. Pathogenesis of gram-negative bacterial infections in warm water fish. Annual Review of Fish Diseases 3:145-185.

Tonguthai, K. 1985. A Preliminary Account of Ulcerative Diseases in the Indo-Pacific Region. National Inland Fisheries Institute, Bangkok, Thailand.

Toranzo, A.E., J.L. Barja, S.A. Poter, R.R. Colwell, F.M. Hetrich, and J.H. Crosa. 1983. Molecular factors associated with virulence of marine vibrios isolated from Chesapeake Bay. Infection and Immunity 39: 1220-1227.

Toranzo, A.E., Y. Santos, and J.L. Barja. 1997. Immunisation with bacterial antigens: vibrio infections. Pages 93-105 *in* R. Gudding, A. Lillehaug, P.J. Midtlyng, and F. Brown, eds. Developments in Biological Standardization.

Torres, J.L., M. Shariff, and A.T. Law. 1990. Identification and virulence screening of *Aeromonas* spp. from healthy and epizootic ulcerative syndrome (EUS)-infected fish. Pages 663-666 *in* R. Hirano, and I. Hanyu, eds. The Second Asian Fisheries Forum. Asian Fisheries Society, Manila, The Philippines.

Wakabayashi, H., and S. Egusa. 1972. Characteristics of a *Pseudomonas* sp. from an epizootic of pond cultured eels (*Anguilla japonica*). Bulletin of the Japanese Society of Scientific Fisheries 38: 577-587.

Wakabayashi, H., K. Kanai, T.C. Hsu, and S. Egusa. 1981. Pathogenic activities of *Aeromonas hydrophila* on fishes. Fish Pathology 15: 319-325.

Wakabayashi, H., S. Egusa, and J.L. Fryer. 1980. Characteristics of filamentous bacteria isolated from the gills of salmonids. Canadian Journal of Fisheries and Aquatic Sciences 37: 1499-1504.

Wakabayashi, H., M. Hikida, and K. Masumura. 1984. *Flexibacter* infection in cultured marine fish in Japan. Helgolander Meeresuntersuchungen 37: 587-593.

Wattanavijarn, W., J. Tangtrongpiros, and S. Wattanodorn, S. 1988. Viruses of ulcerative diseased fish in Thailand, Burma and Lao P.D.R. Page 121 *in* Abstracts First International Conference on the Impact of Viral Diseases on Development of Asian Countries, World Health Organization, New York, New York.

Willoughby, L.G., R.J. Roberts, and S. Chinabut. 1995. *Aphanomysis invaderis* spp. nov.: The fungal pathogen of freshwater tropical fish affected by epizootic ulcerative syndrome. Journal of Fish Diseases 18: 273-275.

Grow-Out Production of Carps in India

S. Ayyappan
J. K. Jena

SUMMARY. Asian aquaculture has been contributing in great measure to the global fish basket. The Indian subcontinent, with a rich biodiversity of fish species, has emerged as an important aquaculture country, particularly in the freshwater environment. Carps form the mainstay of culture practices in the country, supported by a strong traditional knowledge base and scientific input in various aspects of biology, environment, nutrition, and health management. New species and culture systems, integration with other farming systems, use of organic material as nutrient inputs, and depuration measures in waste-fed culture systems are being considered for enhancing aquaculture productivity. This article presents the status of grow-out production of carps in India and traces the growth of the practices over the decades as well as the potential of the sector. *[Article copies available for a fee from The Haworth Document Delivery Service: 1-800-HAWORTH. E-mail address: <getinfo@haworthpressinc.com> Website: <http://www.HaworthPress.com>*

KEYWORDS. Carp, grow-out culture, India

S. Ayyappan, Central Institute of Fisheries Education, Seven Bunglows, Versova, Mumbai-400 061, India.

J. K. Jena, Central Institute of Freshwater Aquaculture, Kausalyaganga, Bhubaneswar-751 002, India.

[Haworth co-indexing entry note]: "Grow-Out Production of Carps in India." Ayyappan, S., and J. K. Jena. Co-published simultaneously in *Journal of Applied Aquaculture* (Food Products Press, an imprint of The Haworth Press, Inc.) Vol. 13, No. 3/4, 2003, pp. 251-282; and: *Sustainable Aquaculture: Global Perspectives* (ed: B. B. Jana, and Carl D. Webster) Food Products Press, an imprint of The Haworth Press, Inc., 2003, pp. 251-282. Single or multiple copies of this article are available for a fee from The Haworth Document Delivery Service [1-800-HAWORTH, 9:00 a.m. - 5:00 p.m. (EST). E-mail address: getinfo@haworthpressinc.com].

INTRODUCTION

Carp culture is based on management practices developed over the years following intensive research utilizing different agro-ecological conditions. The production of carps through semi-intensive and intensive polyculture systems results from stocking the appropriate size of fingerlings at desirable densities and species combinations, along with other factors, such as fertilization, supplementary feeding, aeration, and water exchange. Fish culture has a long history in India, and the species of Indian major carps, such as the catla, *Catla catla*; rohu, *Labeo rohita*; and mrigal, *Cirrhinus mrigala*, in pond culture have been known to the farmers for many years (Chaudhuri et al. 1974). However, the production from these systems remained significantly low at 600 kg/ha/year (Banerjea 1967; Jhingran 1969) until the introduction of scientific composite carp culture technology during the 1970s. Further, the experiences gained over the years through experimentation have led to the gradual increase of fish production, particularly in ponds, with contributions of both Indian major carps and exotic carps. In 1996, national production reached 2 tons/ha/year (GOI 1996), showing enormous potential.

Carps form the major component of aquaculture production in the country and contribute as much as 87% of the total aquaculture production of 1.768 million tons. Among the different groups of fish species cultured worldwide, carps and other cyprinids form the most dominant group, with production levels of 11.5 million tons during 1996, comprising over 50.6% of the total aquaculture production of finfish and shellfish (FAO 1998). Besides the three Indian major carp species, other important medium and minor carp species being cultured in certain parts of the country, though on a smaller scale, are kalbasu, *Labeo calbasu*; bata, *L. bata*; fringed lipped carp, *L. fimbriatus*; reba *Cirrhinus reba*; and mola, *Amblypharyngodon mola*. Further, the country also possesses several other potential candidate carp species, endemic to different regions of the country, that could be cultured; the most important two species are pulchellus, *Puntius pulchelus*, and Cauvery carp, *Cirrhinus cirrhosa*.

The exotic species, such as silver carp, *Hypophthalmichthys molitrix*, and grass carp, *Ctenopharyngodon idella*, introduced during 1959 (Alikunhi and Sukumaran 1964) to fill the two important niches of consuming phytoplankton and macrovegetation, respectively, along with the indigenous carp species, have also contributed significantly to enhancing the yield rates from fish ponds. Further, the Bangkok-strain of common carp, *Cyprinus carpio* var. *communis*, introduced during 1957, has influenced aquaculture in India. The two components of Indian and exotic carps have established a compatible and wholesome aquaculture system, and the combinations are used in diverse habitats and practices such as ponds/tanks, paddy fields, sewage-fed waters, cages, pens, and flow-through and other industrial aquaculture systems (Ayyappan and Jena 1998).

Almost all freshwater aquaculture production in India is pond-based, and the different systems of culture practices include extensive and semi-intensive carp polyculture, sewage-fed fish culture, and integrated fish farming. Much literature is available on various aspects of carp culture in the country, especially on static pond culture. However, the information on other non-conventional carp culture systems, such as culture in cages, pens, raceways, and flow-through systems, are sparse. This paper reviews the available literature on carp farming systems and practices of the country that provides an insight into the present status of carp farming.

CARP CULTURE IN STATIC PONDS/TANKS

Stocking Density, Stocking Combination, and Biomass Production of Carps

Aquatic resources in the form of ponds and tanks spread over 2.36 million ha in India, of which only 0.8-0.9 million ha are presently utilized for aquaculture. Freshwater aquaculture, mainly confined to carp culture, has evolved from the stage of a household activity in small, backyard ponds in the eastern Indian states of West Bengal and Orissa to that of an industry worth about US$2 billion in recent years. Stocking densities and species combinations are important issues to be decided, depending on the level or intensity of operation and the carrying capacity of the system. The rates of stocking are further decided based on the expected growth increments of individual fish and survival rates. The standard recommendation for carp culture in India involves either three species of Indian major carps (catla, rohu, and mrigal) or six carp species comprising both Indian major carps and exotic carps (silver carp, grass carp, and common carp). A stocking density of 4,000-10,000 fingerlings/ha is generally advocated in such a system (Nandeesha 1993). Preliminary studies conducted with Indian major carps in grow-out systems used stocking levels of 1,800-12,355 fish/ha (Alikunhi

1957; Hora and Pillay 1962). Experiments on mixed fish culture with Indian major carps and exotic carps conducted by Alikunhi et al. (1971) during 1962-63 recorded production levels ranging from 1,000-4,900 kg/ha/year, with the stocking density of 690-15,000/ha. They further suggested that for a production level of 3,000-3,500 kg/ha/year under regular manuring and/or artificial feeding, the stocking density is to be maintained at 3,000-3,500 fingerlings/ha. Later, experiments conducted at different laboratories with both Indian major carps and exotic carps recommended stocking densities of 4,000-6,000 fingerlings/ha for optimum production (Lakshmanan et al. 1971; Sukumaran et al. 1972). Various stocking densities ranging from 3,000 fingerlings/ha (Alikhuni et al. 1971) to 13,320 fingerlings/ha (Chaudhuri et al. 1978) were tried in different trials for maximizing the production levels, though not always resulting in higher rates of production.

In composite fish farming (Tables 1 and 2) wherein a number of compatible species are grown together, it is necessary to group the species initially according to identical or overlapping feeding habits and thereafter to decide the best ratio for each group (Lakshmanan et al. 1971). For mixed farming in India, the stocking combination of catla, rohu, and mrigal of 4:3:3 has generally been accepted as the most advantageous. While suggesting a tentative ratio of 3:3:4, Alikunhi (1957) remarked that fish farmers could use their discretion in choosing different combinations of Indian major carps and exotic carps; the ratios sometimes are decided based on consumers' preference or market demand. For example, in the state of Andhra Pradesh, the Indian major carps are the principal species cultivated at densities of 1,500-2,500/ha stocked at a ratio of 2:7:1, with a predominance of rohu (Jhingran 1991).

Experimenting with catla, rohu, and mrigal (3.5:3.5:3.0) at a stocking density of 7,500/ha, Das et al. (1977) recorded gross and net yields of 5,565 kg/ha and 5,174 kg/ha in a culture period of 16 months, with provision of diet inputs for only 12 months. The supplementary diet provided was limited to a mixture of rice bran and mustard oil cake fed at 1-2% of standing biomass daily. Taking only supplementary diet into consideration, the conversion ratio of diet to fish was 2.6:1. Further, with these three species, Das et al. (1979) demonstrated similar production levels (5.2 tons/ha/13 months) in a pond in lower Sundarbans, West Bengal. However, stocking with six species in combination (catla, rohu, mrigal, silver carp, grass carp, and common carp) at a ratio of 1:3:1.25:2.5:1:1.25, Das et al. (1975) recorded higher growth of all carp species at a stocking density of 6,600/ha in the Nilganj center of West Bengal. Fed with a mixture of rice bran and mustard oil cake in an equal ratio as a supplementary diet, the gross and net production levels obtained were 5,253 kg and 5,142 kg/ha/year, respectively. Stocking six carp species, Rao and Raju (1989) demonstrated production levels of 4,185-5,009 kg/ha/year in 3.5 ha ponds, where fish were fed a supplementary diet of equal parts of peanut oil cake and rice bran along with regular fertilization with both organic and inorganic fertilizers.

Experiments conducted at Kulia Fish Farm, Kalyani, by Sinha and Saha (1980) showed gross and net production levels of 4,636 and 4,483 kg/ha/year, respectively, when ponds were stocked with seven carp species (catla, 9.35%; rohu, 14.1%; mrigal, 23.55%; silver carp, 4.75%; grass carp, 13.95%; common carp, 28.9%; silver barb, *Puntius gonionotus*, 6.5%) at a density of 5,350 fingerlings/ha. Though the fish were fed exclusively on a commercial diet, the recorded feed conversion ratio was quite high at 3.9:1. However, the cost of that diet was much cheaper (INR 0.57/kg) than a mixture of mustard oil cake and rice bran (INR 1.05/kg). Experiments (Sinha and Saha 1980) showed higher growth of all exotic carp species over those of Indian major carps.

Studies carried out by Singh et al. (1972) with only exotic carp species suggested that the ratios of silver carp, grass carp, and common carp should be 3:1:2 or 4:2:3 for grow-out and 4:3:3 for fingerlings at optimum production levels. Stocking these three carp species in the ratio of 4:2:3 resulted in gross and net production levels of 2,090 and 2,723 kg/ha/year, respectively, when stocked at a combined density of 3,700 fingerlings/ha. Similarly, other trials with the above three species at ratio of 3:1:2 at 5,000/ha resulted in gross and net production rates of 3,098 and 3,022 kg/ha/year, respectively. The contributions of silver carp in these two experimental trials, however, were as high as 67% and 71% when stocked at 40% and 30% of the total fish composition, respectively. Though encouraging production in the former trials was obtained without feeding a diet other than aquatic weeds for grass carp, no significant increase in production was recorded in later trials where fish were fed a supplementary diet, attributed to a low percentage of recovery of grass carp. Similar results of low survival levels in grass carp as compared to other species were also reported by several other researchers (Hickling 1971; Lakshmanan et al. 1971; Aravindakshan et al. 1997; Jena et al. 1998a).

Composite carp culture trials undertaken at Kulia Experimental Fish Farm at Cuttack over a duration of two years demonstrated production levels of 3,889-5,600 kg/ha/year during the first year, which rose to an average of 8,200 kg/ha/year with a maximum of 9,389 kg/ha/year in the second year (Chaudhuri et al. 1975). Further, experiments conducted by

TABLE 1. Experiments on carp polyculture trials with fertilization and supplementary feeding.

Culture practices	Stocking details	Management measures	Production details	Reference
Composite carp culture with both Indian major carps and exotic carp species	Stocking density: 5,000-6,250 fingerling/ha Stocking ratio: catla 1.0-1.5, rohu 2.0-3.0, mrigal 1.25, silver carp 2.5-3.0, grass carp 0.75-1.0 and common carp 1.25	Culture period: One year Fertilization: Cattle dung at 20.5-25.0 tons with ammonium sulphate, super phosphate and calcium ammonium nitrate at combined doses of 975-1,725 kg/ha/year Supplementary feeding: Mustard oil cake and rice bran at 1:1 ratio by weight	Survival: 74.3-88.8% Gross production (GP): 2.58- 4.21 tons/ha/year Net production (NP): 2.35-3.97 tons/ha/year	Lakshmanan et al. (1971)
Composite carp culture with both Indian major carps and exotic carp species	Stocking density: 5,473/ha Stocking ratio: catla 1.76: rohu 2.74: mrigal 1.15: silver carp 2.29: grass carp 0.91: common carp 1.15	Culture period: 6 months Fertilization: Cattle dung 5 tons with ammonium sulphate 450 kg, single super phosphate 250 kg, and muriate of potash 40 kg/ha/year Supplementary feeding: peanut oil cake and rice bran at 1:1 ratio by weight	Mean survival: 91.5% GP: 3.23 tons/ha/6 month NP: 2.92 tons/ha/6 months	Sinha et al. (1973)
Composite carp culture with six species	Stocking density: 10,540 /ha Stocking ratio: catla 1: rohu 3: mrigal 1: silver carp 2: grass carp 1: common carp 2	Culture period: One year Fertilization: Cattle dung 14.4 tons/ha/year along with urea and triple super phosphate at 1,530 kg/ha/year Supplementary feeding: peanut oil cake and rice bran at 1:1 ratio by weight	GP: 5,734-7,500 kg/ha/year NP: 5,652-7,343 kg/ha/year	Chaudhuri et al. (1974)

Composite carp culture with six species	<u>Experiment I</u> Stocking density: 5,000/ha Stocking ratio: catla 1: rohu 3: mrigal 2: silver carp 2: grass carp 1: common carp 1 Certain number of carp hybrids and *Mugil cephalus* introduced <u>Experiment II</u> Stocking density: 7,500/ha Stocking ratio: catla 1: rohu 2.5: mrigal 1: silver carp 2.5: grass carp 1: common carp 2 Certain number of carp hybrids, *Mugil cephalus*, and *Notopterus chitala* introduced	<u>Experiment I</u> Culture period: One year Fertilization: Urea 72 kg, triple super phosphate 55 kg and cattle dung 3.7-4.6 tons/ha/year Supplementary feeding: peanut oil cake and rice bran at 1:1 ratio <u>Experiment II</u> Culture period: 287 days-one year Fertilization: Urea, triple super phosphate and ammonium sulphate at combined dose of 396-738 kg/ha/year and cattle dung at 3-10 tons/ha/year Supplementary feeding: peanut oil cake and rice bran at 1:1 ratio by weight	<u>Experiment I</u> GP: 3,889-5,604 kg/ha/year NP: 3,770-5,442 kg/ha/year <u>Experiment II</u> GP: 7,409 kg/ha/287 days–9,389 kg/ha/year NP: 7,186 kg/ha/287 days–9,088 kg/ha/year	Chaudhuri et al. (1975)
Carp polyculture with seven species	Stocking density: 5,350/ha Stocking ratio: catla 9.4: rohu 14.1: mrigal 23.3: silver carp 4.7: grass carp 14: common carp 28: *Puntius gonionotus* 6.5	Culture period: One year Fertilization: Urea 200 kg, SSP 250 kg, and cattle dung 15 tons/ha/year Supplementary feeding: Commercial epic fish feed	Survival: 96.8% GP: 4,636 kg/ha/year NP: 4,483 kg/ha/year FCR: 3.9:1	Sinha and Saha (1980)
Carp polyculture	Stocking density: 5,000/ha <u>Treatment I:</u> Stocking ratio: catla 29: rohu 36: mrigal 14: common carp 21 <u>Treatment II:</u> Stocking ratio: catla 12: rohu 24: mrigal 12: silver carp 27 : grass carp 12: common carp 13	Culture period: One year Fertilization: Poultry manure 2.7 tons/ha, ammonium sulphate 761 kg, and single super phosphate 340 kg/ha/year Supplementary feed: Rice bran and peanut oil cake at 1:1 ratio by weight	<u>Treatment I:</u> Survival: 60.4% GP: 4,186 kg/ha/year NP: 4,185 kg/ha/year <u>Treatment !I:</u> Survival: 62.7% GP: 5,011 kg/ha/year NP: 5,010 kg/ha/year	Rao and Raju (1989)

TABLE 1 (continued)

Culture practices	Stocking details	Management measures	Production details	Reference
Carp polyculture with three and six species combination	<u>Treatment I:</u> Stocking density: 5,000/ha Mean stocking size: 7.8 g Stocking ratio: catla 3: rohu 4: mrigal 3 <u>Treatment II:</u> Stocking density: 5,000/ha Stocking ratio: catla 2: rohu 2.5: mrigal 2: silver carp 1.5: grass carp 0.5: common carp 1.5 <u>Treatment III:</u> Stocking density: 10,000/ha Stocking ratio: Same as Treatment I <u>Treatment IV:</u> Stocking density: 10,000/ha Stocking ratio: Same as Treatment II	Culture period: 6 months Fertilization: Cattle dung 15 tons/ha/year; urea 200 kg and single super phosphate 300 kg/ha/year Supplementary feeding: Mixture of peanut oil cake and rice bran at 1:1 ratio at 2% of fish biomass/day in all treatments. Provision of aquatic weed in Treatments II and IV Water depth: 1.4-1.5 m	<u>Treatment I:</u> Survival: 84.7% Mean size of harvest: 432 g GP: 1,831 kg/ha/6 months NP: 1,791 kg/ha/6 months FCR: 1.49 <u>Treatment II:</u> Survival: 86.2% Mean size of harvest: 567 g GP: 2,442 kg/ha/6 months NP: 2,385 kg/ha/6 months FCR: 1.47 <u>Treatment III:</u> Survival: 82.1% Mean size of harvest: 318 g GP: 2,611 kg/ha/6 months NP: 2,532 kg/ha/6 months FCR: 1.69 <u>Treatment IV:</u> Survival: 80.4% Mean size of harvest: 444 g GP: 3,565 kg/ha/6 months NP: 3,472 kg/ha/6 months FCR: 1.67	Jena (1998)

Composite carp culture with both Indian major carps and exotic carp species	Stocking density: 10,000/ha Stocking ratio: catla 2: rohu 2: mrigal 2: silver carp 2: grass carp 0.5: common carp 1.5	Culture period: One year Three treatments were single cropping for one year (Treatment I), single stocking and multiple harvesting (Treatment II) and multiple cropping, i.e., two crops of six months each in one year (Treatment III) Fertilization: Cattle dung 15 tons/ha, urea 200 kg and single super phosphate 300 kg/ha/year Supplementary feeding: peanut oil cake and rice bran at 1:1 ratio by weight	<u>Treatment I</u> Mean growth: 680 g Mean survival: 88.4% GP: 6,011kg/ha/year NP: 5,844 kg/ha/year FCR: 3.16 <u>Treatment II</u> Mean growth: 738 g Mean survival: 88.0% GP: 6,488 kg/ha/year NP: 6,320 kg/ha/year FCR: 2.53 <u>Treatment III</u> Mean growth: 858 g Mean survival: 85.5% GP: 7,195 kg/ha/year NP: 6,828 kg/ha/year FCR: 1.67	Jena et al. (2000)
Intensive carp polyculture	<u>Experiment I:</u> Stocking density: 25,000/ha Mean stocking size: 98-125 g in different ponds Stocking ratio: catla 3.5: rohu 3: mrigal 3: grass carp 0.5 <u>Experiment II:</u> Stocking density: 25,000/h Mean stocking size: 111-122 g in different ponds Stocking ratio: catla 2.5: rohu 2.5: mrigal 2.5: silver carp 2: grass carp 0.4: kalbasu 0.1	Culture period: One year Fertilization: *Azolla* at 40 tons/ha/year as bifertilizers at weekly intervals; lime (CaO) at 200 kg/ha/year, and single super phosphate at 100 kg/ha/year Supplementary feeding: feed mixture comprised rice bran 38%, peanut oil cake 35%, soybean meal 20%, fish meal 5%, and vitamin and mineral mixture 2% Provision of aquatic weed for grass carp at *ad libitum* Aeration: 4-6 hours/day.	<u>Experiment I:</u> Mean survival: 88.0% in different ponds Mean size of harvest: 548 g GP: 15,666 kg/ha/year NP: 11,987 kg/ha/year FCR: 2.38 <u>Experiment II:</u> Mean survival: 93.5% Mean size of harvest: 661 g GP: 15,463 kg/ha/year NP: 12,544 kg/ha/year FCR: 2.58	Tripathi et al. (2000)

TABLE 2. Experiments on carp culture trials without provision of supplementary diet.

Culture practices	Stocking details	Management measures	Production details	Reference
Carp polyculture with seven species	Stocking density: 4,450 fingerlings/ha Stocking ratio: catla, 8.5%; rohu, 28%; mrigal, 5.5%; silver carp, 22.5%; grass carp, 17%; common carp, 17%; and *Osphronemus gorami*, 1.5%	Culture period: One year Fertilization: Cattle dung 25 tons/ha/year, and super phosphate 113 kg/ha/year Aquatic weed: 4.4-14.4 tons/ha/year	Survival: 74.3% GP: 2,228-3,638 kg/ha/year NP: 2,028-2,436 kg/ha/year	Lakshmanan et al. (1971)
Polyculture of silver carp, grass carp, and common carp	Stocking density: 3,700 fingerlings/ha Stocking size: 24-69 g Stocking ratio: silver carp, 40%; common carp, 30%; and grass carp, 30%	Culture period: One year Fertilization: Ammonium sulphate 692 kg/ha, and single super phosphate 650 kg/ha Aquatic weed: 17 tons/ha/year	Survival: 88.4% GP: 2,909 kg/ha/year NP: 2,723 kg/ha/year	Singh et al. (1972)
Composite culture of both Indian major carps and exotic carp species	Stocking density: 6,000 fingerlings/ha	Culture period: One year Fertilization with both organic (cattle dung) and inorganic fertilizers (urea, triple super phosphate, and calcium ammonium nitrate)	Experiment I: Survival: 67.3% Size of harvest: 508-1,081 g GP: 3,341 kg/ha/year Experiment II: Survival: 64% Size of harvest: 522-1,918 g GP: 2,588 kg/ha/year	Chakrabarty et al. (1979)

Composite culture of both Indian major carps and exotic carp species	Stocking density: 7,500 fingerlings/ha	Culture period: One year Treatment I: Application of chemical fertilizers (urea, super phosphate, and potassium chloride) at 1,020 kg/ha/year Treatment II: Application of chemical fertilizers (as Treatment I) at 650 kg/ha/year and organic manure (cattle dung) at 10 tons/ha/year	Treatment I: Survival: 78.3% GP: 4,297 kg/ha/year NP: 4,221 kg/ha/year Treatment II: Survival: 60.3% GP: 3,352 kg/ha/year NP: 3,276 kg/ha/year	Saha et al. (1979)
Composite carp culture with six species	Stocking density: 5,000 fingerlings/ha Stocking size: 1-14.51 g Stocking ratio: catla, 10%; rohu, 10%; mrigal, 10%; silver carp, 30%; grass carp, 20%; and common carp, 20%	Culture period: 9 months Treatment I: Fertilization: Super phosphate 713-756 kg/ha Aquatic weed for grass carp 119 tons/ha Treatment II: Fertilization: Urea 225-270 kg/ha; cattle dung 6,000-6,720 kg/ha Aquatic weed for grass carp: 119 tons/ha	Treatment I: Survival: 95.4-98.0% Size of harvest: 235-1,278 g GP: 2,609-333kg/ha/9 months Treatment II: Survival: 95.0-97.0% Size of harvest: 330-1,609 g GP: 3,642-3,966 kg/ha/9 months	Tripathi and Mishra (1986)
Carp polyculture with catla, rohu, mrigal, and silver carp	Stocking density: 10,000-30,000 fingerlings/ha Mean size of stocking: 10 g Stocking ratio: catla, 20%; rohu, 30%; mrigal, 30%; and silver carp, 20%	Culture period: One year Fertilization: Cattle dung 15 tons/ha/year; urea 100 kg N (217 kg); and single super phosphate 50 kg P (312.5 kg) Water depth: 1.4-1.5 m	Survival: 62-82% Size of harvest: 125-244 g GP: 2,010-2,225 kg/ha/year NP: 1,900-1,925 kg/ha/year	Jena (1998)

Chaudhuri showed production levels of 10,390 kg/ha in a year in Burma by culturing the Indian major carps together with small numbers of carp hybrids and gourami, *Osphronemus gorami* (FAO 1971). Studies on intensive carp culture carried out for a period of five years, 1989-1994, through a series of experimental trials at the Central Institute of Freshwater Aquaculture, Bhubaneswar, India, demonstrated production levels of over 10 tons/ha/year in all the experiments (CIFA 1998) and recorded the highest national production, 17.3 tons/ha/year (Tripathi et al. 2000). The stocking densities during the studies were 15,000-35,000 fingerlings/ha. While production levels of over 15 tons/ha/year were registered at a stocking density of 25,000/ha in three successive experimental trials, a higher stocking density of 35,000/ha resulted decrease in production (CIFA 1998). Further, the other important aspect of the study was the higher percentage of Indian major carps in the species stocked, 75-95%.

Lin (1955) has given details of polyculture practices with several carp species in China. It is observed that Chinese carp culture is mainly organic-based. Zhu et al. (1990) reported an increased net yield of 10.2 kg/ha/day in Chinese integrated farm ponds as a result of manure application, while Schoreder (1978) reported yields of up to 30 kg/ha/day in Israel in polyculture of common carp with tilapia through intensive manuring. Introduction of Asiatic carps has made a breakthrough through in many countries. In Vietnam, it has led in average production of 2-3 tons/ha/year in family ponds and 7-8 tons/ha/year in well-managed ponds (Singh 1990). Though Chinese carp culture has been practiced in Malaysia and Thailand traditionally, the use of grass carp with tawes, *Punitus gonionotus*, and Mossmbique tilapia, in subsequent years was successful (Sinha 1979). Carp culture in Bangladesh obviously uses a similar ecosystem. Islam and Dewan (1986) in their study of pond fish culture recorded annual production ranging between 1,700 and 3,889 kg/ha, obtained through semi-intensive culture of Indian major carps, as well as Chinese and common carps. However, Hasan (1990) reported that the average production of Bangladesh is as low as 940 kg/ha.

Biogas slurry has been observed to be one of the potential manurial inputs for enhancing growth of plankton, rather than raw cattle dung, and resulted in higher carp production levels (Kalyani and Shetty 1987). Tripathi et al. (1994) evaluated the use of water hyacinth-based biogas slurry on carp production and observed a production of 2,227-3,270 kg/ha in a 10-month growing period with an application rate of 15-20 tons/ha/year. Further, evaluation of biogas slurry at different rates showed an application level of 30 tons/ha/year to be optimal (Tripathi et al.

1991, 1992) with fish production levels of 1,377-2,487 kg/ha/year. Experiments conducted with sole application of biogas slurry at 5-15 tons/ha/year has recorded yields of 1,956-2,096 kg/ha/year without supplementary diet, but with supplementary diet, production was 5,470-7,230 kg/ha/year (Prasad et al. 1993).

It has been observed that a large share of the available freshwater area of India is infested with aquatic weeds, which are posing great concerns for judicious utilization of pond sites. Studies have shown that some of the submerged weeds such as *Hydrilla*, *Najas*, and *Ceratophyllum*, and duckweed species such as *Spirodela*, *Lemna*, and *Wolffia*, can be effectively utilized as diet for fish species such as grass carp, *Tilapia zilli*, and *Puntius gonionotus*. Under a new system of carp culture known as weed-based culture, yields of over 4,000 kg/ha/year have been achieved when fish are stocked at a density of 4,000 fingerlings/ha with as much as 50% grass carp and the other 50% comprising five other carp species. No fertilizer was used except aquatic/terrestrial vegetation applied for grass carp (Tripathi and Mishra 1986). Under similar experimental trials carried out with grass carp as the major species, Aravindakshan et al. (1999) recorded production of 2,407-2,517 kg/ha/year. Stocked at a density of 4,000 fingerlings/ha with grass carp at 40-50%, and the remaining percentage comprising other carp species, the experiments did not utilize any other inputs except aquatic weeds like *Ceratophyllum* and *Najas*, provided at regular intervals as food for grass carp; fertilization was limited to single super phosphate at a rate of 100 kg/ha/year and lime (CaO) at 250 kg/ha/year at periodical intervals.

The technology of scientific carp culture in India was developed at the Pond Culture Division of the Central Inland Fisheries Research Institute, Cuttack, Orissa; this technology was disseminated to different agro-climatic zones and virtually refined through the work carried out at different centers of the All India Co-Ordinated Research Project (AICRP) on Composite Culture of Indian and Exotic carps (redesigned later as AICRP on Composite Fish Culture and Fish Seed Production) initiated in 1971 by the Indian Council of Agricultural Research, New Delhi. In the initial experiments conducted at Kulia Fish Farm, West Bengal, the gross and net yields of fish were as much as 3,232 kg and 2,917 kg/ha/6 months, respectively (Sinha et al. 1973). Further, studies from all six AICRP centers revealed initial production ranging from 2,436-6,522 kg/ha/year through fertilization and supplementary feeding at various agro-climatic zones (Jhingran 1991). Later, experiments conducted in these centers achieved varied levels of production showing the potential of composite carp culture technology.

At the Pune Center in Maharashtra, culture with a six-species combination recorded production levels of 2,199-2,242 kg/ha/8 months in 1972, to as high as 10,148 kg/ha/year in 1979 (Mathew et al. 1979). The center had the distinction of achieving production levels of over 10 tons/ha/year for five consecutive years. The highest national production of 10,164-10,673 kg/ha/year was also achieved at this center (Rao and Singh 1982, 1984). Even at low stocking of 5,000-6,000 fingerlings/ha, the center could demonstrate high production levels of 4,197-4,355 kg/ha in 8 months and 4,053-5,327 kg/ha in 6 months in ponds using six-species combinations. Further, trials that stocked of carp fry instead of fingerlings at 6,000/ha also resulted in a spectacular production level of over 10 tons/ha/year, with all species attaining an average growth rate of over 1 kg, except common carp. Krishnamurthy et al. (1976) reported production levels of 3,448-5,894 kg/ha/6 months and 6,191-7,337 kg/ha/6 months with six-species composition from the Karnal Center at Haryana. High rates of production achieved at the Pune Center at Maharashtra and the Karnal Center at Haryana under AICRP were correlated with water intake (Tripathi 1982, 1984). Due to high seepage at Karnal, about 20% of the water loss were replaced every day through a canal and/or a tubewell. This created almost a flow-through water environment and gave consistently high yields, resulting in a record production of over 8,200 kg/ha in 8 months (Shah and Tyagi 1984). Krishnamurthy and Aravindakshan (1979) recorded gross and net production levels of 5.7 tons/ha/year and 5.2 tons/ha/year in running water ponds, provided with artificial feed.

Studies on carp polyculture with Indian major carp and exotic carps conducted at varied stocking densities and ratios, generally showed higher growth rates for silver carp, grass carp, and common carp than those for Indian major carp species (Sukumaran et al. 1972; Chaudhuri et al. 1974, 1975, 1978; Sinha and Gupta 1975; Sinha and Saha 1980; Jena 1998; Jena et al. 1998a; Tripathi et al. 2000).

Supplementary Feeding

The importance of feeding a supplementary diet for enhancing fish production has been reported by several workers (Ling 1967; Hickling 1971; Chaudhuri et al. 1975; Chakbrabarty et al. 1979a; Nandeesha 1993). Using nitrogenous fertilizers, urea and ammonia sulphate, at 186 and 386 kg/ha without diet, Murty et al. (1978) obtained a net production of 2,275 kg/ha/year. With the addition of a supplementary diet, in addition to fertilizers at the above rates, a net production of 3,859 kg/ha/year was recorded. Khan et al. (1979) conducted detailed experiments on the role of fertilizer and/or diet and registered production levels of 1,053-1,491 kg/ha/year in ponds without any treatment; 1,397-2,303 kg/ha/year in pond which received fertilizers alone; 2,901-4,470 kg/ha/year in ponds which received diet alone; and 4,414-6,535 kg/ha/year in ponds with the addition of both fertilizers and diet. Further, Singh and Singh (1975) found that the weight of fish increased three- to fourfold in 6 months when ponds were provided with both fertilizers and supplementary diet, as compared to the weights attained in ponds fertilized with cattle dung alone. Jena (1998) reported higher production of carps (stocked with catla, rohita, mrigala, and *H. molitrix)* ranging from 3,588 to 4,100 kg/ha/year with supplementary diet compared to 2,010 to 2,225 kg/ha/year in treatments without supplementary diet. The importance of a nutritious diet for carp culture has been stated, and several formulations have been used for carp fry and fingerlings (Sen et al. 1978; Singh et al. 1980; Mohanty et al. 1990; Jafri et al. 1991; Jena et al. 1996, 1998c, 1999). However, the conventional mixture of rice bran and peanut/mustard oil cake at equal proportions in dough form is still used extensively as the only form of supplementary diet in grow-out production of carps in India (Tripathi 1990). Further, since the cost of oil cake and rice bran has been increasing rapidly, farmers in the state of Andhra Pradesh are using combinations of these two ingredients based on fish growth rate and pond productivity. During the summer, when growth rates are high, fish are provided with a diet mixture consisting of 30-40% oil cake and 60-70% rice bran *ad libitum*. However, when the growth of fish is slower, particularly during the monsoon and winter seasons, fish are fed with rice bran alone or mixed with a small percentage of oil cake (Nandeesha 1993). Other ingredients are also used based on their availability and cost, mostly in addition to rice bran and peanut cake or as a substitute for the latter.

Jayaram and Shetty (1981) demonstrated that a balanced diet, even without fish meal, could induce good growth in carps. Nandeesha et al. (1986) demonstrated successful replacement of fish meal with cheaper slaughter-house waste or silkworm fecal matter in diets while culturing carps in earthen ponds. However, in the case of commercial carp culture in India, the use of animal ingredients, especially in grow-out systems, is almost non-existent, except for the use of 15-20% fish meal in the diet for broodstock production (Nandeesha 1993). Studies on intensive carp culture carried out at the Central Institute of Freshwater Aquaculture, Bhubaneswar, over a period of five years, have utilized formulated diets composed of rice bran, peanut oil cake, soybean meal, fish meal, and vitamin-mineral mixture (with the following proximate composition: moisture–9.06%; crude protein–30.27%; crude lipid–8.95%; fiber–9.55%; gross energy–3.49 kcal/g), and have reported production levels of over, 15 tons/ha/year (CIFA 1998).

In polyculture systems, supplementary diets are often given arbitrarily, though it is essential to provide an optimum amount of diet to obtain maximum conversion of diet and growth of fish species. Thus, an understanding of the relationship between the rate of feeding and the rate of conversion of the diet is of the utmost importance, as it avoids wastes.

Aeration

Retardation of growth at higher stocking densities in carps and other fish species has been explained by several researchers (Refstei and Kittelsen 1976; Trzebiatowski et al. 1980; Vijayan and Leatherland 1988; Kjartansson et al. 1988; Holm et al. 1990; Degani 1991; Bjoernesson 1994; Martinez et al. 1997; Jena et al. 1998b). The requirement of aeration in cases of higher densities has been emphasized in several reports (Boyd 1982; Mohanty 1995; Aravindakshan et al. 1997). While increasing production with the provision of aeration has been demonstrated by several workers in species like Nile tilapia, *Oreochromis niloticus*, and channel catfish, *Ictalurus punctatus* (Marek and Sarig 1971; Loyacano 1974; Rappaport and Sarig 1975), information on its effect on carp culture has been limited (Vijayan and Varghese 1986; George et al. 1990; Rout 1995; Mohanty 1995; Aravindakshan et al. 1997; Jena 1998). Raising carp juveniles for a period of 90 days, Aravindakshan et al. (1997) recorded gross production levels of 3.07-3.58 tons/ha in treatments with aeration compared to 2.45 tons/ha in the control treatment.

Intensification of culture systems requires the use of aeration (Boyd and Ahmad 1987), and the necessity of aerators at stocking densities beyond 15,000/ha has been emphasized by the Central Institute of Freshwater Aquaculture (CIFA) for increasing production levels (CIFA 1998). There are five types of mechanical aerators commonly employed in aquaculture ponds: vertical pumps, pump sprayers, paddlewheels, propeller-aspirator pumps, and diffused air systems (Boyd and Watten 1989). Of these, propeller-aspirator, pump aerator, and paddlewheel aerators are commonly used in aquaculture (Ruttangagosright et al. 1991). Rappaport et al. (1976) evaluated the efficiency of different methods of aeration with respect to fish yields and recorded highest fish yield with the spray type of surface aerator. Ruttanagosrigit et al. (1991) registered higher aeration efficiency values for the propeller-aspirator-pump aerators than for the paddle wheel aerators. Studies conducted by Boyd and Ahmad (1987) and Ahmad and Boyd (1988) showed that even the same paddlewheel aerators designed and developed at Auburn University showed 1.7 time greater aeration efficiency than the Taiwan-style paddlewheel aerators, which suggests that efficiency of aerators depends greatly on their design. Aeration in aquaculture ponds is usually provided during night hours when the oxygen contents in ponds remain low. Further, daytime aeration is usually avoided to conserve the commonly observed super-saturation of dissolved oxygen (Szyper and Lin 1990).

The obvious role of aeration is to supply oxygen to fish. In addition, aeration of water may affect a variety of other biological systems in ponds, such as microbial transformations of nitrogen or organic carbon degradation. The indirect effect of artificial aeration is the mixing of water, an effect that depends to a large extent on the device used to aerate the pond (Avinmelech et al. 1992). Aeration with minimal mixing may suffice to create an aerated zone that supports fish survival in the pond. However, only a fraction of the pond volume is aerated, while a part of the pond may remain anaerobic. On the other hand, ponds that are effectively stirred are uniformly aerated. Aerobic conditions, provided by the aeration of fish ponds are also essential for the process of nitrification, especially in the case of intensive production. A temporary build-up of nitrite may be a problem, especially in salt-free water (Colt et al. 1981; Hasan and McIntosh 1986). In non-aerated conditions, nitrification process is affected, and therefore mineralized nitrogen may be accumulated as ammonium ions (Avinmelech et al. 1992). Mixing of waters by mechanical aeration in intensive ponds minimizes the existence of non-desirable anaerobic zones in the pond, besides reducing ammonium accumulation in water.

CARP CULTURE IN CAGES AND PENS

Farming of fish in cages has been a practice for over a century in Asia, especially in countries such as Cambodia, Indonesia, and China. Further, in recent years it has been considered as a highly specialized and sophisticated modern aquaculture technique, gaining importance for intensive exploitation of water bodies, especially larger in nature, all over the world. In India, cage culture was attempted for the first time in the case of air-breathing fishes in swamps (Dehadrai et al. 1974). The studies on cage culture of different carp species carried out in India are presented in Table 3.

Attempts have also been made at replacing ground nurseries for raising fry. Initial studies conducted by Natarajan et al. (1979) and Menon (1983) showed encouraging results, with survival levels ranging 25-85% during the rearing of Indian major carp fry in floating cages.

The production from grow-out cages varies greatly depending on the type of management of inputs. Further, the number of fish that can be stocked in the cages depends on carrying capacity, water exchange, species of fish, and quality and quantity of supplemental feed input. With this high-tech system of high stocking and enriched formulated feeds, production levels of as much 35, 37.5 and 25 kg/m^3/month have been recorded with common carp in Japan, Germany, and the Netherlands, respectively. There are several types of cages including surface cages resting on the bottom, submerged, and floating (Cohche 1979). However, the floating cage has been generally accepted as the most appropriate type for fish. The size of fish cages depends on the scale of operation, species and size of fish, and infrastructural/financial/managerial resources. The size of cages varies from 150 m^3 in Indonesia to 60-180 m^3 and 40-625 m^3, respectively, in Cambodia and Vietnam (Pantulu 1979). However, the cages employed in India are small, in the ranges of 1 to 36 m^3, both cylindrical and rectangular and suitable for manual operation.

Fish culture in pens had its origin in Japan where yellowtail, *Seriola quinqueradiata* is cultured extensively, the method of which later spread to Southeast Asian countries. In India, reports on fish culture in pens either in fresh water or brackish water are scanty (Lalmohan 1983; Laal et al. 1990; Routray and Routray 1998). Further, all these studies are restricted to only production of fry and fingerlings. However, the captive culture systems such as pens possess great potential where land for construction of ponds is expensive or not available and suitable watersheds are not being fully utilized. Considering the existence of vast expanses of reservoirs, with a water area of 3 million ha, along with the requirement of larger size juveniles as stocking material for cultured-based capture fisheries, the technology of pen culture would be of great use for the development of large water bodies in the country.

TABLE 3. Grow-out production of carps through cage culture in India.

Species	Stocking details	Management measures	Production details	Reference
Catla	Cage size: 15.0 m^3 Stocking density: 49 fingerlings/m^3 Stocking size: 13 g	Culture period: 8 months Supplementary feeding: mixture of peanut oil cake and rice bran (1:1) provided at 20% of fish biomass daily initially, which was progressively reduced to 5%	Harvested size: 544 g GP: 2.7 kg/m^2/month (1.8 kg/m^3/month) FCR: 5.6	Sukumaran et al. (1986)
Catla	Cage size: 15.75 m^3 Stocking density: 13 fingerlings/m^3 Stocking size: 50 g	Culture period: 6 months Supplementary feeding: mixture of peanut oil cake, and rice bran (1:1) provided at 5-10% of fish biomass/day	Harvested size: 772 g GP: 1.41 kg/m^3/month FCR: 6.03	Govind et al. (1988)
Silver carp	Cage size: 10 m^3 Stocking density: 15 fingerlings/m^3 Stocking size: 61 g	Culture period: 157 days Supplementary feeding: mixture of rice bran, peanut oil cake, and deoiled silkworm pupae (3:2:1) provided at 3-5% of fish biomass/day	Harvested size: 472 g NP: 0.7 kg/m^3/month FCR: 3.18	Kumaraiah et al. (1991)
Grass carp	Cage size: 3 m^3 Stocking density: 33-67 fingerlings/m^3 Stocking size: 7-10 g	Culture period: 6 months Supplementary feeding: provision of only aquatic weed	Harvested size: 450 g GP: 2.0-3.3 kg/m^3/month	Bandyopadhyay et al. (1991)

INTEGRATED FISH FARMING

Integrated farming refers to a combination of practices incorporating the recycling of wastes and resources from one farming system to the other, in order to optimize production efficiencies and achieve maximum biomass harvest from unit area, with due consideration to environmental upkeep. Under the integrated fish-farming system, the waste from different component systems, such as livestock, poultry, and agriculture byproducts, is recycled as input for production of fish. Thus, integrated fish-farming involves two or more production technologies to function together. Fish-livestock farming combined with growing agricultural crops is a workable model of such integration and has been demonstrated by Chinese small-scale farming systems. The system has also been successfully demonstrated in Malaysia, Thailand, Taiwan, Hungary and several other countries (Buck et al. 1979; Delmendo 1980; Tan and Huat 1980; Chen and Yen 1980; Sharma and Olah 1986). However, little attention is paid to this type of system in India, though initial trials on the integration of pig, duck, poultry, and fish culture have given encouraging results (Sharma et al. 1979a, 1979b; Jhingran and Sharma 1980; Toor et al. 1991).

In India, cattle dung has been used extensively as the major source of organic manure for pond fertilization in carp polyculture, along with other inorganic fertilizers (Lakshmanan et al. 1971; Singh et al. 1972; Sukumaran et al. 1972; Sinha et al. 1973). In composite fish culture, while it is used normally at application level of 5,000-10,000 kg/ha/year in less productive ponds (Sukuarman et al. 1972), the quantity could be as high as 25 tons/ha/year (Lakshmanan et al. 1971). Under the system of poultry-cum-fish farming, production of 4,500 to 5,000 kg of fish, more than 70,000 eggs, and about 1,250 kg (live weight) chicken meat were produced from a one-ha pond area in one year, without the use of any supplementary diet or chemical fertilizers (Jhingran and Sharma 1980; Sharma et al. 1985). Evaluating production efficiencies with fish-duck and fish-poultry integration, Sharma and Das (1988) demonstrated higher production levels of 4,665 kg/ha/year in the latter as against 4,340 kg/ha/year in the former, when stocked with six carp species at a combined density of 6,000 fingerlings/ha. It is estimated that one ton of deep litter is produced by 25-30 birds, thus requiring 500-600 poultry birds to serve the need of one ha of aquaculture pond (Sharma and Das 1988; Jhingran 1991; Tripathi 1991).

Duck-fish integration is the most common integration, practiced mainly in countries such as China, Hungary, East Germany, Poland,

Russia, and to some extent, India, utilizing the mutually beneficial relationship between fish and ducks. The raising of ducks over fish ponds integrates very well with fish polyculture. Under the system of duck-cum-fish culture, it has been possible to obtain 3,500-4,000 kg fish, 15,000 eggs, and 500 kg duck meat from one ha of pond area in one year (Jhingran 1991). Ducks feed on organisms, such as larvae of aquatic insects, tadpoles, molluscs, and aquatic weeds, which may not be the food of stocked fish. Jhingran and Sharma (1980), Sharma et al. (1985), and Sharma and Das (1988) have dealt in detail with the technologies developed for Indian conditions. In India, mainly the ducks of egg-laying species are grown, the most important breeds being local variety 'Indian runner' and exotic Khaki campbell. It has been found out that 200-300 ducks are sufficient to produce manure adequate enough to fertilize one ha of water under polyculture (Sharma and Das 1988; Tripathi 1990).

In fish-cum-pig farming, pig farming is integrated with fish culture, where the housing units are constructed on pond embankments in such a way that the wastes and washings are drained into the fish ponds. The dung obtained from the pigs not only acts as fertilizer, but also contains 70% of digestible food for the fish, which is consumed directly. Sustained research efforts of this farming system have made this system most suitable for small and marginal farmers and have resulted in an economically viable system (Sharma et al. 1979a; Jhingran and Sharma 1980; Sharma and Olah 1986; Sharma and Das 1988; Sharma et al. 1988). According to Jhingran (1991), each fully grown pig voids between 500-600 kg feces in a year; the excreta released by 30-40 pigs is adequate to fertilize one ha of water, resulting is a yield of 6-7 tons of fish/ha/year. Stocking fingerlings of six carp species at combined density of 8,500/ha, Sharma and Das (1988) recorded production levels of 6,792 kg/ha in a culture period of one year. Evaluating the management parameters in integrated fish-cum-pig farming, Sharma and Olah (1986) recorded production rates of 18.4 kg/ha/day by recycling the pig manure in fish polyculture pond, without addition of supplementary feeding and inorganic fertilization. It may be mentioned that the fish yield rates obtained by different experimental trials through recycling of excreta of pigs, ducks, and poultry without supplementary feed and inorganic fertilizer are indeed quite impressive, which can be favorable compared with the yields obtained from fish-livestock farming in United States, Malaysia, Thailand, Philippines, Hungary and Taiwan (Buck et al. 1979; Cruz and Shehadeh 1980; Woynarovich 1980; Chen and Yen 1980).

SEWAGE-FED FISH CULTURE

The practices of the recycling sewage through agriculture, horticulture, and aquaculture have been in vogue traditionally in several countries (Edwards 1990, 1993 and 1996). Sewage-fed fish culture in *bheries* of West Bengal is also an age-old practice (Ghosh et al. 1985; Jana and Datta 1996; Jana 1998). Though the area of coverage is gradually reducing, about 5,700 ha are still utilized for raising 7,000 tons of fish annually by adding raw sewage to ponds (Saha 1970). It may be mentioned that use of sewage in fish culture is not permitted in many countries due to public health concerns. However, in many south Asian countries, this has been used as a resource alternative to conventional organic fertilizers. Ghosh et al. (1974) recorded gross production of 7,676 kg/ha in 7 months in sewage-fed ponds stocked with catla, rohu, mrigal, and silver carp at a stocking density of 50,000/ha. Sreenivasan (1968) reported maximum fish production from sewage-fed moats at Webster was 5,486 kg/ha/year. Investigations carried out in the state fish farm near Calcutta resulted in a maximum production of 3,119 kg/ha (Saha 1970). In ponds stocked with bata, *Labeo bata*, Datta et al. (1994) registered an average production of 1,270 kg/ha in 6-10 months at stocking density of 50,000/ha. Studies have shown that average fish production rates of 2,000-3,000 kg/ha/year can be achieved by using sewage alone. The important species being cultured in the system include Indian major carp species such as catla, rohu, mrigal, and kalbasu, *Labeo calbasu*, besides other medium and minor carp species such as bata; rebae, *Cirrhinus reba*; and mola, *Amblypharyngodon mola*. Emphasis in these practices has been on the recovery of nutrients from waste water and raising protein-rich fish from sewage. However, it is advocated that only primary treated sewage be used instead of raw form. Further, depuration measures for a period of at least 15 days after final harvest of fish grown in sewage water are suggested to be mandatory in order to avoid any possible risk of human health hazard.

REFERENCES

Ahmad, T., and C. E. Boyd. 1988. Design and performance of peddle wheel aerators. Aquaculture Engineering 7:39-62.

Alikunhi, K. H. 1957. Fish Culture in India. Farm Bulletin 20, Indian Council of Agricultural Research, New Delhi, India.

Alikunhi, K. H., and K. K. Sukumaran. 1964. Preliminary observations on Chinese carps in India. Proceedings of the Indian Academy of Sciences 60:171-188.

Alikunhi, K. H., K. K. Sukumaran, and S. Parameswaran. 1971. Studies on composite fish culture: Production by composite combinations of Indian and Chinese carps. Journal of Indian Fisheries Association 1:26-57.

Aravindakshan, P. K., J. K. Jena, S. Ayyappan, H. K. Muduli, and S. Chandra. 1997. Evaluation of aeration intensities for rearing of carp fingerlings. Journal of Aquaculture 5:63-69.

Aravindakshan, P. K., J. K. Jena, S. Ayyappan, P. Routray, H. K. Muduli, S. Chandra, and S. D. Tripathi. 1999. Evaluation of production trials with grass carp as a major component in carp polyculture. Journal of the Inland Fisheries Society of India 31:64-68.

Arce, G. R., and C. E. Boyd. 1975. Effect of agricultural limestone on water chemistry, phytoplankton productivity and fish production in softwater ponds. Transactions of the American Fisheries Society 104:308-312.

Avnimelech, Y., N. Mozes, and B. Weber. 1992. Effect of aeration and mixing on nitrogen and organic matter transformations in simulated fish ponds. Aquaculture Engineering 11:157-169.

Ayyappan, S., and J. K. Jena. 1998. Carp culture in India–a sustainable farming practice. Pages 125-153 *in* P. Natarajan, K. Dhevendaran, C. M. Aracindhan, and S. D. Rita Kumari, eds. Advances in Aquatic Biology and Fisheries. Vol. II. Department of Aquatic Biology and Fisheries, University of Kerala, Trivandrum, India.

Bandyopadhyay, M. K., S. K. Singh, and B. Sarkar. 1991. Growth and production of *Ctenopharyngodon idilla* reared in plastic cages. Pages 35-37 *in* Proceedings of the National Symposium on New Horizons in Freshwater Aquaculture, Association of Aquaculturists and Central Institute of Freshwater Aquaculture, Bhubaneswar, India.

Banerjea, S. M. 1967. Water quality and soil condition of fish ponds in some states of India in relation to fish production. Indian Journal of Fisheries 14:115-144.

Bjoernsson, B. 1994. Effect of stocking density on growth rate of halibut (*Hippoglossus hippoglossus* L.) reared in large circular tanks for three years. Aquaculture 123:259-270.

Boyd, C. E. 1982. Water Quality Management for Pond Fish Culture. Elsevier, Amsterdam, the Netherlands.

Boyd, C. E., and T. Ahmad. 1987. Evaluation of Aerators for Channel Catfish Farming. Bulletin 584, Alabama Agricultural Experiment Station, Auburn University, Alabama.

Boyd, C. E., and B. J. Watten. 1989. Aeration system in aquaculture. CRC Critical Review Aquatic Science 1: 425-472.

Buck, D. H., R. J. Baur, and C. R. Rose. 1979. Experiment in recycling swine manure fish ponds. Pages 489-492 *in* T. V. R. Pillay and W. A. Dill, eds. Advances in Aquaculture. Fishing News Books Ltd., Farnham, Surrey, England.

Chakrabarty, R. D., N. G. S. Rao, and P. R. Sen. 1979. Culture of fish in ponds with fertilization. Pages 5-6 *in* Abstract Proceedings of the Seminar on Inland Aquaculture. Central Inland Fisheries Research Institute, Barrackpore, India.

Chaudhuri, H., R.D. Chakrabarty, N. G. S. Rao, K. Janakiram, D. K. Chatterjee, and S. Jena. 1974. Record fish production with intensive culture of Indian and exotic carps. Current Science 43:303-304.

Chaudhuri, H., R. D. Chakrabarty, P. R. Sen, N. G. S. Rao, and S. Jena. 1975. A new high in fish production in India with record yields by composite fish culture in freshwater ponds. Aquaculture 6:343-355.

Chaudhuri, H., N. G. S. Rao, M. Rout, and D. R. Kanaujia. 1978. Record fish production through intensive fish culture. Journal of Inland Fisheries Society of India 10:19-27.

Chen, T. P., and Li Pin Yen. 1980. Integrated agriculture-aquaculture studied in Taiwan. Pages 239-241 *in* R. S. V. Pullin, and Z. H. Shehadeh, eds. Proceedings of the ICLARM-SEARCA Conference on Integrated Agri-Aquaculture Farming System, International Center for Living Aquatic Resources Management, Manila and the Southeast Asian Regional Center for Graduate Study and Research in Agriculture, Los Banos, Laguna, Philippines.

CIFA. 1998. Intensive Carp Culture Utilizing Biotechnological Tools for Achieving Fish Production Levels of 25 tons/ha/annum. Project Completion Report. India: Central Institute of Freshwater Aquaculture, Bhubaneswar, India.

Coche, A. G. 1979. A review of cage fish culture and its application in Africa. Pages 428-441 *in* T. V. R. Pillay and W. A. Dill, eds. Advances in Aquaculture. Fishing News Books Ltd., Farnham, Surrey, England.

Colt, J., R. Ludwig, G. Tchobanoglous, and J. J. Cech. 1981. The effects of nitrite on the short term growth and survival of channel catfish, *Ictalurus punctatus*. Aquaculture 24:111-122.

Cruz, E. M., and Z. H. Shehadeh. 1980. Preliminary results of integrated pig-fish and duck-fish production tests. Pages 225-238 *in* R. S. V. Pullin, and Z. H. Shehadeh, eds. Proceedings of the ICLARM-SEARCA Conference on Integrated Agri-aquaculture Farming System, International Center for Living Aquatic Resources Management, Manila and the Southeast Asian Regional Center for Graduate Study and Research in Agriculture, Los Banos, Laguna, Philippines.

Das, P., D. Kumar, and M. K. Guha Roy. 1975. National demonstration on composite fish culture in West Bengal. Journal of the Inland Fisheries Society of India 7:112-115.

Das, P., M. Sinha, D. Kumar, D. P. Chakraborty, and M. K. Guha Roy. 1977. Culture of Indian major carps with record yield in a demonstration pond. Journal of the Inland Fisheries Society of India 9:105-110.

Das, P., J. G. Chatterjee, A. B. Mondal, and D. P. Chakraborty. 1979. Prospects of carp culture in lower Sundarbans. Page 59 *in* Abstract Proceedings of the Symposium on Inland Aquaculture. Central Inland Fisheries Research Institute, Barrackpore, India.

Das, P., D. Kumar, A. K. Ghosh, D. P. Chakraborty, and M. L. Bhaumik. 1980. High yield of Indian major carps against encountered hazards in a demonstration pond. Journal of the Inland Fisheries Society of India 12:70-78.

Datta, A. K., M. L. Bhowmilk, S. C. Mandal, and S. D. Tripathi. 1994. Rearing of bata in sewage-fed fish culture ponds. Page 59 *in* Abstract Proceedings of the Symposium on Inland Aquaculture. Central Inland Fisheries Research Institute, Barrackpore, India.

Degani, G. 1991. Effect of diet, population density and temperature on growth of larvae and juveniles of *Trichogaster trichopterus* (Bloch and Schneider 1801). Journal of Aquaculture in the Tropics 6:135-141.

Dehadrai, P. V., R. N. Pal, M. Choudhury, and D. N. Singh. 1974. Observations on cage culture of air-breathing fishes in swamps of Assam. Journal of the Inland Fisheries Society of India 6: 89-92.

Delmond, M. N. 1980. A review of integrated livestock-fowl farming systems. Pages 59-72 *in* R. S. V. Pullin, and Z. H. Shehadeh, eds. Proceeding of the ICLARM-SEARCA Conference on Integrated Agri-Aquaculture Systems, International Center for Living Aquatic Resources Management, Manila and the Southeast Asian Regional Center for Graduate Study and Research in Agriculture, Los Banos, Laguna, Philippines.

Edwards. P. 1990. An alternative excreta-reuse strategy for aquaculture: the production of high-protein animal feed. Pages 209-221 *in* P. Edwards, and R. S. V. Pullin, eds. Wastewater-fed Aquaculture. Asian Institute of Technology, Bangkok, Thailand.

Edwards, P. 1993. Reuse of human wastes in aquaculture, Part-II. Encology 7(9):1-15.

Edwards, P. 1996. A note on wastewater-fed aquaculture systems: status and prospects. Naga, International Center for Living Aquatic Resources Management 19: 33-35.

FAO. 1971. Report to the Government of Burma on Fish Culture Development. Based on Work of H. Chaudhuri, Inland Fishery Biologist (Fish culture), FAO (UNDP) Report 2954, Food and Agriculture Organization of the United Nations, Rome, Italy.

FAO. 1998. Aquaculture Production Statistics. Food and Agriculture Organization of the United Nations, Rome, Italy.

George, J. P., K. Venkateshvaran, and G. Venugopal. 1990. Effect of artificial aeration on the growth and survival of *Labeo rohita* fry of Sathanpur fish farm. Pages 76-81 *in* P. Keshavanath, and K. V. Radhakrishnan, eds. Proceedings of the Workshop on Carp Seed Production Technology. Asian Fisheries Society, Indian Branch, Mangalore, India.

Ghosh, A., L. H. Rao, and S. C. Banerjee. 1974. Studies on the hydrobiological conditions of a sewage-fed pond with a note on their role in fish culture. Journal of the Inland Fisheries Society of India 6:51-61.

Ghosh, A., S. K. Saha, A. K. Roy, and P. K. Chakrabarty. 1985. Package of practices for using domestic sewage in carp production. Pages 1-19 *in* Central Inland Fisheries Research Institute, Aquaculture Extension Manual. New Series 8, Barrackpore, India.

GOI. 1996. Hand Book of Fisheries Statistics. Ministry of Agriculture, Government of India, New Delhi, India.

Govind, B. V., S. Ayyappan, S. L. Raghavan, and M. F. Rahman. 1988. Culture of catla in floating net cages. Mysore Journal of Agricultural Sciences 22:517-522.

Hampson, B. L. 1976. Ammonia concentration in relation to ammonia toxicity during a rainbow trout rearing experiment in closed freshwater-seawater system. Aquaculture 9:61-70.

Hasan, M. R. 1990. Aquaculture in Bangladesh. Pages 105-139 *in* M. Mohan Joseph, ed. Aquaculture in Asia. Asian Fisheries Society, Indian Branch, Mangalore, India.

Hasan, M. R., and D. J. McIntosh. 1986. Effect of chloride concentration on the acute toxicity of nitrite to common carp. Aquaculture and Fisheries Management 17:19-30.

Hickling, C. F. 1971. Fish Culture. Faber and Faber, London, England.

Holm, J. C., T. Refstie, and S. Bo. 1990. The effect of fish density and feeding regimes on individual growth rate and mortality in rainbow trout (*Oncorhynchus mykiss*). Aquaculture 89:225-232.

Hora, S. L., and T. V. R. Pillay. 1962. Handbook on Fish Culture in the Indo-Pacific Fisheries Region. FAO Fisheries Biology Technical Paper 14, Food and Agriculture Organization of the United Nations, Rome, Italy.

Islam, M. S., and S. Dewan. 1986. Resource use and economic returns in pond fish culture. Bangladesh Journal of Agricultural Economics 9:141-150.

Jafri, A. K., D. K. Khan, M. A. Anwar, M. F. Hassan, and Erfanullah. 1991. Growth and survival of Indian major carp spawn fed artificial diets. Pages 485-490 *in* V. R. P. Sinha, and H. C. Srivastava, eds. Aquaculture Productivity. Oxford and IBH Publishing Co. Private Ltd., New Delhi, India.

Jana, B. B. 1998. Sewage-fed aquaculture: the central model. Ecological Engineering 11:73-85.

Jana, B. B., and S. Datta. 1996. Nutrient variability in six sewage-fed tropical fish ponds. Environmental Research Forum 5 & 6:379-382.

Jayaram, M. G., and H. P. C. Shetty. 1981. Formulation, processing, and water stability of two new pelleted feeds. Aquaculture 23:355-359.

Jena, J. K. 1998. Input Management in Carp Culture for Optimization of Production Levels. Doctoral dissertation, Orissa University of Agriculture and Technology, Bhubaneswar, India.

Jena, J. K., P. K. Mukhopadhyay, S. Sarkar, P. K. Aravindakshan, and H. K. Muduli. 1996. Evaluation of a formulated diet for nursery rearing of Indian major carp under field condition. Journal of Aquaculture in the Tropics 11:299-305.

Jena, J. K., P. K. Aravindakshan, S. Chandra, H. K. Muduli, and S. Ayyappan. 1998a. Comparative evaluation of growth and survival of Indian major carps and exotic carps in raising fingerlings. Journal of Aquaculture in the Tropics 13:143-150.

Jena, J. K., P. K. Aravindakshan, and W. J. Singh. 1998b. Nursery rearing of Indian major carp fry under different stocking densities. Indian Journal of Fisheries 45:163-168.

Jena, J. K., P. K. Mukhopadhyay, and P. K. Aravindakshan. 1998c. Dietary incorporation of meat meal as a substitute for fish meal in carp fry rearing. Indian Journal of Fisheries 45:43-49.

Jena, J. K., P. K. Mukhopadhyay, and P. K. Aravindakshan. 1999. Evaluation of formulated diet for raising carp fingerlings in field condition. Journal of Applied Ichthyology 15:188-192.

Jena, J. K., S. Ayyappan, and P. K. Aravindakshan. 2000. Comparative evaluation of production performance in varied cropping patterns of carp polyculture systems. Page 155 *in* Abstract Proceedings of the Fifth Indian Fisheries Forum, Asian Fisheries Society, Indian Branch, Mangalore and Association of Aquaculturists, Bhubaneswar, India.

Jhingran, V. G. 1969. Potential of inland fisheries. Indian Farming 19:22-25.

Jhingran, V. G. 1991. Fish and Fisheries of India. Hindustan Publishing Corporation, New Delhi, India.

Jhingran, V. G., and B. K. Sharma. 1980. Integrated livestock-fish farming. Pages 135-142 *in* R. S. V. Pullin, and Z. H. Shehadeh, eds. Proceedings of the ICLARM-SEARCA Conference on Integrated Agri-Aquaculture Farming System, International Center for Living Aquatic Resources Management, Manila and the Southeast Asian Regional Center for Graduate Study and Research in Agriculture, Los Banos, Laguna, Philippines.

Kalyani, R., and H. P. C. Shetty. 1987. Use of biogass slurry as a fertilizer for production of carps. Page 24 *in* Abstract Proceedings of the First Indian Fisheries Forum, Asian Fisheries Society, Indian Branch, Mangalore, India.

Khan, H. A., D. N. Mishra, and B. C. Tyagi. 1979. Investigations on composite fish culture with and without supplementary feeding of fish and fertilization of pond. Pages 35-36 *in* Abstract Proceedings of the Symposium on Inland Aquaculture, Central Inland Fisheries Research Institute, Barrackpore, India.

Kjartansson, H., S. Fivelstad, J. M. Thomassen, and M. J. Smith 1988. Effects of different stocking densities on physiological parameters and growth of adult Atlantic salmon (*Salmo salar* L.) reared in circular tanks. Aquaculture 73:261-274.

Krishnamurthy, K. N., and P. K. Aravindakshan. 1979. A note on composite fish culture in a running water pond at Bhavanisagar, Coimbatore District, Tamil Nadu. Page 20 *in* Abstract Proceedings of the Symposium on Inland Aquaculture, Central Inland Fisheries Research Institute, Barrackpore, India.

Krishnamurthy, K. N., K. Sukumaran, D. V. Pahwa, and B. C. Tyagi. 1976. Progress Report of the Karnal sub-center of Haryana for the period of November 1972 to August 1975. Pages 1-12 *in* Third Workshop on All India Co-ordinated Research Project on Composite Fish Culture and Fish Seed Production, Central Inland Fisheries Research Institute, Barrackporę, India.

Kumaraiah, P., S. Parameswaran, and N. M. Chakraborty. 1991. New perceptions in cage fish culture with reference to the growth and production of silver carp in cages. Pages 246-249 *in* Proceedings of the National Symposium on New Horizons in Freshwater Aquaculture, Association of Aquaculturists and Central Institute of Freshwater Aquaculture, Bhubaneswar, India.

Laal, A. K., S. K. Sarkar, A. Sarkar, B. L. Pandey, and K. P. Singh. 1990. Culture possibilities of Indian major carps in sewage fed zone of a channel of river Ganga. Journal of Aquaculture in the Tropics 5:53-59.

Lakshmanan, M. A. V., K. K. Sukumaranan, D. S. Murty, D. P. Chakraborty, and M. T. Philipose. 1971. Preliminary observations on intensive fish farming in freshwater ponds by the composite culture of Indian and exotic species. Journal of the Inland Fisheries Society of India 3:1-21.

Lalmohan, R. S. 1983. Experimental culture of *Chanos chanos* in fish pens in a coastal lagoon at Mandapam. Indian Journal of Fisheries 30:287-295.

Lin, S. Y. 1955. Chinese system of pond stocking. Proceedings of Indo-Pacific Fisheries Council 5:113-125.

Ling, S. W. 1967. Feed and Feeding of Warm Water Fishes in Ponds in Asia and Far East. FAO Fisheries Report 44:291-309.

Loyacano, H. A. 1974. Effect of aeration in earthen ponds on water quality and production of white catfish. Aquaculture 3: 261-271.

Marek, M., and S. Sarig. 1971. Preliminary observation of super-intensive fish culture in the Beth-swan Valley in 1969-1970. Bamidgeh 23:93-99.

Martinez-Cordova, L. R., H. Villarreal-Colmenares, A. Prochas-Cornejo, J. Naranjo-Paramo, and A. Aragon-Noriega. 1997. Effect of aeration rate on growth, survival and yield of white shrimp *Penaeus vannamei* in low water exchange ponds. Aquaculture Engineering 16:85-90.

Mathew, P. M., B. K. Singh, and D. P. Chakraborty. 1979. Stocking fry in composite fish culture. Pages 45-46 *in* Abstract Proceedings of the Symposium on Inland Aquaculture, Central Inland Fisheries Research Institute, Barrackpore, India.

Menon, V. R. 1983. On the result of rearing carp fry in floating nurseries inn Tamil Nadu. Pages 95-98 *in* Proceedings of the National Seminar on Cage and Pen Culture. Fisheries College, Tamil Nadu Agricultural University, Tuticorin, India.

Mohanty, S. N., D. N. Swamy, and S. D. Tripathi. 1990. Protein utilization in Indian major carp fry *Catla catla* (Ham.), *Labeo rohita* (Ham.) and *Cirrhinus mrigala* (Ham.) fed four protein diets. Journal of Aquaculture in the Tropics 5:173-179.

Mohanty, U. K. 1995. Comparative evaluation of growth and survival of Indian major carp fry in aerated *vis-a-vis* non-aerated ponds under different stocking densities. Master's thesis, Orissa University of Agriculture and Technology, Bhubaneswar, India.

Murty, D. S., G. N. Saha, C. Selvaraj, P. V. G. K. Reddy, and R. K. Dey. 1978. Studies on increased fish production in composite fish culture through nitrogenous fertilization with and without supplementary feeding. Journal of the Inland Fisheries Society of India 10:39-45.

Nandeesha, M. C. 1993. Aquafeeds and feeding strategies in India. Pages 213-54 *in* M. B. New, A. G. J. Tacon, and I. Csavas, eds. Farm-Made Aquafeeds. Regional Office for Asia and the Pacific, Food and Agriculture Organization of the United Nations, Bangkok, Thailand and ASEAN-EEC Aquaculture Development and Coordination Programme, Bangkok, Thailand.

Nandeesha, M. C., K. V. Devaraj, and N. S. Sudhakara. 1986. Growth response of four species of carps to different protein sources in pelleted feeds. Pages 603-608 *in* J. L. Maclean, L. B. Dizon, and L. V. Hossilos, eds. Proceedings of the First Asian Fisheries Forum. Asian Fisheries Society, Manila, Philippines.

Natarajan, A. V., R. K. Saxena, and N. K. Srivastava. 1979. Experiments in raising quality fish seed in floating nurseries and its role in aquaculture in India. Pages 45-49 *in* International Workshop on Cage and Pen Culture, Tigbaun, Philippines.

Pantulu, V. R. 1979. Floating cage culture of fish in the lower Mekong river basin. Pages 423-427 *in* T. V. R. Pillay, and W. A. Dill, eds. Advances in Aquaculture. Fishing News Books Limited, Farnham, Surrey, England.

Prasad, G. S., A. Qureshi, P. S. Rao, and B. C. Devi, 1993. Studies on the growth of Indian major carps by utilizing biogas slurry. Page 2 *in* Abstract Proceedings of the Third Indian Fisheries Forum, Asian Fisheries Society, Indian Branch, Mangalore, India.

Rao, K. J., and T. S. R. Raju. 1989. Observations on polyculture of carps in large freshwater ponds of Kolleru lake. Journal of Aquaculture in the Tropics 4:157-164.

Rao, P. L. N., and B. K. Singh. 1982. Report on the progress of work from 1 September 1980 to 31 March 1982 at Hadapsar Fish Farm, Pune sub-center (Maharashtra). Pages 135-139 *in* Sixth Workshop on All India Co-Ordinated Research Project on Composite Fish Culture and Fish Seed Production, Central Inland Fisheries Research Institute, Barrackpore, India.

Rao, P. L. N., and B. K. Singh. 1984. Report on the progress of work from July 1982 to January 1984 at Pune sub-center (Maharashtra). Pages 66-74 *in* Seventh Workshop of All India Co-Ordinated Research Project on Composite Fish Culture and Fish Seed Production, Central Inland Fisheries Research Institute, Barrackpore, India.

Rappaport, U., and S. Sarig. 1975. The results of tests in intensive growth of fish at the Gonosar (Israel) station ponds in 1974. Bamidgeh 27: 75-82.

Rappaport, U., S. Sarig, and M. Marck. 1976. The results of tests of various aeration systems on the oxygen regime in Genosar experimental ponds and growth of fishes there in 1975. Bamidgeh 28: 35-49.

Refstie, T., and A. Kittelsen. 1976. Effect of density on growth and survival of artificially reared Atlantic salmon. Aquaculture 8:319-326.

Rout, P. R. 1995. Aeration and its effect on the rearing of Indian major carp fingerlings. Master's thesis, Orissa University of Agriculture and Technology, Bhubaneswar, India.

Routray, P., and M. D. Routray. 1998. Culture possibilities of grass carp, *Ctenopharyngodon idella*, Val., in pens of low saline brackishwater areas of Chilka lake. Journal of Aquaculture 6:13-17.

Ruttanagosright, W., Y. Musig, C. E. Boyd, and L. Sukchareon. 1991. Effect of salinity on oxygen transfer by propeller-aspirator-pump and paddle wheel aerators used in shrimp farming. Aquaculture Engineering 10:121-131.

Saha, G. N., K. Raman, D. K. Chatterjee, and S. R. Ghosh. 1975. Relative response of three nitrogenous fertilizers in different pond soils in relation to primary productivity, plankton and survival and growth of *Labeo rohita* spawn. Journal of the Inland Fisheries Society of India 7:62-172.

Saha, G. N., D. K. Chatterjee, C. Selvaraj, and N. N. Mazumdar. 1979. Evaluation of chemical fertilizers in enhancing fish production in composite fish culture in tropical freshwater ponds. Page 48 *in* Abstract Proceedings of the Symposium on Inland Aquaculture, Central Inland Fisheries Research Institute, Barrackpore, India.

Saha, K. C. 1970. Fisheries of West Bengal. Government Press, Calcutta, West Bengal, India.

Schroeder, G. L. 1978. Autotrophic and heterotrophic production of microorganisms in intensively-managed fish ponds and related fish yields. Aquaculture 14:303-325.

Sen, P. R., N. G. S, Rao, S. R. Ghosh, and M. Rout. 1978. Observation on the protein and carbohydrate requirements of carps. Aquaculture 13:245-255.

Sen, P. R., N. G. S. Rao, and A. N. Mohanty. 1980. Relationship between rate of feeding, growth and conversion in major Indian carps. Journal of Fisheries 27:201-208.

Shah, K. L., and B. C. Tyagi. 1984. Report on the progress of work from 1 April 1982 to February 1984 at Karnal sub-center (Haryana). Pages 42-65 *in* Seventh Workshop of All India Co-Ordinated Research Project on Composite Fish Culture and Fish Seed Production, Central Inland Fisheries Research Institute, Barrackpore, India.

Sharma, B. K., and M. K. Das. 1988. Studies on integrated fish-livestock crop farming system. Pages 27-30 *in* M. M. Joseph, ed. Proceeding of the First Indian Fisheries Forum, Asian Fisheries Society, Indian Branch, Mangalore, India.

Sharma, B. K., and J. Olah. 1986. Integrated fish-pig farming in India and Hungary. Aquaculture 54:135-139.

Sharma, B. K., D. Kumar, M. K. Das, S. R. Das, and D. P. Chakraborty. 1979a. Observations on swine dung recycling in composite fish culture. Page 99 *in* Abstract Proceeding of the Symposium on Inland Aquaculture, Central Inland Fisheries Research Institute, Barrackpore, India.

Sharma, B. K., M. K. Das, S. R. Das, and S. K. Neogi. 1979b. Observations on fish-duck rearing. Page 95 *in* Abstract Proceedings of the Symposium on Inland Aquaculture, Central Inland Fisheries Research Institute, Barrackpore, India.

Sharma, B. K., M. K. Das, and D. P. Chakraborty. 1985. Package of practices for increasing production in fish-cum-livestock farming system, Central Inland Fisheries Research Institute, Aquaculture Extension Manual, New Series 5, Barrackpore, India.

Singh, C. S., and K. P. Singh. 1975. Feeding experiments on Indian major carps in Tarai ponds. Journal of the Inland Fisheries Society of India 7:212-215.

Singh, S. B. 1990. Status of aquaculture in the Socialist Republic of Vietnam. Pages 371-384 *in* M. M. Joseph, ed. Aquaculture in Asia, Asian Fisheries Society, Indian Branch, Mangalore, India.

Singh, S. B., K. K. Sukumaran, P. C. Chakrabarti, and M. M. Bagchi. 1972. Observations of composite culture of exotic carps. Journal of the Inland Fisheries Society of India 4: 38-50.

Singh, S. B., S. R. Ghosh, P. V. G. K. Reddy, R. K. Dey, and B. K. Mishra. 1980. Effect of aeration on feed utilization by common carp fingerlings. Journal of the Inland Fisheries Society of India 12:64-69.

Sinha, M., and P. K. Saha. 1980. Efficacy of a commercial fish feed for composite fish culture. Journal of the Inland Fisheries Society of India 12:51-55.

Sinha, V. R. P. 1979. New trends in fish farm management. Pages 123-126 *in* T. V. R. Pillay, and W. A. Dill, eds. Advances in Aquaculture, Food and Agriculture Organization of the United Nations, Rome, Italy.

Sinha, V. R. P., and M. V. Gupta. 1975. On the growth of grass carp *Ctenopharyngodon idella* Val. in composite fish culture at Kalyani, West Bengal (India). Aquaculture 5:283-290.

Sinha, V. R. P., M. V. Gupta, M. K. Banerjee, and D. Kumar. 1973. Composite fish culture in Kalyani. Journal of the Inland Fisheries Society of India 5: 201-208.

Sukumaran, K. K., H. A. Khan, P. U. Verghese, and P. M. Mathew. 1972. Progress Report of Jaunpur Sub-center, Uttar Pradesh, for the period of October 1971 to October 1972. Page 11 *in* Second Workshop on All India Co-Ordinated Research Project on Composite Fish Culture of Indian and Exotic Fishes, Central Inland Fisheries Research Institute, Barrackpore, India.

Sukumaran, P. K., S. L. Raghavan, M. F. Rahman, and S. Ayyappan. 1986. Cage culture of the carp, *Catla catla* (Hamilton) in a freshwater tank in Bangalore. Pages 145-147 *in* Proceedings of the National Symposium on Fish and Environment, Haridwar, India.

Szyper, J. P., and C. K. Lin. 1990. Techniques for assessment of stratification and effects of mechanical mixing in tropical fish ponds. Aquaculture Engineering 9:151-165.

Tan, E. S. P., and K. K. Huat. 1980. The integration of fish farming with agriculture in Malaysia. Pages 175-188 *in* R. S. V. Pullin and Z. H. Shehadeh, eds. Proceedings of the ICLARM-SEARCA Conference on Integrated Agri-Aquaculture Farming Systems, International Center for Living Aquatic Resources Management, Manila and the Southeast Asian Regional Center for Graduate Study and Research in Agriculture, Los Banos, Laguna, Philippines.

Tripathi, S. D. 1982. Report of the project co-ordinator. Pages 167-177 *in* Sixth Workshop of All India Co-Ordinated Research Project on Composite Fish Culture and Fish Seed Production, Central Inland Fisheries Research Institute, Barrackpore, India.

Tripathi, S. D. 1984. Report of the project co-ordinator. Pages 155-213 *in* Seventh Workshop of the All India Co-Ordinated Research Project on Composite Fish Culture and Fish Seed Production, Central Inland Fisheries Research Institute, Barrackpore, India.

Tripathi, S. D. 1990. Freshwater Aquaculture in India. Pages 191-222 *in* M. M. Joseph, ed. Aquaculture in Asia, Asian Fisheries Society, Indian Branch, Mangalore, India.

Tripathi, S. D., and D. N. Mishra. 1986. Synergistic approach in carp polyculture with grass carp as a major component. Aquaculture 54:157-160.

Tripathi, S. D., S. Bhandari, M. Das, and S. Ayyappan. 1991. Evaluation of manurial application levels of biogas slurry in carp farming. Pages 157-160 *in* Proceedings of the National Symposium on New Horizons in Freshwater Aquaculture, Association of Aquaculturists, Bhubaneswar and Central Institute of Freshwater Aquaculture, Bhubaneswar, India.

Tripathi, S. D., S. Ayyappan, M. Das, S. Bhandari, and V. S. Basheer. 1992. Production plan for manurial management of fish pond using biogas slurry. Pages 17-21 *in* S. D. Tripathi, M. Ranadhir, and C. S. Purushothaman, eds. Aquaculture Economics. Proceedings of the Workshop on Aquaculture Economics, Special Publication 7, Asian Fisheries Society, Indian Branch, Mangalore, India.

Tripathi, S. D., P. K. Aravindakshan, S. Ayyappan, R. Singh, A. N. Mohanty, and H. K. Muduli. 1994. Carp polyculture with cattle dung and water hyacinth based biogas slurry as manures. Pages 34-38 *in* S. Ayyappan, B. Seshagiri, M. Ranadhir, and S. D. Tripathi, eds. Proceedings of the National Seminar on Forty Years of Freshwater Aquaculture in India, Association of Aquaculturists, Bhubaneswar and Central Institute of Freshwater Aquaculture, Bhubaneswar, India.

Tripathi, S. D., P. K. Aravindakshan, S. Ayyappan, J. K. Jena, H. K. Muduli, S. Chandra, and K. C. Pani. 2000. New high in carp production in India through intensive polyculture. Journal of Aquaculture in the Tropics 15:119-128.

Trizebiatowski, J., J. Filipiak, and R. Jakubowski. 1980. On the effect of stocking density on growth and survival of rainbow trout. Aquaculture 22:289-295.

Vijayan, M. M., and J. F. Leatherland. 1988. Effect of stocking density on the growth and stress-response in brook charr, *Salvelinus fontinalis*. Aquaculture 75:159-170.

Vijayan, M. M., and T. J. Varghese. 1986. Effect of artificial aeration on growth and survival of Indian major carps. Proceedings of the Indian Academy of Sciences (Animal Science) 95:371-378.

Woynarovich, E. 1980. Utilization of piggery wastes in fish ponds. Pages 125-128 *in* R. S. V. Pullin, and Z. H. Shehadeh, eds. Proceeding of ICLARM-SEARCA Conference of the Integrated Agri-Aquaculture Farming Systems, International Center for Living Aquatic Resources Management, Manila and the Southeast Asian Regional Center for Graduate Study and Research in Agriculture, Los Banos, Laguna, Philippines.

Zhu, Y., Y. Yang, J. Wan, D. Huo, and J. A. Mathias. 1990. The effect of manure application rate and frequency upon fish yield in integrated fish farm ponds. Aquaculture 91:233-251.

The Potential and Sustainability of Aquaculture in India

B. B. Jana
Santana Jana

SUMMARY. India is a very populous country with more than one billion people. In order to provide food for this growing population, serious environmental problems may result. Despite many benefits from the green, blue, and silver revolutions adopted in India, there has been much concern resulting from intensive agricultural practices that led to environmental problems in both terrestrial and aquatic ecosystems. Increasing demand for aquatic resources also caused inland fisheries to decrease over the past few decades. The location of aquaculture projects, landscape destruction, soil and water pollution by pond effluents, over-exploitation of important fish stocks, depletion in biodiversity, conflicts over agriculture and aquaculture among various stakeholder groups over resource and space allocation, and international fish trade controversies have threatened the long-term sustainability of fisheries and aquaculture industries. The subject of sustainable aquaculture has not been adequately projected in terms of current aquaculture practices aimed to boost a rural economy. This review briefly describes the key issues of aquaculture unsustainability in terms of intensive aquaculture, nutrient enrichment syndrome, soil and groundwater salinization, destruction of mangroves, loss of biodiversity, marine pollution and loss of fish stock,

B. B. Jana, Aquaculture and Applied Limnology Research Unit, Department of Zoology, University of Kalyani, Kalyani-741 235, West Bengal, India.

Santana Jana, Department of Fisheries Economics and Statistics, West Bengal University of Animal and Fishery Sciences, Mohanpur-741 252, West Bengal, India.

[Haworth co-indexing entry note]: "The Potential and Sustainability of Aquaculture in India." Jana, B. B., and Santana Jana. Co-published simultaneously in *Journal of Applied Aquaculture* (Food Products Press, an imprint of The Haworth Press, Inc.) Vol. 13, No. 3/4, 2003, pp. 283-316; and: *Sustainable Aquaculture: Global Perspectives* (ed: B. B. Jana, and Carl D. Webster) Food Products Press, an imprint of The Haworth Press, Inc., 2003, pp. 283-316. Single or multiple copies of this article are available for a fee from The Haworth Document Delivery Service [1-800-HAWORTH, 9:00 a.m. - 5:00 p.m. (EST). E-mail address: getinfo@haworthpressinc.com].

use of aquachemicals and therapeutics, hormone residues, etc. The strategies for sustainability have been highlighted with respect to rice-cum-fish culture, carp polyculture, integrated farming with livestock, rural aquaculture, intensification of small farms, wastewater-fed aquaculture, crop rotation, probiotics, feed quality, socioeconomic considerations, environmental regulations and fisheries acts, transboundary aquatic ecosystems, impact of alien species, ethical aspects of intensive aquaculture, responsible fisheries, and environmental impact assessment. A suggested model outlines the feedback mechanisms for achieving long-term sustainability through improved farm management practices, integrated farming, use of selective aquachemicals and probiotics, conservation of natural resources, regulatory mechanism, and policy instruments.

KEYWORDS. Sustainable aquaculture, integrated resource management, aquachemicals, socio-economic factors, responsible fisheries

INTRODUCTION

The issue of malnutrition has become important to the growing population in the Third World. The world population has been projected to increase from 4 billion in 1980 to 12 billion in 2030, with an estimated 500 million people from developing countries lying below the poverty line who will get less protein than they need. Retention of dietary protein and energy in fish farming is approximately twice that in chicken and swine production, with correspondingly lower waste production (Asgard and Austreng 1995). Because of high food conversion efficiency, economic potential, and suitability for tropical waters along with fast growth at high ambient temperatures, aquaculture utilizing three-dimensional pond resources was duly recognized to promote the development of environmentally sound and sustainable aquaculture into rural, agricultural, and coastal development.

To make the country self-sufficient in food production, India has made remarkable strides in crop production. Despite much benefit from the green and blue revolutions adopted in India, there have been much environmental problems resulting from intensive agricultural practices that led to environmental problems in both terrestrial and aquatic ecosystems. Increasing demand for aquatic resources also caused inland fisheries to decrease over the past few decades (Anon 1998). The location of aquaculture projects, landscape destruction, soil and water pollution by pond effluents, over-exploitation of important fish stocks, depletion in biodiversity, conflicts over agriculture and aquaculture among various stakeholder groups over resource and space allocation, and international fish trade controversies have threatened the long-term sustainability of fisheries and aquaculture industries. Human population, as an essential part of the ecosystem can, therefore, contribute a more balanced view of sustainability in the long run.

Because the present state-of-the-art of aquaculture in India is not always in the right direction to ensure sustainability, the Green Bench of the Supreme Court of India, in December 1996, ordered the closure of all semi-intensive and intensive shrimp farms within 500 m of the high-tide line, banned shrimp farms from all public lands, and required farms that closed down to compensate their workers with 6 years of wages in a move to protect the environment and to rehabilitate the local people and resources. Recently, the aquaculture activities pertaining to the coastal regulation zone (CRZ) have come into prominence in India in view of the economic importance of aquaculture on the one hand and clean environmental space on the other.

The ill effects of intensive farming are well documented, though not much investigation has been carried out. The subject of sustainable aquaculture has not been adequately projected in terms of current aquaculture practices in the developing countries with poverty-driven socio-economic conditions, the major constraint to sustainable aquaculture development. This study briefly describes ecological and socio-economic consequences of ill-planned and unscientific aquaculture projects, citing some examples of case studies, and recommends measures to achieve long-term sustainability through improved farm management practices, integrated farming, use of selective aquachemicals and probiotics, conservation of natural resources, regulatory mechanism, and policy instruments.

PRODUCTION SCENARIO

The total world production from fisheries of finfish and shellfish reached 112 million tons in 1995. There was steady rise in total aquaculture production of finfish and shellfish during the nineties (Figure 1). The total world fisheries production in 1996 is estimated to be 121 mil-

FIGURE 1. Global trends in aquaculture production. A steady rise is noticeable during the nineties.

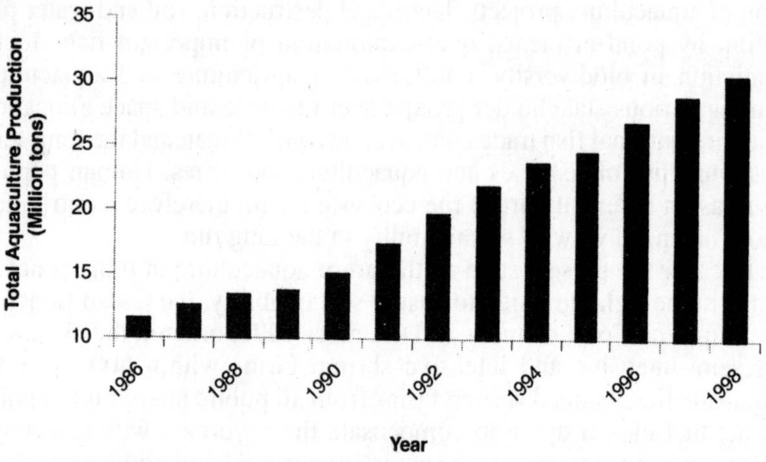

lion tons, whereas the production from inland waters was more than 23 million tons with contributions of 7.5 and 15.5 million tons from capture fisheries and aquaculture, respectively (Barg et al. 1999). This suggests that the total world fisheries production from inland water was ~19%, of which ~13% came from inland aquaculture and ~6% from mariculture.

The total annual production of cultured fish, molluscs, crustaceans, and aquatic plants is projected to exceed 40 million tons by 2010. World aquaculture production has evolved from some 2 million tons in the beginning of the 1960s to 34.1 million tons in 1996 (Wiefels 1999). Global aquaculture production has grown at an average annual rate of 10% since 1984, compared with 3% for livestock meat and 1.6% for capture fisheries production (Rana 1997). It is further estimated that total world aquaculture production was ~22% of the total world fisheries production in 1996, compared to ~13% in 1990 (FAO 1999). During the same period, the proportionate contribution of capture fisheries declined by about 8.5%. Of special importance, more than 85% of total world aquaculture production came from developing nations and particularly from low-income southeast Asian countries (Khoo and Tan 1980). Clearly as an economically viable source, shrimp farming in these countries has stepped into an important area for export. Seventy-five percent of cultured shrimp came from Asia, with major contributions from Thailand, Indonesia, China, and India.

In India, total fish production increased from 4.2 million tons in 1992 to 4.9, 5.1, and 5.6 million tons in 1995, 1996 and 1999, respectively (Anon 1999a). During 1999, the share from marine and inland sources was 3.03 and 2.57 million tons, respectively. This shows an average annual growth rate of 5.2%, which is much lower than the world average. India produces about 1.2 million tons of freshwater fish from about 670,000 ha of freshwater ponds and about ~0.04 million tons of shrimp from an area of about 70,000 ha of brackish water (Anon 1997a). The shrimp production peaked during 1994-95, recording ~0.083 million tons in an area of 100,700 ha, followed by a declining trend (~0.067 million tons) in 1997-98, though the area under culture (141,519 ha) increased by 14% (Anon 1999a).

In 1996, the total exploited quantity from marine sources is 2.69 million tons or about 69% of estimated total marine fishery resource (3.9 million tons) in India, which has vast coastline of 8,129 km extending over 9 maritime states. The islands of Lakshadweep, Andamans, and Nicobar have an exclusive economic zone (EEZ) of 2.02 million km^2 and a continental shelf of 0.512 million km^2 (Nayar et al. 1999). According to Datta and Chakrabarti (1997), Indian marine fish production has reached the maximum sustainable yield level, which is defined as 50% of the resources. Obviously, there are reported cases of over-exploitation, depletion of stocks of certain species (ASCI 1996), and marine biodiversity (Perez and Mendoza 1998).

In a perspective plan document prepared by Central Marine Fisheries Research Institute, India's domestic demand for fish by 2020 is projected to be 7.2 million tons and 0.60 million tons for additional exports (Anon 1999a). The current per capita availability of fish in India is 9.85 kg, which is far from the requirement on the global scale. The ninth plan (1997-2002) proposes to raise the per capita consumption to 11.24 kg. In order to make the modest provision of 9.85 kg per person, 5 million tons of fish production are needed in the domestic market alone by 2020. The overall requirement of fish has been estimated to be 4.0 million tons from marine fisheries and 3.8 million tons from aquaculture sectors to meet the demand in domestic and international markets (Anon 1999a).

Despite aquaculture making notable contributions during the past decade, it still remains a small sub-sector of agriculture in India. It is a matter of concern that vast inland water resources remain either under-utilized or un-utilized; present aquaculture production utilizes only 0.142 million ha compared to its potential of 1.19 million ha. Adequate scientific planning and management of a large number of village ponds would make a significant contribution to the improvement of rural economy and provide fish protein to the people.

AQUACULTURE SUSTAINABILITY

There is often a conflict between aquaculturists on the one hand and local fishermen, agriculturists, and environmentalists on the other. In recent years, coastal aquaculture has suffered a major setback in India, due to unplanned and unregulated development resulting in environmental degradation and disease problems within the pond system, and in environmental hazards at the production site. The situation is further aggravated because of the fact that the source of water itself, i.e., sea or creek water used for shrimp farming, is highly contaminated by its own operations, industrial effluents, and domestic sewage. There is an urgent need for food security that has led to utter negligence of environmental protection in the Third World. Enormous population growth and poverty coupled with unplanned development culminated into unsustainable development, disregarding environmental laws and leaving bleak prospects for future generations.

Major conflicts arise due to: large-scale diversion of farmland for converting paddy fields into brackishwater shrimp farms; intensive farming of finfish and shellfish; unscientific management of shrimp ponds; depletion of the water table, resulting from excess extraction of water for shrimp farming; slumming of small farms in utter disregard of the minimum distance criterion; lack of proper drainage facilities; lack of adequate maintenance of ecosystem health in terms of specific oxygen levels and other water quality parameters; poor pond preparation techniques; improper emulation process and production cycles; and release of pond effluent without pre-treatment.

Sustainable development has been considered and defined in many ways. At least four types of sustainability (environmental, social, cultural, and economic) were recognized at the Earth Summit (Hempel 1998). The Brundtland Commission has defined sustainable development as such development that meets the needs of the present without compromising the ability of future generations to meet their own needs (WCED 1987). Sustainable development has been defined by FAO (1988) as the management and conservation of the natural resource base and the orientation of technological and institutional change in such a manner as to ensure the attainment and continued satisfaction of human needs for present and future generations. It represents a compromise between the forces of economic growth and those of environmental protection (Hempel 1998). Jacobs et al. (1987) called sustainable development a reaction against the *laissez faire* economic theory that considered living resources as free goods, external to the development process, essentially infinite and inexhaustible. Sustainability implied not only long-term economic viability but also environmental safety, biodiversity, conservation of natural resources, social harmony, and socioeconomic development.

Though sustainability in agriculture refers to increasing production while maintaining the production base of soil fertility, this may also be a suitable conceptual framework for aquaculture where ecosystem health is of prime importance (FAO 1990). Complex interactions between aquatic organisms and the physical environment result in exchanges of mass and energy through network systems of food chain and food web (Shell 1993). In developing countries, sustainable aquaculture development has been coupled with the aquaculture activities that are oriented towards large-scale, low-cost fish production to meet to demand of economically poor and growing populations. It may be useful to introduce the term "protein crops" versus "cash crops" in order to signify a basic value criterion related to various methods of production. "Protein crop" is used when production aims at achieving as much animal protein as possible within the regional ecosystems to meet the protein needs of low-income populations, mostly in the developing world. Asian integrated fish-cum-rice production is an example. In contrast, "cash crop" aims at high-value products for distant markets to meet the demand of high-income groups, mostly in the developed world. Typically, the farmed species chosen are high-value species such as salmon, trout, and shrimp. Production is capital intensive and depends on input from outside the ecosystem of the production area.

In India, a wide range of aquaculture practices are followed to produce both protein and cash crops. The production of protein crops by means of extensive traditional farming or integrated rice-cum-fish culture has been the major source of fish protein (Jana 1998a). Shrimp farming, an example of a cash crop, has emerged recently as an important aquaculture practice for earning foreign exchange in the global market. In essence, the principle of sustainable development needs to be increasingly applied to reduce the possibility of any future harm to the environment from the industrial scale of aquaculture. In other words, aquaculture systems need to be productive within the carrying capacity of the ecosystem, as well as cost-effective, environmentally sound, and socially acceptable in the long run.

KEY ISSUES OF AQUACULTURE UNSUSTAINABILITY

Intensive Aquaculture

While efficient, extensive culture has developed gradually over the years, intensive farming has gained prominence in recent years, which has of course brought much benefit and bounty to the country. However, the environmental issues of intensive shrimp farming have created considerable debate because of the generation of a substantial amount of particulate organic wastes and soluble inorganic excretory wastes. It is estimated that intensive shrimp farming can generate 1,500 kg of total nitrogen (N) and 400 kg of total phosphorus (P)/ha of pond per year (FAO 1996) and thus pose a serious threat via environmental pollution.

Aquaculture practices are closely associated with environmental issues. Increase in the concentration of suspended solids and dissolved nutrients in water bodies is known as hypernutrification, whereas the consequences of algal bloom development is called eutrophication. Eutrophication may induce the bloom of toxic algae, while hypernutrification may bring a change in species composition. Evidently, the negative impacts of such hypernutrification and eutrophication increase with an increase in stocking density of cultured species, especially with shrimp.

Studies (Pullin 1989; ESCAP/FAO/UNIDO 1991; Bergheim and Asgard 1996) reveal that application of rotenone, tea-seed cake, and quicklime, which lose their toxicity in a short time after application, may cause minimal adverse effects on an environment. On the other hand, fertilizers, feed, and fish feces largely affect the survival, growth, and reproduction of fish and other biota, as well as the species diversity of the habitats. According to an estimate (ESCAP/FAO/UNIDO 1991), the optimal pond fertilization rate of 4 kg N and 1 kg of P/ha/day used to achieve high yields is as high as that used on the most intensively fertilized field crops. In culture practice, excessive organic loading through aquaculture activities leads to increased oxygen consumption, giving rise to decreased dissolved oxygen levels and increased biological oxygen demand (BOD) levels in water, along with anoxic conditions stimulating hydrogen sulphide production in sediments with a reduced diversity of benthic organisms under intensive culture (Pullin 1989; Bergheim and Asgard 1996). The degree of pollution in shrimp farming is highly dependent upon the type of farming adopted. Shrimp grow-out systems are grouped as (characterized by natural food and tidal flushing), semi-intensive (supplemental diets and occasional pumping of water), and intensive (complete dependence on formulated diets, water pumping, and circulation/aeration), with stocking rates of 1-3, 3-10, and 10-50 shrimp/m^2, respectively. Additionally, the effluent quality from an aquaculture farm is greatly influenced by the rate of production per unit volume of water and water retention time on the farm, apart from other factors such as depth and temperature of the water and cleaning operations on the farm (Pillay 1994). It is estimated that intensive systems can generate 7 to 31 times more N load than semi-intensive systems (Edwards 1993). The contribution of nutrients from a cage of salmon in northern Europe was estimated to be generally on the order of 10 to 20 kg P and 75 to 95 kg N (Penczak et al. 1982).

The contribution of fish feces increases the concentration of ammonium nitrogen and phosphate-phosphorus in the surrounding water column up to two and four times, respectively (Larsson 1984). Notwithstanding, the N loading from aquaculture operations, ranging from 75 to 211 kg/ton fish produced, is shown to be relatively low (4-13%) compared to agriculture and other point-source activities.

There are reasons to believe that the aquaculture industry is relatively safe environmentally, though intensive farming has raised considerable debate in recent years. However, in order to make it free from controversy and sustainable, a more holistic approach needs to be applied. When the environmental and socioeconomic consequences of shrimp culture are listed, negative factors appear to outnumber the positive (Primavera 1998).

Soil and Groundwater Salinization

Though most aquaculture activities are highly complementary to agriculture, intensive shrimp culture was reported to reduce agricultural yields in certain localities where soil conditions allowed saline water to seep through embankments and pond bottoms into adjacent agricultural fields. In addition, excessive abstraction of groundwater for agriculture, domestic water supply, industrial activities, and in some cases shrimp culture are causing seawater intrusion into coastal aquifers. However, in a study in India conducted by the National Environmental Engineering Research Institute and Marine Products Exports Development Authority revealed that aquaculture farm effluents do not alter the salinity of the soil/water beyond 25 m.

The water quality of pond effluents at the time of harvest and cleaning of ponds under an intensive farming system may become more polluting with total N level ranging from 1,900 to 2,600 mg/L and 40-110 mg/L P. Such effluents should not be discharged directly into coastal waters but into sedimentation or treatment ponds. Hence, a shrimp farm over 8 ha should be equipped with wastewater treatment or sedimentation ponds covering not less than 10% of the total area. Because suspended particles settle more rapidly from brackish water than from fresh water (Yoo and Boyd 1994), the benefit accrued from sedimentation ponds in shrimp culture is obvious.

Destruction of Mangrove Swamps

Mangrove swamps are increasingly used for the construction of shrimp farms in tropical countries. It is reported that between 27% and 50% of the total mangrove areas have been destroyed for aquaculture activities in some countries in Southeast Asia, whereas it is only 13% in Equador (Pillay 1994). There has been much controversy regarding destruction of mangrove vegetation for shrimp farming. In India, utilization of mangrove areas for shrimp farming is minimal (Ninawe 1999a). From the point of view of suitability for siting aquaculture farms, mangrove swamps do not rank very high, mainly because of their acid sulphate soils that prove detrimental to shrimp growth. Soils under red mangroves, *Rhizophora* spp., have a peaty texture because of the extensive network of minute rootlets. Therefore, mangrove destruction may not be closely linked with well-planned and well-managed shrimp farms.

However, destruction of seed stocks and numerous marine fish species may be a problem. Thousands of women and children from fishing families along the coast are engaged in the collection of wild shrimp juveniles from the sea and rivers. They select the wild shrimp juveniles out of mix taken from the sea, and they destroy fish spawn by throwing them onto dry land. As a result, massive destruction of fish spawn along the coasts of India is seriously depleting the fish populations of the sea, having a direct economic impact on fishermen. This issue has come to prominence in the context of loss of biodiversity due to negligence. Therefore, there is an urgent need to educate fishermen about the loss of biodiversity due to existing practices.

Marine pollution has also been a problem for India. Until recently, the ocean has been regarded as a buffer that can receive huge quantities of nutrients without being polluted. Though dissolved nutrients are easily diluted in the water column, heavy particles such as uneaten diets and feces may fall to the bottom and may become a source of pollution. Complaints from fishermen that many fish catches from coastal water fishing zones in the Bay of Bengal are being considerably depleted due to discharge of effluents from shrimp farming ponds have been received by the Fisheries Department. It is suggested that optimal diet composition, feeding regimens, and managerial efficacy, may reduce pollution risk significantly.

Environmental Impacts of Aquachemicals and Therapeutics

While disease outbreaks may be managed by proper animal husbandry and environmental management, it is not be possible to completely eradicate disease incidence without the application of antibiotics. The drugs used in aquaculture are topical disinfectants, organophosphates, antimicrobials (sulphonamide, tetracyclines, quinolones, nitrofurans, chloramphenicol, etc.), and a wide range of anthelminthes and parasiticides (de Kinkelin and Michel 1992). Also the chemicals used in aquaculture systems are diet additives and hormones. There are several major environmental and human health hazards associated with the use of aquachemicals and therapeutics.

Persistence in Aquatic Habitats

Although many chemicals used in aquaculture degrade rapidly, some persist for several days and even months. Formalin, a widely used parasiticide and fungicide, has a half-life of 36 hours in water. Dichlorovos, a parasiticide, may persist for over a week in sea water. Metal-based compounds such as organotin can, however, persist for several months in sediment. Antibiotics, such as oxytetracycline, oxonilic acid, and fumequine have been detected in sediments even after six months following their application (Sze 2000). Studies have shown that the antimicrobial activities of oxytetracycline persisted in bottom sediments up to 12 weeks after its application, causing much alterations in the bacterial flora and sediment processes (Sze 2000).

Residues in Cultured Animals

Residues of persistent chemicals in aquaculture products are a matter of serious concern. Developed countries are increasingly imposing restrictions on the occurrence of such residues in cultured organisms, and the concept of maximum residue limit has been focused by food safety experts. As a result, fish exporting countries are compelled to create facilities for monitoring products before shipment to international markets.

Persistent organic pollutants (POPs) pose a danger to human health and the environment on a global scale. POPs are toxic, last for a long time in the environment, travel long distances from the source of emission, accumulate in the fatty tissue of living organisms, and increase in concentration at successive trophic levels.

The United Nations Environment Program mandate for the global treaty being negotiated is to reduce and/or eliminate the production, emission, and uses of 12 specific POPs. These are the pesticides aldrin, chlodane, DDT, dieldrin, endrin, heptachlor, mirex, and toxaphene; the industrial chemicals such as polychlorinated biphenyls (PCBs) and hexaxchlorobenzene; and the combustion byproducts dioxins and furans (Durkee 1999). Significant advancements have been made in recent years towards rational and wise use of pesticides and antibiotics. For example, from 1987 to 1996 the use of antibiotics in Norwegian salmon culture decreased from 1.065 g to 0.003 g/kg fish produced (Asgard et al. 1998).

In order to minimize public health risks some countries have stipulated withdrawal periods for drugs administered to fish. As disintegration half-lives are dependent on temperature and other factors, the general tentative withdrawal period has been recommended as 80 days and 40 days after final application in waters with temperature below 10°C and above 10°C, respectively (Rusmussen 1988). Information is inadequate; there is a need to synthesize and disseminate information on the use of management of aquachemicals (Barg and Lavilla-Pitogo 1996).

Pesticides such as organophosphates and ectoparasiticides, when used, result in the release of significant quantities of toxic materials in the ambient water and can adversely affect larval stages of crustaceans and other organisms.

Drug Resistance

Antibacterial agents are extensively used, orally or by immersion, to treat bacterial diseases, in addition to prophylactic measures. In the process, about 20-30% of the antibiotic administered orally are actually taken up by the fish, and the rest reach the environment directly and eventually undergo degradation, which can lead to selective resistance of native bacteria to drugs (Pillay 1994). Such applications have resulted in increased resistance in obligate fish pathogens such as *Aeromonas salmonicida*, as well as in opportunistic pathogens, including *Vibrio* sp. and the motile aeromonads. Non-pathogenic bacteria in the marine environment may eventually pose a threat to humans by transferring resistance to human pathogens mediated through plasmid relocation. The frequency of resistance increases by the presence of antibacterial agents whose concentrations are not sufficient to kill bacteria. With a view to solving the problem of drug resistance, Chythanya et al. (1999) suggested developing new classes of antimicrobial drugs and a managerial approach that may help to avoid the development of resistance to existing antibiotics. The antibiotics should be used at slightly higher doses so that the microbial population level in the aquatic system is reduced before mutants have a chance to appear. However, no well-founded data are available on these aspects.

The organic matter accumulated in the pond bottom is decomposed under aerobic conditions and releases carbon dioxide and nitrate. Presence of antibacterial residues inhibits microbial activity and aerobic degradation, leading to accumulation of organic matter and, thereby, anoxic conditions in the sediment. Subsequent anaerobic degradation can result in increased production of toxic products such as sulphides, methane, and ammonia, which are detrimental to the fish population under culture.

Hormone Residues

Different kinds of third-generation hormones are used in aquaculture. Experience gained so far has clearly shown the need for caution in the use of hormones for induced spawning and growth promotion and of the steroids such as methyltestosterone for obtaining mono-sex progeny. The possible effect of hormones and drugs residues of treated finfish and shellfish on human health has to be weighed against the benefits that may be derived from their use in aquaculture.

STRATEGIES FOR SUSTAINABILITY

Aquaculture has an initial cost: an investment of time, effort, and resources. In general, the more the intensive the culture system, the greater the capital investment. Important linkages for efficient fish production are scientific knowledge, organization, managerial efficacy, etc. Nevertheless, there is considerable debate on the issue. Can efficient aquaculture be environmentally sound and still contribute to sustainable food production? In order to achieve sustainability, integrated resource management has emerged as the key factor for significant aquaculture production, especially in the developing countries.

Integrated fish farming has, therefore, become an important tool for rational use of various resources. The rationale behind raising fish on animal manure becomes apparent when it is found that 72% to 79% of N, 61% to 87% of P and 81% to 92% of potassium (K) in the diet rations fed to animals are recovered in their excreta. As a result, a farm integrating agriculture, horticulture, fish culture, dairy and poultry has become a popular model of farming in different parts of the world, with varying degrees of success. In India, the management of solid waste for biogas and vegetable production and of waste water for fish production through recycling of organic wastes has assumed greater significance in view of its immense potential in the improvement of environmental sanitation and the development of rural economy.

The main objective of integrating fisheries and agriculture is to maximize the synergistic and to minimize the antagonistic interactions between the two sectors. The synergistic actions are mainly due to recycling of nutrients arising in the course of the agriculture-livestock fish production process, from integrated pest management, and from optimal use of water resources. Antagonistic interactions arise from the application of aquachemicals that harm aquatic living organisms, eutrophication, soil erosion, drainage of wetlands and swamps, and the obstruction of fish migration routes. Direct interactions between fisheries and agriculture occur where these two sectors compete for same kind of resource, especially land and water, and where measures are aimed at higher agricultural production. Agricultural activities such as damming, wetland reclamation, drainage, water abstraction, and transfer for irrigation have resulted in a considerable decline of inland fisheries over the past few decades (Anon 1997).

Integrated pest management practices, fish culture, and rice farming are complementary to each other, as cultured fish can reduce pest populations and enrich paddy fields through feces. Selection of suitable pesticides with high efficiency but low toxicity to fish is most important in the management of rice-cum-fish culture. During pesticide application, the water level in paddy fields is lowered to force fish to move to the sumps to avoid toxicity. Different kinds of culture practices such as raising fry and fingerlings and production of marketable size fish are followed (Gupta et al. 1998), and modifications in design criteria have been made over the years. With savings on pesticides and additional earnings from fish sales, the net profit on rice-fish farms is reported to be significantly higher than on rice monoculture farms by a margin ranging from 7% to 65% (Halwart 1995).

Culture Practices

In areas where paddy fields retain water for 3 to 9 months in a year, rice-cum-fish culture can provide an additional supply of fish. Depending on the intensities of cultural practices involved, fish culture in rice fields may be grouped as follows: secondary crop of paddy after paddy; along with paddy during the period of cultivation; and continual fish culture, transferring the fish to specially prepared ditches or channels during the harvesting period or when fields are drained. Over 80 million ha of land produces the world's supply of rice, and rice-cum-fish culture yields approximately 3 kg of fish per ha for an inundated period of 3-8 months (Jhingran 1995). There was a twelve-fold increase in economic benefits from integrated rice-fish systems combined with vegetables or fruit crops grown on the bunds, as compared with traditional rice farming (Anon 1999a).

In India, low-lying, waterlogged tracts covering an area of above 2.3 million ha is the highest in the world. Though six million ha are under rice cultivation, only 0.03% of this area is now used for rice-fish culture. Some other agricultural crops such as maize, bananas, coconuts, etc., are also integrated with fish culture. The most common practice used at present is culture of fingerlings to market-sized fish in rice fields concurrently with rice crops. Practices in rice-cum-fish culture revealed the use of two crops of rice (a tall variety of rice in Kharif and a high yielding variety of rice in Rabi) and a single crop of fish. Experiments performed using traditional and non-traditional species of fish in different states of India indicated a wide range of fish yield, from 17 kg/ha to 240 kg/ha within a growing period of 2 to 9 months (Ghosh et al. 1985; Jhingran 1995).

Introduction of certain traditional and non-traditional fish species (bata, *Labeo bata*; catla, *Catla catla*; mrigal, *Cirrhinus mrigala*; murrel, *Channa striatus*; Pearl spot, *Etroplus suratensis*; mourala, *Amblypharyngodon mola*; common carp, *Cyprinus carpio*; and Mosambique tilapia, *Oreochromis mossambicus*) in rice-cum-fish culture is another way of reducing the cost of supplementary diet and inorganic fertilizers, by recycling the organic wastes from both plant and animal origin (Ninawe 1999b).

Culture of brackishwater prawns and fishes is also common in the states of West Bengal and Kerala in India. The important species of fish cultured in paddy fields are mullets, *Mugil parsia*, *M. tade*, *Rhinomugil corsula*; sea bass, *Lates calcarifer*; catfish, *Mystus gulio*; freshwater prawns, *Palaemon carçinus*, *P. stylifirus*, *Macrobrachium rude*, *Metapenaeus monoceros*, *M. dobsoni*, *M. brevicornis*; brakishwater shrimps, *Penaeus semisulcatus*, *P. indicus*, *Penaeus monodon*; and crabs *Caridina gracilirostris* and *Acetes* sp. However, the present trend of leaning heavily towards multiple cropping in the course of a year has become a problem for development (Table 1).

INTEGRATED AQUACULTURE

In general, cultured fish are selected on the basis that resource input should be as low as possible. Carp polyculture has gained considerable importance as a basis for rational utilization of resources in the pond ecosystem and for contributing much towards sustainability (De et al. 2000). Originated in China and India, polyculture is a traditional production system of aquaculture in many countries of Asia involving traditional management protocols or integration with different sectors of animal husbandry. Basically, it consists of raising compatible species that occupy different ecological niches and feed on different food resources in the farm at optimal level for maximum fish production. It de-

TABLE 1. List of fish and prawn used in rice-cum-fish culture in India.

Freshwater fishes	Brackish water fishes	Prawns and crabs
Rohu (*Labeo rohita*)	Mullet	Freshwater prawn
Catla (*Catla catla*)	(*Mugil parsia*)	(*Macrobrachium rude*)
Mrigal (*Cirrhinus mrigala*)	(*Mugil tade*)	
Sal (*Channa striatus*)	(*Rhinomugil corsula*)	Non-penaeid prawn
		(*Palaemon carcinus*)
Pearl spot	Sea bass	(*Acetes* sp.)
(*Etroplus suratensis*)	(*Lates calcarifer*)	
		Penaeid prawn
Commom carp	Catfish	(*Penaeus monodon*)
(*Cyprinus carpio*)	(*Mystus gulio*)	(*Penaeus indicus*)
		(*Penaeus semisulcatus*)
Mossambique tilipia		(*Metapenaeus brevicornis*)
(*Oreochromis mossambicus*)		(*Metapenaeus monoceros*)
		Crab
		(*Caridina gracilirostris*)

velops a symbiotic relationship between the species and thereby creates an ecologically a benign environment minimizing aquaculture-dependent environmental degradation.

In practice, 4-8-species combination with exotic and indigenous carps, air breathing fish, freshwater prawn, and other suitable fish of different feeding habits is followed. Fish production under different conditions was found to range from 1,504 kg/ha/yr to 9,300 kg/ha/year (Jana 1997).

Low cost fish farming using manure from ruminants, buffalo, cattle, pig, and other domestic animals has been used in many developing countries. The economic viability of pigs and use of their manure is not generally acceptable. On the other hand, cow manure is often used in fish culture ponds in India. However, among the various livestock integrated with fish farming, production of ducks has been found to be economically the most efficient (Gavina 1994) and has been shown to be viable (Sharma and Olah 1986; Sharma 1989).

Fish farming integrated with chicken production also appears to be promising. Many fish ponds are used as receptacles for domestic sewage. In some countries of southeast Asia, live-stock, crops, human wastes, and fish are integrated in either two- or three-tier systems, making it more integrated and productive (Chakravarty and Jana 1984). In a long-term study conducted by the Asian Institute of Technology for promoting sustainable aquaculture systems for small-scale farmers, Edwards et al. (1994) observed that the manure from 1.7 buffaloes would produce the same fish yield in a 200 m^2 pond as integration with 30 feedlot ducks, assuming that fish yield is proportional to manure N content. Daily buffalo manure loading rates of 0, 2, 4, and 6 kg dry matter (DM)/pond/day were equivalent to daily manure outputs from 0, 1, 2 and 3 buffaloes, respectively. The response curve relating to buffalo manure input to net fish yield (NFY) (Edwards et al. 1994) indicated a steep rise to an NFY of 20 kg/200 m^2 pond with a daily input of 1 kg buffalo manure in the 6-month experiment, but then only a gradual doubling of fish output in response to a tripled buffalo manure input, suggesting that buffalo manure is nutritionally poor. A study (Edwards et al. 1994) indicated that buffalo manure inhibited phytoplankton growth due to its low soluble N and P inputs and by marked reduction of light penetration into the pond water column (Shevgoor et al. 1994) which stained pond water dark brown due to release of tannins and flavonoids. As a result, phytoplankton biomass seldom exceeded 15 mg/L in buffalo-manured ponds, compared with 35 mg/L in duck-manured ponds.

Regression analysis indicated that a more efficient use would be a daily input of a maximum of 1 kg DM per 200 m^2 pond (Edwards et al. 1996). The buffalo manure conversion efficiency would almost double from 1.0 to 1.9% with a recommended reduced buffalo manure loading rate from 5 to 2 kg DM leading to further increase in fish yield. This suggests a limited value of buffalo manure as a major pond fertilizer, due to competition from vertically integrated agro-industry. Attempts to introduce feed-lot livestock/fish integrated technology in a top-down mode have failed in Africa and Asia (Pullin and Prein 1995), perhaps due to lack of technique standardization. However, many regions of the world do not allow the use of human or animal wastes in any food production system, and the practice of using human or animal wastes may be of limited value.

Intensification of fish production on a small-scale farm is possible. A minimum harvest of 100 kg fish/year can be produced on a small-scale farm, the total area of which may be only about 10,000 m^2, from a small pond typically measured with an area of tens or hundreds of square meters, if NFY is relatively high. High NFY can usually be attained from a small-scale pond by significant amounts of off-farm, agro-industrial inputs. Doubling or tripling of the extrapolated NFY of 3 tons/ha has been achieved (Knud-Hansen and Batterson 1994).

Wastewater-fed aquaculture is assuming greater importance in developing countries due to the fact that the quantity of waste water generated by the ever-increasing human population is beyond the capacity of conventional treatment plants (Edwards 1980). In India about 90 tons N, 32 tons P and 55 tons K valued at Rs. 61 million (US $1 = Rs. 46) could be recovered from the country's domestic sewage of 15 billion L daily (Anon 1996a). This is possible through recycling of wastes into fish biomass and thereby providing excellent employment opportunities for millions of people in developing countries.

Practice guidelines are now available for sewage-fed aquaculture. In general, waste water is treated for one day in a stabilization pond (anaerobic pond), followed by a five-day treatment in facultative ponds prior to discharge into a culture pond. Various species of carp stocked at 10,000/ha or higher is used. A production range between 2 and 8 tons/ha/year is reported from well-managed semi-intensive carp polyculture ponds in Asia. It is essential that the fish be microbiologically safe for consumption, provided the count of fecal coliforms in the pond water is $< 10^3$/100 mL (WHO 1989). However, in many developed countries, even biowaste is considered unsafe, and thus fish are not allowed to grow in an environment with human waste.

In India, wastewater-fed aquaculture systems started in the state of West Bengal in 1925. Since then, the facility has become one of the largest fish farm in the world, covering an area of 12,000 hectares and producing on an average of 1,200 kg of fish/ha/year. A record production of 9,534 kg/ha/year from some of the sewage-fed ponds was reported in 14 months (Banerjee 1981; Jana 1998b). Fish culture in treated domestic sewage yielded a net production of 3,782 kg/ha/year without adding any supplementary diet of exogenous fertilizer (Pataik 1987; Orino et al. 1987; Verma 1989; Sharma and Olah 1986; Sharma 1989; Mahadevaswamy and Venkataraman 1998). Early investigations have demonstrated that carp production can be increased to 10 and 15 tons/ha/year even in non-aerated ponds (William et al. 1991).

In some fish farms in southern India, intensive aquaculture using red tilapia has been used to reduce the level of pollution by over 99%; the BOD values were reported reduced to 30 mg/L (from more than 100 mg/L) in the case of distillery effluents and to less than 100 mg/L from 2,000 mg/L of pre-treated effluent waters from breweries (Banerjee and Srinivasan 1988; Balasubramanian et al. 1990; Rangaswami 1993). Some fish farms in India used waste water from over 800 sago/starch factories for fish culture. In these farms, the waste water was treated, using anaerobic digestion for methane production and treated further to reduce the BOD level below 250 mg/L so as to make it useful for fish culture.

In recent years, there has been a major shift in the reclamation strategies of waste water in India from high-cost, traditional engineering technology to environmentally sound, low-cost, sustainable ecotechnology integrating some aquatic macrophytes as living biofilters. In essence, the macrophyte-based ecoengineering reclamation approach is a low-cost, environmentally sound, and sustainable method for controlling nutrient enrichment and keeping heavy metal pollutants from waste water. Floating, submerged, and emergent macrophytes have shown immense potentials for the reclamation of wastewater (Welch and Lindell 1996; Jana et al. 1996; Haq and Ghosal 2000; Saha and Jana in press a, b; Saha et al. 2000). Such reclaimed water can be used as a resource for fish farming (Saha 2000). While water hyacinth, *Eicchornia crassipes* is widely used for removing nutrients and heavy metals from waste water (Jana and Das 1999; Shome and Banerjee 2000), duckweed in fish ponds appears to provide a complete, balanced diet for some carp that consume it directly. The feces of fish consuming duckweed are consumed directly by detritus feeders used in fish culture (Ninawe 1999b).

Aquaculture wastes are biodegradable and thus pose relatively less threat to the environment than industrial wastes. The pond sludge used as a fertilizer for agriculture. In conventional sewage systems such as activated sludge, trickling filters and stabilization ponds are designed to treat waste water and remove nutrients. In a typical sewage-fed aquaculture system, the ponds are dried up, and the sludge that accumulates at the bottom of ponds is removed before filling a pond with sewage effluents.

ENVIRONMENTAL ASPECTS

Concurrent and rotational cultivation practices of finfish and shellfish with rice are gaining interest currently in Italy, Japan, the U.S., Spain, and several African countries, though is originated in Asia. Some shrimp farmers in Indonesia obtained relatively good harvests of shrimp after their ponds were used to produce one crop of tilapia. Some farmers in Tamil Nadu in India experimented with crop rotation and successfully cultured *P. monodon* during the dry season and freshwater prawn during the wet season (Srinivasan et al. 1997). In general, the greater the phylogenic differences between the culture organisms selected in crop rotation, the better the sanitary effects (Francis and Clegg 1990).

Recent studies suggest that the use of antibiotics and other chemicals in controlling shrimp pathogens becomes ineffective as the strains grow more resistant to these chemicals. Moreover, the bacterial pathogen, *Vibrio harveyi*, produces a biofilm coating that protects it from drying, and chemical disinfection procedures that are followed during pond preparation. As a result, biological control has become an alternative means of preventing shrimp disease outbreaks. The main principle of the biological control is to enhance the growth of beneficial microorganisms (probiotics) that serve as an antagonosis of target pathogens. Secondly, the environment is mitigated in such a way as to promote the growth of beneficial microorganisms. In practice, this is often done using probiotics and crop rotation, respectively (Paclibare et al. 1998). Nevertheless, experiences on the use of probiotics in shrimp culture have yielded conflicting results. Moreover, the use of probiotics may still not be cost-effective, because high amounts of the costly probiotic products must be added to the ponds frequently. It appears that there is great promise in this particular area of research, and further research is needed to standardize the techniques.

Improvement in diet quality in the form of high-energy diets and the use of different feeding regimens can reduce the solid-waste and nutrient load (especially N) (Kibria et al. 1998) and fish production cost. Extended diets can also reduce nitrogen and phosphorus discharge from aquaculture operations. Recently it is claimed that as much as 50% of the animal protein in salmon diets has been replaced by single-cell proteins produced by using methane gas (Berg et al. 1995).

The wetlands, coastal wetlands, estuaries, and seasonal flood plains serve as repositories of aquatic biodiversity (Willman et al. 1998). Many of the world's large and small river basins have undergone major anthropogenic changes in their hydrological regimens over the past few decades. While the construction of dams, reservoirs, embankments, barrages, and channels has brought large benefits for fisheries, in many cases modifications in hydrological systems have caused drastic declines in natural fish populations, fish catch, loss of spawning grounds, migratory routes, etc. As a result, some indigenous fish species have become extinct in India.

Efficient Use of Water Resources

Apart from the benefit of fish production, enhanced fish culture in reservoirs and channels often helps to maintain water quality and the physical functions of these water bodies. Stocking with grass carp controls aquatic weeds in irrigation channels, thereby facilitating water flow and reducing evaporation rates during water transport. Stocking and fish culture can also reduce human health hazards caused by mosquitoes and other insects. A close link between fisheries and irrigation departments (Blehle 1998) would enable the ponds to function as water-holding facilities allowing improved irrigation management and consequently increasing vegetable yields. Such an integrated management approach would contribute much towards environmental sustainability.

Socioeconomic Considerations

Economic viability is another important aspect of sustainability. The economic viability of a farm depends on some economic factors such as scale of operation, finance and credit condition, cost and return from the project, market demand, and government support measures. A fishing industry cannot be economically viable without efficient commercial networks, modern processing units, and other infrastructures. A minimum economic size for farms is of considerable importance for sustainable and efficient management. For sustainability, small-scale aquaculture has been proved efficient. In many areas it has been found the rate of growth in large-scale aquatic facilities is less than that of the small-scale farm. For poverty alleviation and food security, the impact of small-scale rural aquaculture projects is of importance in developing countries (Gupta and Dey 1999). The impact indicators for rural aquaculture projects could be broadly categorized as socioeconomic, ecological, and institutional. On the other hand, the development of export-oriented aquaculture, especially of salmonids and shrimp, is important for large-scale enterprises.

In large-scale aquatic farming, to sustain production, the entrepreneurs set up different units for different activities with suitable horizontal linkages. In small-scale aquatic farming, vertical integration is possible. The construction and maintenance of aqua-farms, seed and food production, together with harvesting and sales may all be integrated in a centralized operational form (Pillay 1994). As a result of economic reform in China, considerable reduction in the size of farms has led to remarkable increase in output. Smaller farms have therefore immense potential in terms of productivity and profitability.

Medium-scale aquaculture enterprises have gained increasing importance in India. The owners of such farms have been successful in mobilizing additional resources from different sectors. The state-sponsored development programs and credit from banks give financial support to the aquaculture farms. Further, the external funding sources such as Food and Agricultural Organisation, United Nations Development Programme, United Nations agencies (Shehadeh and Orzeszko 1997), also provide finance to aquaculture farms. The amount contributed by the development banks accounted for as much as 69% of the funding and 40% of the projects, while the contributions from bilateral sources represented 17% of funding and 6% of the projects. Multilateral sources contributed 34% of the projects and 7% of funding. The major beneficiaries of external aid to aquaculture farms from developmental banks during 1988-95 are India, China, Bangladesh, and Mexico (Shehadeh and Orzeszko 1997).

Among the various types of culture fisheries resources in inland sectors of India, freshwater ponds are the greatest potential source for sustainable and substantial increase in fish production. In India, ponds are categorized into three classes according to size. Small ponds with surface areas below 0.1 ha belonging to poor, marginal farmers are mostly seasonal, semi-derelict, and highly silt-affected because of lack of financial resources for reclamation, resulting in very poor yields (0.2 tons/ha). Medium-size ponds, mainly representing the traditional culture system, contribute about 0.8 tons/ha. The third category, bigger ponds, show a productivity of 1-2 tons/ha and 2.5 tons/ha. The fisheries sector in India at present faces some problems. All three sizes of ponds continue to remain underutilized. Conversion of agricultural lands into aquaculture farms in an unwise manner, multi-ownership problems, conflicts between aquaculture and agriculture have led to unsustainable development of the aquaculture industry in India. It is suggested that better use of these ponds via financial resources, low cost technology, extension service including training programs particularly for women, and marketing facilities should be conducted.

It is estimated that freshwater aquaculture provides full time employment for about 1.2 million laborers and part-time employment for 250,000 people, while brackishwater aquaculture employs about 67,000 and 15,000 people full-time and part-time, respectively (Sinha 1999). In 1993, the number of fish farmers was 5.8 million, which was 0.70% of the total population of India. Another 3-3.5 million people are believed to be engaged in fish processing and ancillary industries.

With economic liberalization of the Indian economy, there has been a good impact on the fisheries export market. It has resulted in a steady rise in the quantity and value of exports The export quantity is found to have increased from 99,306 tons in 1988 to 29,4264 tons in 1995, i.e., a growth of 196% during the period, while in value, (in Rs) exports have increased from 420 to 1,07 million tons during the same period.

Aquacultural exports continued to be an extreme focus sector during the ninth plan of India. Indian fisheries have strong competitive advantage, not only regarding huge natural resources, but also because of abundant skilled and unskilled labor at comparatively low wages (ASCI 1996).

Sustainability in aquaculture fish production will give an adequate exportable surplus of fish in the coming years. At present the aquaculture sector accounts for 43% of the total output of shrimp in the country and 66.8% of the total earnings from shrimp exports.

Environmental Regulation and Fisheries Act

Society is required to accomplish its goals through regulation. Absence of environmental regulation and tax enforcement of laws are the main reasons for the unplanned development of aquaculture industries in India. Recently, the government of India has given aquaculture one of the highest priorities in terms of liberal economic policy, strategy, and planning. At the national level, the Local Agenda 21 (LA 21) is of considerable importance for sustainable development at the local government level (GACGC 1999). As a result, of often irresponsible ecological activities, the Supreme Court of India has permitted eco-friendly, improved technology in the traditional system and improved traditional (selective stocking of species) systems within the coastal regulation zone (CRZ) to increase production to the level of 1.5 tons/ha/crop to ensure optimum utilization of the resources. Though the Indian Fisheries Act (Act IV of 1897) came into force during the British regime in 1897, the act has not strictly implemented the prohibition of poisoning and dynamiting, regulation of mesh, establishing a fishing season, etc. As a result, many aquaculture-generated unsustainable regimens and environmental hazards have emerged. However, there is hardly any documentation on environmental impact assessment. Because of the great economic impact of aquaculture on the rural economy, on the other hand, and of environmental degradation on the other, aquaculture activities need to be dealt with cautiously, re-addressing the coastal zone regulations in view of the present economic scenario.

According to Pillay (1992), the rules published in the Federal Registrar (Environmental Protection Agency 1974) on permit application to discharge effluents from aquaculture production systems require: identification of the quality and quantity of pollutants to be used in the aquaculture project; information on conversion efficiency of pollutants; data on culture organisms; data on water quality parameters; health effects of the proposed aquaculture projects; identification of pollutants produced by the species under culture; identification of the disposal method to be used. However, the EPA insists on tests only for settleable solids, and there are not any rigid water quality standards except in the case of marine siting regulations. Seim et al. (1997) stressed the need for more stringent requirements before granting discharge permits. Such regulations are particularly important in India in order to promote sustainability in aquaculture.

Many rivers and lake basins and their respective catchment areas are being shared by two or more countries, which is very common in Europe. Some rivers such as the Brahmaputra, the Padma, the Teesta, the Mathabhanga Churni, the Ichamati and others are shared by India and Bangladesh. Several aquaculture practices such as the introduction of non-native species, genetically altered stocks, siting and management, etc., have adverse effects on these transboundary ecosystems. An exotic fish such as *Oreochromis nilotica* is not a legally introduced species in India; nevertheless it is a very common cultured species in West Bengal, through transportation by fish traders from neighboring Bangladesh. Similarly, African catfish and other species find their way into India through fish traders.

There are numerous international agreements aimed at the environmental protection of transboundary aquatic ecosystems (58 of FAO Aquaculture development) (Caddy and Griffiths 1995). Similar agreements should be made for species introduction in the national interest.

Alien species have received international attention in recent years, due to their potential to assist in human welfare. While much of the recent attention focused on their adverse impacts, not all alien species are reported to be bad. Rainbow trout, *Oncorhynchus mykiss* is one of the important exotic species that is well established in the upland waters of India (Gopalakrishnan et al. 1999). A good number of exotic carps such as common carp, silver carp, and grass carp, introduced into India during the sixties, have not only adapted and become well established under Indian conditions but also have shown great promise as species compatible with Indian major carps in carp polyculture. The list is increasing to include more species introduced either legally or illegally through fish traders.

Ethical Aspects of Intensive Aquaculture

While irresponsible coastal aquaculture can have adverse environmental and socioeconomic consequences, fish and shrimp farming conducted in a sustainable and eco-friendly manner can bring many positive benefits such as food production, poverty alleviation, employment generation, and rural development (Singh 1999). Responsible fisheries may bridge the gap between fisheries regulation acts on one hand and the benefit of increased production on the other for nature and resource conservation. The technical guidelines for responsible fisheries (Anon 1997) provide annotations to the Principles of Article 9 of the Code of Conduct for Responsible Fisheries and for sustainable aquaculture development. The Code of Conduct for Responsible Fisheries has become the single-most important guiding principle to ensure long-term sustainability development in the fisheries sector (Feidi 1999). The priority issues for implementation of the code consisted of two articles and two annexes. The last seven articles discuss the general principles for fisheries management aquaculture development, and fisheries research.

TOWARDS A BALANCED MODEL

In pond aquaculture, sustainability should be used as a point of view rather than as a list of necessary steps to be taken in the production process. As a general guideline for the enhancement of inland capture fisheries production, Born (1999) listed a number of measures such as introduction of new species, stocking, fertilization, environmental mitigation, elimination of unwanted species, introduction of cage culture, aquaculture, and genetic modification. Such recommendations may be applicable under the humid tropical conditions of India. The government of India has taken the following programs for fisheries and aquaculture development: enhancing the fish production contributed by different sectors; generating employment and higher income in fisheries sector; improving the socioeconomic conditions of traditional fishermen and their families and of fish farmers; augmenting the export of marine, brackish and freshwater fin and shellfish and other aquatic species; increasing the per capita availability and consumption of fish to about 11 kg per annum; adopting an integrated approach towards sustainable marine and inland fisheries and aquaculture; and conservation of aquatic resources and genetic diversity. For implementing these programs, the Aquaculture Authority of India (Saktivel 1999) has made the following practical guidelines:

- Low stocking density should be aimed at the production of 2.0 to 4.0 tons/ha in two crops.
- Aquaculture farms should be located away from agricultural lands and mangrove areas.
- Fish farms above 5 ha should be provided with facilities for waste treatment.
- Beneficial microbes in the form of probiotics should be used to reduce load of pathogens.
- Feed quality and feeding techniques to reduce nutrient enrichment (Eikebrokk et al. 1991; Jensen 1991) by the use of high energy diets (Johnsen et al. 1993) and extended diets should be improved. Also, it is wise to practice semi-intensive farming coupled with the use of low pollution feed.
- The count of faecal coliforms in the pond water should be $< 10^3$ per 100 mL (WHO 1989), in order to make fish microbiologically safe for human consumption.
- Use of biodegradable pesticides should be encouraged.
- Remote sensing should be used for identification of potential aquaculture sites.
- A buffer zone of fresh water must be provided in the aquafarm in order to avoid ecosystem damage in the neighbouring areas.
- Beneficial coexistence of coastal aquaculture and agriculture should be promoted by the use of rotational systems of rice-fish or rice-shrimp culture.
- Low cost fish farming integrated with live stock should be promoted.

The major factors those are responsible for sustainable aquaculture have been outlined in a model (Figure 2). Quantifying optimal sustain-

FIGURE 2. A proposed model for sustainable aquaculture development in India. Sustainability is inversely proportional to the intensity of aquaculture activity. Solid line indicates the real activity, broken lines the possible sustainable benefit which can be achieved by reducing intensification. Other socioeconomic factors are also shown.

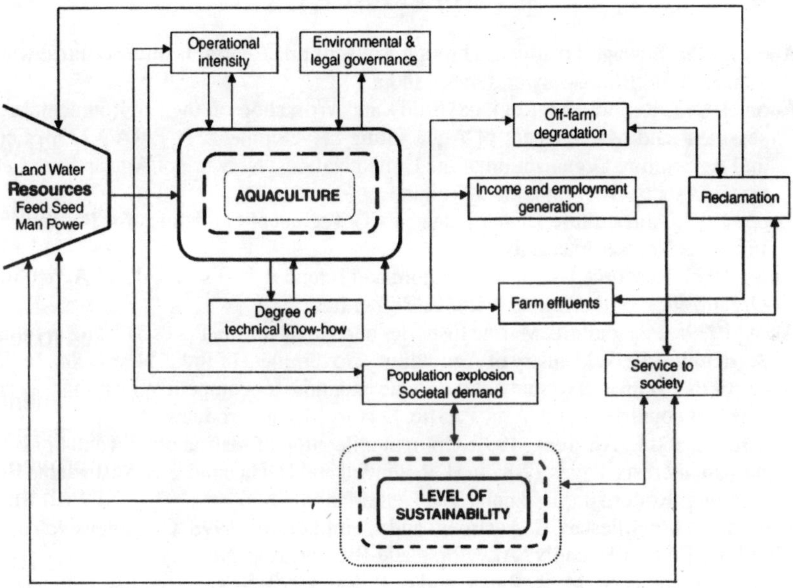

ability as nearly half of the resource exploited, the level of sustainability is again under the influence of multidisciplinary forces of the society, the more the demand without understanding the less the sustainability. The fundamental or potential sustainability thus can be achieved through functioning of interdependent factors like intensity of operation, use of technical know how and finally the ability of the society to compromise with nature.

REFERENCES

Anon. 1996a. Sewage Treatment Through Aquaculture. Central Institute of Freshwater Aquaculture, Bhubaneswar, Orissa, India.

Anon. 1997a. Report on a Regional Study and Workshop on the Environmental Assessment and Management of Aquaculture Development, TCP/RAS 2253, Food and Agriculture Organization of the United Nations, Network of Aquaculture Centres in Asia Pacific, Bangkok, Thailand.

Anon. 1997. Aquaculture Development. FAO Technical Guidelines for Responsible Fisheries No. 5 Rome, Italy.

Anon. 1998. Resource Use in Aquaculture and Inland fisheries. Food and Agriculture Organization of the United Nations, Rome, Italy.

Anon. 1999a. Pisciculture, Marine Fisheries and Aquaculture. Pages 517-567 *in* Indian Agriculture 1999, Ministry of Agriculture, Government of India, New Delhi, India.

ASCI 1996. Comparative study of corporate and industry competitiveness in India and selected countries in the Asia-Pacific, Sector: Marine Products.

Asgard, T., and E. Austreng. 1995. Optimal utilization of marine protein and lipids for human interests. Pages 79-87 *in* H. Rosenthal and H. Haaland, eds. Sustainable Fish Farming. Rotterdam, Balkema, The Netherlands.

Asgard, T., M. Hillestad, E. Austreng, and I. Holmefjord. 1998. Can Intensive Aquaculture Be Eco-Friendly? Akvaforsk and Biomar, Aas, Norway.

Balasubramanian, S., M. R. Rajan, and S. P. Raj. 1990. Sewage fed fish culture with reference to different BOD levels. Pages 266-274 *in* Proceedings of National Workshop on Animal Biotechnology, Chennai, India.

Banerjee, R. K. 1981. Utilization of Agricultural and Industrial Wastes in Integrated farming. Summer Institute of Farming System Integrating Agriculture, Livestock and Fish Culture, Central Institute of Capture Fisheries Research Institute, Barrackpore, India.

Banerjee, R. K., and K. V. Srinivasan. 1988. Recycling, reuse of *Penicillium mycelium* as fish pond manure. Biological Wastes 23: 107-116.

Barg, U., and C. R. Lavilla-Pitogo. 1996. The use of chemicals in aquaculture. FAO Aquaculture Newsletter No. 14:12-14.

Barg, U., J. Kapetsky, M. Pedini, B. Satia, U. Wijkstrom, and R. Willmann. 1999. Intergrared resource management for sustainable inland fish production. FAO Aquaculture Newsletter No. 23: 4-8.

Barica, J. 1992. Sustainable management of urban lakes: A new environmental challenge. Central Canadian Symposium on Water Pollution Research, Ontario, Canada.

Berg, G. M., T. Storebakken and G. Baeverfjord. 1995. A new bacterial protein source for Atlantic salmon. Pages 104-105 *in* N. Svennevig, and A. Krogdahl, eds. Quality in Aquaculture. European Aquaculture Society, Special Publication, 23, Ghent, Belgium.

Bergheim, A., and T. Asgard. 1996. Waste production from aquaculture. Pages 52-80 *in* D. J. Baird, M. C. M. Beveridge, L. A. Kelly, and J. F. Muir, eds. Aquaculture and Water Resource Management, Blackwell Science Publications, Oxford, England.

Blehle, C. 1998. Integrated aquaculture and irrigation in Zambia. FAO Aquaculture Newsletter 19:14-18.

Born, B. 1999. Overview of inland fishery enhancements from a global perspective. FAO Aquaculture Newsletter 21:10-18.

Caddy, J. F., and R. C. Griffiths. 1995. Living Marine Resources and Their Sustainable Development: Some Environmental and Institutional Perspectives. FAO Fisheries Technical Paper 353, Rome, Italy.

Chakravarty, N. M., and B. B. Jana. 1984. Utilization of organic wastes in fish culture: A brief review. Pages 72-86 *in* M. C. Dash, B. K. Senapati, and P. C. Mishra, eds. Proceedings of the National Seminar on Organic Wastes Utilization and Vermicomposting. Sambalpur University, Orissa, India.

Chythanya, R., D. K. Nayak, and M. N. Venugopal. 1999. Antibiotic resistance in aquaculture. Infofish International 18(3): 30-32.

Datta, S., and M. Chakrabarti. 1997. A Management Perspectives for Sustainable Fisheries Under Globalised Market Conditions: The Indian Scenario. Final Project Report, Indian Institute of Management, Ahmedabad, India.

de Kinkelin, P., and C. Michel. 1992. The use of drugs in aquaculture. Infofish International 11(4):45-49.

De, U. K., M. Basu, and G. Kundu. 2000. The ecological basis of carp polyculture towards sustainable fish production. Pages 169-173 *in* B. B. Jana, R. D. Banerjee, B. Guterstam, J. Heeb, eds. Waste Recycling and Resource Management in the Developing World. University of Kalyani, India and Ecological Engineering Society, Wolhusen, Switzerland.

Durkee, L. C. 1999. Persistent organic pollutants pose serious challenges to urban areas. UNEP IETC Newsletter, 4-5, May 1999.

Edwards, P. 1980. A review of recycling organic wastes into fish: Emphasis on the tropics. Aquaculture 21: 261-279.

Edwards, P. 1993. Environmental issues in integrated agriculture-aquaculture and wastewater-fed fish culture systems Pages 139-170 *in* R. S. V. Pullin, H. Rosenthal, and J. L. Maclean, eds. Environment and aquaculture in the Developing Countries. International Center for Living Aquatic Resources Management (I CLARM) Conference Proceedings 31, Manila, Philippines.

Edwards, P., K. Kaewpaitoon, D. C. Little, and N. Siripandh. 1994. An assessment of the role of buffalo manure for pond culture of tilapia. I. On-station experiment. Aquaculture 126: 83-95.

Edwards, P., H. Demaine, N. Innes-Taylor, and D. Turongruang. 1996. Sustainable aquaculture for small-scale farmers: Need for a balanced model. Outlook on Agriculture 25(1):19-26.

Eikebrokk, B., H. Flogstad, A. Hergheim, and T. Asgard. 1991. Prospects and perspectives for the development of green production technologies in Northern Seas agriculture. Pages 85-100 *in* J. M. Debois, and B. Bratten, eds. Seminar on Agriculture and Aquaculture. 1st ENS Conference, Stavanger, Norway.

ESCAP, FAO, UNIDO. 1991. Comparative Economic Indicators of the Fertilizer Sector. Agrochemical News in Brief, Special Issue, November 91. Fertilizer Advisory, Development and Information Network and Asia and the Pacific/ESCAP, Bangkok, Thailand.

Feidi, I. H. 1999. The implementation of the code of conduct for responsible fisheries. Infofish International 18(6): 56-61.

Food and Agriculture Organization (FAO). 1988. Aspects of FAO's Policies, Programmes, Budget and Activities Aimed at Contributing to Sustainable Development. Document to the 94th session of FAO Council, FAO, Rome, Italy.

Food and Agriculture Organization (FAO). 1990. FAO Activities Related to Environment and Sustainable Development. FAO Council Document. CL/98/6, 1990, Rome, Italy.

Food and Agriculture Organization (FAO). 1996. Integration of Fisheries into Coastal Area Management. 3. FAO Technical Guidelines for Responsible Fisheries. Food and Agriculture Organization of the United Nations, Rome, Italy.

Food and Agriculture Organization (FAO). 1999. The State of World Fisheries and Aquaculture 1998. Food and Agriculture Organization of the United Nations, Rome, Italy.

Francis, C. A., and M. D. Clegg. 1990. Crop Rotations in Sustainable Production Systems. *In* Sustainable Agricultural Systems. Soil and Water conservation Society, Iowa.

GACGC (German Advisory Council on Global Change). 1999. World in Transition: Ways Towards Sustainable Management of Freshwater Resources. Springer-Verlag, Berlin.

Gavina, L. D. 1994. Pig-duck-fish-*Azolla* integration on La Union, Philippines. International Center for Living Aquatic Resources Management (ICLARM) 17:18-20, Manila, Philippines.

Ghosh, A., S. K. Saha, R. K. Banerjee, A. B. Mukherjee, and K. R. Naskar. 1985. Package of practices for increased production in rice-cum-fish farming system. Aquaculture Extension Manual 4, Central Institute of Inland Fisheries Research Institute, Barrackpore, India.

Gupta, M. V., J. D. Sollows, M. A. Mazid, A. Rahman, M. G. Hussain, and M. M. Dey. 1998. Integrating Aquaculture with Rice Farming in Bangladesh: Feasibility and Economic Viability, Its Adopttion and Impact. ICLARM Technical Report 55, Manila, Philippines.

Gupta, M. V., and M. M. Dey. 1999. A framework for assessing the impact of small scale rural aquaculture projects on poverty alleviation and food security. FAO Aquaculture Newsletter No. 23:22-25.

Gopalakrishnan, A., K. K. Lal, and A. G. Ponniah. 1999. Conservation of Nilgiri rainbow trout in India. Naga, The ICLARM Quarterly 22(3):16-19.

Halwart, M. 1995. Fish as biocontrol agents in rice: The potential of common carp *Cyprinus carpio* and Nile tilapia (*Oreochromis nilotica*), Weikersheim. Margraf.

Haq, A. H. M. R., and T. K. Ghosal. 2000. Wastewater reclamation using duckweed. Pages 495-499 *in* B. B. Jana, R. D. Banerjee, B. Guterstam and J. Heeb, eds. Waste Recycling and Resource Management in the Developing World, University of Kalyani, India and Ecological Engineering Society, Wolhusen, Switzerland.

Hempel, L. C. 1998. Environmental Governance: The Global Challenge. Affiliated East-West Press Pvt. Ltd., New Delhi, India.

Jacobs, P., J. Garner, and D. A. Munro. 1987. Sustainable and equitable development. *in* P. Jacobs and D. A. Munro, eds. Conservation with Equity. IUCN, Cambridge, England.

Jana, B. B. 1997. Text Book of Fish Culture. Ananda Publishers, Calcutta, India (in Bengali).

Jana, B. B. 1998a. Sustainable aquaculture towards protein crop and cash crop production in India. Pages 151-159 *in* A. Moser, ed. The Green Book of Eco-Tech, Sustain, Graz, Austria.

Jana, B. B. 1998b. Sewage-fed aquaculture: The Calcutta model. Ecological Engineering 11:73-85.

Jana, B. B., and S. Das. 1999. Eichhornia induced depuration of cadmium by tilapia. Journal of Aquaculture in the Tropics 14:201-207.

Jana, B. B., S. Datta, and S. Saha. 1996. Reclamation of eutrophic waters: Nutrient removal by water hyacinth in response to qualitative and quantitative variations of fertilizers. Pages 331-336 *in* J. Staudenmann, S. Schonborn, and C. Etnier, eds. Recycling the Resource: Ecological Engineering for Wastewater Treatment. Environmental Research Forum, 5, Transtec Publications, Switzerland.

Jensen, J. B. 1991. Environmental regulations of freshwater fish farms. Pages 51-63 *in* C. B. Cowey, and C. Y. Cho, eds. Nutritional Strategies and Aquaculture Waste. Proceedings of the first International Symposium on Nutritional Strategies in Management of Aquaculture Waste. University of Guelph, Ontario, Canada.

Johnsen, F., M. Hillestad, and E. Ausreng. 1993. High energy diets for Atlantic salmon. Effect on pollution. Pages 391-401 *in* S. J. Kaushik, and P. Luquet, eds. Fish Nutrition in Practice, INRA, Paris, France.

Jhingran, V. G. 1995. Fish and Fisheries of India. Hindustan Publishing Corporation (India), Delhi, India.

Khoo, K. H., and Tan, E. S. P. 1980. Review of rice-fish culture in Southeast Asia. Pages 1-14 *in* R. S. V. Pullin, and Z. H. Shehadeh, eds. Integrated Agriculture-Aquaculture Farming Systems. ICLARM Conference Proceedings 4.

Kibria, G., D. Nugegoda, R. Fairclough, and P. Lam. 1998. Can nitrogen pollution from aquaculture be reduced ? Naga, The ICLARM Quarterly 21(1):17-25.

Knud-Hansen, C. F., and T. R. Batterson. 1994. Effect of fertilization frequency on the production of Nile tilapia (*Oreochromis niloticus*). Aquaculture 123:271-280.

Larsson, A. M. 1984. Hydrological and Chemical Observations in a Coastal Area with Mussel Farming. W. Sweden University of Gothenberg, Department of Oceanography, Report 46.

Mahadevaswamy, M. C., and I. V. Venktaraman. 1998. Integrated utilization of rabbit dropping for biogas and fish production. Biological Wastes 25:249-256.

Nayar, S., T. M. Anil, and M. Joseph. 1999. *Quo vadis*, marine fisheries in India? Infofish International 18(4):62-67.

Ninawe, A. S. 1999a. Coastal aquaculture versus environmen: Pros and cons. Infofish International 18(2):43-47.

Ninawe, A. S. 1999b. Need to promote sewage fed fish culture as ecofriendly production technology. Indian Journal of Environment and Ecoplanning 2:75-82.

Orino, C., C. B. Cowey, and J. Y. Chiou. 1987. Leaf protein concentrate as a protein source in diet for carp and rainbow trout. Bulletin of Japanese Society for Scientific Fisheries 44: 49-52.

Paclibare, J. O., M. C. J. Verdegem, W. B. van Muiswinkel, and B. E. A. Huisman. 1998. The potential for crop rotation in controlling diseases in shrimp culture. Naga, The ICLARM Quarterly 21(4): 22-24.

Pataik, S. 1987. Utilization of aquatic weeds. USAID Training programme of freshwater aquaculture. Central Institute of Freshwater Aquaculture, Bhubaneswar, Orissa, India 2: 231-232.

Penczak, T., W. Galicka, M. Molinski, E. Kusto, and M. Zalewski. 1982. The enrichment of mesotrophic lake by carbon, phosphorus and nitrogen from the cage aquaculture of rainbow trout *Salmo gairdneri.* Journal of Applied Ecology 19: 371-393.

Perez, J. F., and J. F. Mendoza. 1998. Marine fisheries, genetic effects and biodiversity. Naga, The ICLARM Quarterly 21(4): 7-14.

Pillay, T. V. R. 1992. Aquaculture and the Environment. Fishing News Books, Blackwell Publications Ltd., Oxford, England.

Pillay, T. V. R. 1994. Aquaculture Development: Progress and Prospects. Fishing News Books, Blackwell Publications Ltd., Oxford, England.

Primavera, J. H. 1998. Tropical shrimp farming and its sustainability. Pages 257-289 *in* S. S. De Silva, ed. Tropical Mariculture, Academic Press, London, England.

Pullin, R. S. V. 1989. Third world aquaculture and the environment. Naga, The ICLARM Quarterly 12(1):10-13.

Pullin, R. S. V., and M. Prein. 1995. Fish ponds facilitate natural resources management on small scale farms in tropical developing countries. Pages 169-186 *in* J. J. Symoens, and J. C. Micha, eds. Proceedings of the Seminar on the Management of Integrated Freshwater Agro-Piscicultural Ecosystems in Tropical Areas. Technical Centre for Agricultural and Rural Cooperation and Royal Academy of Overseas Sciences, Brussels, Belgium.

Rana, K. 1997. Recent Trends in Global Aquaculture Production:1984-1995. FAO, No. 16, 14-19.

Rangaswami, G. 1993. Fish culture in treated waste waters. Infofish International 25:54-57.

Rusmussen, F. 1988. Therapeutics Used in Fish Production: Pharmokinetics, Residues and Withdrawal Periods. EIFAC/XV/88Inf.,13, FAO, Rome, Italy.

Saha, S. 2000. Reclamation of Eutrophic and Wastewater Resources for Aquaculture: Management Through Certain Macrophytes and Fishes. Doctoral dissertation, University of Kalyani, West Bengal, India.

Saha, S., and B. B. Jana. In Press-a. Determination of water hyacinth biomass/water area ratio for effective reclamation in simulated model of wastewater at two nutrient levels. International Journal of Environmental Studies.

Saha, S., and B. B. Jana. In Press-b. Nutrient removal potential of emergent (*Scripus articulatus*) and floating (*Lemna major*) macrophytes: a comparison in situ hypertrophic mesocosms. International Journal of Environmental Studies.

Saha, A. K., M. S. Shah, and Y. A. Zamadar. 2000. Nutrient recovery and treatment of fish processing plant effluent by water hyacinth. Pages 511-504 *in* B. B. Jana, R. D. Banerjee, B. Guterstam, and J. Heeb, eds. Waste Recycling and Resource Management in the Developing World. University of Kalyani, India and Ecological Engineering Society, Wolhusen, Switzerland.

Saktivel, M. 1999. Potential of coastal aquaculture-constraints, options and developmental strategies. Indian Society of Fisheries Professionals Newsletter 1(3):4-6.

Seim, W. K., C. E. Boyd, and J. S. Diana. 1997. Environmental considerations. Pages 163-182 *in* S. Egna and C. E. Boyd, eds. Dynamics of Pond Aquaculture, CRC Press, Boca Raton, Florida.

Sharma, B. K., and J. Olah. 1986. Integrated fish pig farming in India and Hungary. Aquaculture 54:135-139.

Sharma, B. K. 1989. Fish Culture Integrated with Various Systems of Livestock Farming. Integrated Farming. KVV and TTC, Central Institute of Freshwater Aquaculture, Bhubaneswar, Orissa, India.

Shehadeh, Z. H., and J. Orzeszko 1997. External assistance to the aquaculture sector in developing countries. FAO Aquaculture Newsletter 16:7-8.

Shell, E. W. 1993. The Development of Aquaculture: An Ecosystem Perspectives. Alabama Agriculture Experimental Station, Auburn University, Alabama.

Shome, J. N., and R. D. Banerjee. 2000. A conceptual wastewater utilization model: A holistic approach towards zero discharge option. Pages 515-528 *in* B. B. Jana, R. D. Banerjee, B. Guterstam, and J. Heeb, eds. Waste Recycling and Resource Management in the Developing World, University of Kalyani, India and International Ecological Engineering Society, Wolhusen, Switzerland.

Singh, T. 1999. Benefits of sustainable shrimp culture. Infofish International 18(3): 25 32.

Sinha, V. R. P. 1999. Rural Aquaculture in India. RAP Publication, 21, Bangkok, Thailand.

Srinivasan, M., S. A. Khan, and S. Rajagopal. 1997. Culture of prawn in rotation with shrimp. Naga, The ICLARM Quarterly, 20:21-23.

Sze, C. P. 2000. Antibiotic use in aquaculture: The Malayasian perspectives. Infofish International 19(2):24-28.

Verma, J. P. 1989. Utilization of Aquatic Weeds as Fish Feed. Integrated Fish Farming. KVV and TTC, Central Institute of Freshwater Aquaculture, Bhubaneswar, Orissa, India.

WCED (World Commission For Environment and Development). 1987. Our Common Future. Oxford University Press, Oxford, England.

Welch, E. B., and T. Lindell. 1996. Ecological Effects of Wastewater–Applied Limnology and Pollutant Effects. Chapman and Hall, London, England.

WHO. 1989. Guidelines for the Use of Wastewater in Aquaculture. Report of WHO Scientific Group, Technical Series No. 778, WHO, Geneva, Switzerland.

Wiefels, R. C. 1999. Trade prospects for aquaculture species in Asia and Latin America. Infofish International 18(4):14-17.

William, K. J., S. Paul, and S. William. 1991. Duckweed Aquaculture–A New Aquatic Farming System for Developing Countries. A Special Publication of the Agricultural Division of the World Bank.
Willman, R., M. Halwarf, and U. Barg. 1998. Integrating fisheries and aquaculture to enhance fish production and food security. Food and Agriculture Organization of the United Nations, Rome, Italy.
Yoo, K. H., and C. E. Boyd. 1994. Hydrology and Water Supply for Pond Aquaculture. Chapman Hall, New York, New York.

Applied Nutrition in Freshwater Prawn, *Macrobrachium rosenbergii*, Culture

P. K. Mukhopadhyay
P. V. Rangacharyulu
Gopa Mitra
B. B. Jana

SUMMARY. The freshwater prawn, *Macrobrachium rosenbergii*, commonly called 'scampi,' is the most important culturable freshwater species of prawn with a fairly high growth rate, wide range of temperature (15-35°C) and salinity tolerance. The nutrients required by this species for growth and associated physiological functions are similar to those required by other crustaceans. Current data suggest that a digestible protein level of above 30% is required for maximum growth and protein efficiency. Ingredients such as mussel meat meal, squid meal, and shrimp meal serve as potential sources of protein in formulated diets. The complete quantitative requirements for all the essential amino acids for this prawn species have not yet been worked out. With respect to non-protein energy supply in the form of carbohydrate, studies indicated that scampi is capable of utilizing various carbohydrate sources efficiently. Dietary carbohydrate in the form of complex polysaccharides appears to be more effective as an energy source. Like other crustaceans, scampi also has limited ability to synthesize sterol from acetate and mevalonic acid. Quantitative estimates of the essential fatty acid require-

P. K. Mukhopadhyay, P. V. Rangacharyulu and Gopa Mitra, Central Institute of Freshwater Aquaculture, Kausalyaganga, Bhubaneswar-751 002, India.

B. B. Jana, Department of Zoology, University of Kalyani, Kalyani-731 234, India.

[Haworth co-indexing entry note]: "Applied Nutrition in Freshwater Prawn, *Macrobrachium rosenbergii*, Culture." Mukhopadhyay, P. K. et al. Co-published simultaneously in *Journal of Applied Aquaculture* (Food Products Press, an imprint of The Haworth Press, Inc.) Vol. 13, No. 3/4, 2003, pp. 317-340; and: *Sustainable Aquaculture: Global Perspectives* (ed: B. B. Jana, and Carl D. Webster) Food Products Press, an imprint of The Haworth Press, Inc., 2003, pp. 317-340. Single or multiple copies of this article are available for a fee from The Haworth Document Delivery Service [1-800-HAWORTH, 9:00 a.m. - 5:00 p.m. (EST). E-mail address: getinfo@haworthpressinc.com].

ments have been made in post-larval scampi. Data on vitamin and mineral requirements are scarce. Recent developments in diet formulation, practical feeding practices and the aquaculture potential in India have been indicated with emphasis on sustainable production.

KEYWORDS. Nutrient requirement, *M. rosenbergii*, culture, freshwater prawn

INTRODUCTION

The culture of freshwater prawn, *Macrobrachium rosenbergii* (commonly called 'scampi'), has received a great deal of attention in India as a preferred crustacean. Under controlled culture in freshwater and low saline ponds in inland, as well as coastal areas, it grows fastest among all freshwater prawns. It shows a wide range of temperature and salinity tolerance, acceptance of a large range of formulated diets, culture compatibility with non-predaceous species of fish, and it has a shorter larval period (Ling and Costello 1976; Behrends et al. 1986; Durairaj et al. 1992). The growth and production of scampi are intricately linked to gross nutrient inputs provided either in the form of endogenously available live food organisms or exogenous diet. The growing interest of freshwater prawn culture in recent years, therefore, deserves attention in view of sustainable production in India using the existing technology and provision of material input such as feed. Enzyme activity measured in the digestive tract of this prawn by Tyagi and Prakash (1967), Murthy (1977), Fair et al. (1980), and Lee et al. (1980) indicated the presence of wide ranges of digestive enzymes including trypsin, amino peptidases, amylases, cellulase, protease, esterases, and lipases. All these are indicative of the capability of this species to digest a relatively large range of complex proteins, carbohydrates, and lipids. The type of feeds used in large-scale, semi-intensive, and intensive prawn farming include components such as prawn-head meal, trash fish meal, chicken offals, clam meat, silk worm pupae, meat and bone meal, mussel meat meal, squid meal, various cereal grains, oil seed cakes, and several other animal husbandry and agro by-products available in India.

Freshwater Prawn Culture in India

Freshwater prawn is mostly grown in natural or managed water bodies in both mono- and polyculture system in India including the land-locked states such as Punjab and Uttar Pradesh. Shallow, self-drainable ponds are ideal for its culture. Promotion of culture of the species incorporating scientific principles has been taken up as a priority sector of aquaculture development in the country as the species is available on both east and west coast. Statewise detail of freshwater prawn production in India has been depicted in Table 1. Natural production based on pond productivity alone is generally low, but under diet-based intensive conditions it has been possible to produce annually 19 tons of prawn/ha/year (Ninawe 1994). This has been attributed to recent advances in nutrition and optimization of feeding regimes.

Traditional prawn farming was dependent on riverine juvenile collection. The age-old practice of collecting juveniles from flooded fields adjacent to the rivers and stocking them in enclosures is still practised in certain states in India. Increased juvenile collection from nature coupled with establishment of hatcheries, has given a lift to freshwater prawn culture while some of the inland states are keen to adopt technologies utilizing the abundantly-available saline water for seed production (Tripathi 1992). While it has specific breeding seasons in different parts of India, prawn can be spawned throughout the year. Generally,

TABLE 1. Production of *M. rosenbergii* in India (1998-1999) (adapted from Santhana Krishnan and Viswakumar 2000).

State	Area under culture (ha)	Production (Mt)	Yield (kg/ha)
Andhra Pradesh	5,200	1,875	360
Gujarat	201	73	363
Karnataka	110	55	500
Kerala	131	100	763
Maharashtra	40	33	825
Orissa	840	150	179
Tamil Nadu	40	23	0.575
West Bengal	3,500	1,594	455
Total	10,062	3,903	390

freshwater prawn attains gonadal maturity within 5-7 months of culture in ponds depending on environmental temperature (around 30°C) and adequate feeding (Rao 2000). Gravid females are either collected from earthen ponds or raised in fiberglass tanks or large plastic pools. For broodstock generation, early juveniles are stocked in (3-4/m^2) in fertilized ponds (Table 2) while the adults are maintained at 2/m^2 in the ratio of one male to four females. After hatching in the incubation tank, the broodstocks are again released to the pond. The captive broodstock has a fecundity of about 500 larvae/g body weight of the female (Rao 2000).

Larval rearing is a critical stage in freshwater prawn culture which needs 28-35 days to complete the larval cycle (1.9-7.5 mm length) with post larval production of 25-45/L. In most of the hatcheries in India a "clear-water larval growing technique" is adopted. Clear-water technique can be used in open or closed systems. In the open system, more than half of the larval culture medium is replaced daily with fresh clean water whereas in the closed system the remaining medium is re-circulated through a sand-gravel filter or bio-filter consisting of inert materials, such as pebbles, sand or granite chips, limestone, and clam shells placed into a container (Chakraborty 1998). Generally, the larvae are grown in two-phase larval culture unit where the early larvae (stage I to V or VI) are grown at high density (500 to 700 larvae/L) in cylindro-conical tanks for a week and in the 2nd phase, the advanced larvae (stage V or VI to PL) are stocked at low density (50/L) until metamor-

TABLE 2. Different pond manuring practices followed in India for *M. rosenbergii* culture.

Manure/Fertilizer	Rate (kg/ha)	Reference
Duck/chicken droppings or cow/pig dung, or rice bran/ground nut oil cake	200-5,000	Dandapat et al. (2000)
Urea	100	
Single super phosphate	200	Biswas et al. (1992)
Mohua (*Bassia latifolia*) oil cake	2,500	
Lime	500	
Quicklime	200	Ahmed and Varghese (1992)
After 15 days, poultry manure	1,000	
Cow dung	2,500	
Ammonium sulphate	100	Durairaj et al. (1992)
Super phosphate	50	

phosis in about 20 days (Rao 2000) with 10-12% dilution of sea water. Hatchery-produced prawn seed (15 mm long and 0.015 g wt.) are grown in nursery ponds (500-1000/m^2) for a period of 45-60 days (stocking density 25 individuals/m^2) to obtain juveniles of 3-5 g. Depending upon the culture system, the stocking density of prawn seed in the nursery phase is limited to 10-15 individuals/m^2 in order to obtain juveniles of 5-10 g (Rao 2000). These juveniles are then stocked into grow-out ponds at a density of 5-8/m^2 to obtain an average harvest-size weight of 50 g with an expected production of 1-2 tons/ha/crop (6 months).

Under monosex culture, male juveniles are stocked at a density of 1.5-2.0/m^2 in order to obtain 1-5 tons with an average harvestable size of 100 g. Being benthophagic, freshwater prawn is a good candidate species for polyculture with surface and column feedings carps (Ahmed and Varghese 1992). For polyculture with Indian major carps the relative proportions of carp: prawn can vary from one farm to another.

In an intensive polyculture system, the fish silver carp, *Hypophthalmicthys molitrix*; catla, *Catla catla*; grass carp, *Ctenopharyngodon idella*; rohu, *Labeo rohita*; common carp, *Cyprinus carpio*; mrigal, *Cirrhinus mrigala*; and chital, *Notopterus chitala* were stocked at 10,000/ha along with male prawn (average weight 0.66 g) in one pond and with female prawn (Av. wt. = 0.5 g) in another pond. Production in the two ponds was 7,090 kg and 84-96 kg/ha, respectively, after 12 months. Male prawn attained an average weight of 295 g with 89% survival and the females an average weight of 230 g with 100% survival.

Integration of prawn culture with crop production is another culture practice followed in India. When freshwater prawn juveniles were stocked together with rohu, common carp, and java carp, *Puntius javanicus*, in pens made of woven bamboo splicings in deep water rice fields, it was reported that rice production was increased by presence of fish and prawn (Das et al. 1991). Culture of freshwater prawn in rotation with tiger prawn, *Penaeus monodon*, followed particularly in Tamil Nadu state is considered a way of overcoming the problems of disease infestation resulting from exclusive freshwater prawn culture (Durairaj et al. 1992). The feasibility of utilizing the extensive water area available under the 'bund channel' system of coconut farming at Kumarakom (Kuttanad), Kerala for freshwater prawn culture has been evaluated with the relative efficiency of manuring and feeding on monoculture. Production ranging from 300-805 kg/ha in 6-7 months has been achieved using a diet comprised of groundnut oil cake, rice bran, and clam meat (Padmakumar et al. 1992).

Freshwater prawn require hideouts during moulting when they are vulnerable to predation, so providing shade and shelter generally promote growth and survival rates and thereby increase production (Tidwell 1997). A specially designed shell-string developed at the Central Institute of Freshwater Aquaculture (Bhubaneswar, India) is hung into the larval growing tanks for shelter and hiding (Mukhopadhyay 1999, unpublished data). Sometime acasia, *Acasia* sp., tree branches (about 3.0% of the water volume) are used for shelter (Rama Rao 1992).

Much variation in growth performances of prawn is noticed under different management practices. Most important is the management of suitable water quality variables (Table 3) and proper feeding practices which will fulfil the nutrient requirement of the species. Techniques for unilateral eyestalk ablation can be used to accelerate growth of juveniles in intensive culture with a reduction in production cost (Sierra 1999).

A variety of scientific culture practices such as indoor to open outdoor, clear or green, saline water to artificial sea water, extensive backyard hatcheries to intensive recirculatory systems with bio-filters has been developed each having its own advantage under the prevailing conditions (Tripathi 1992). Different pond manuring practices followed in India is presented in Table 2. Production of freshwater prawn in India

TABLE 3. Certain water quality parameters recorded in some pond culture trials of *M. rosenbergii* at CIFA farm, Bhubaneswar, India (Dandapat et al. 2000; Rao and Ayyappan 2000).

Water temperature (°C)	15-35
pH	7.0-8.5
Transparency (Sechi disk, cm)	22-35
Dissolved oxygen (ppm)	3.0-8.0
Nitrite (ppm)	< 0.1
Nitrate (ppm)	< 20
Un-ionized ammonia (ppm)	< 0.1
Ionized ammonia (ppm)	< 1.5
Alkalinity (ppm)	50-100
Iron (ppm)	< 0.3

during different years (1989-98) has been depicted in Figure 1 and Table 4.

NUTRITIONAL REQUIREMENTS

Requirement for Protein and Amino Acids

Requirement of protein has been studied in post-larval and juvenile scampi. Like other crustaceans, under controlled laboratory conditions the optimum dietary crude protein level ranges from 30-45% (Rangacharyulu 1999). The utilization of dietary protein is mainly affected by its amino acid composition, level of protein intake, calorie content of the diet, digestibility of the protein, physiological state of the species, water temperature, and size of prawn. Various protein sources have been used in experimental and practical diet formulations in freshwater prawn. Pu-

FIGURE 1. Production of *M. rosenbergii* in India (FAO 2000)

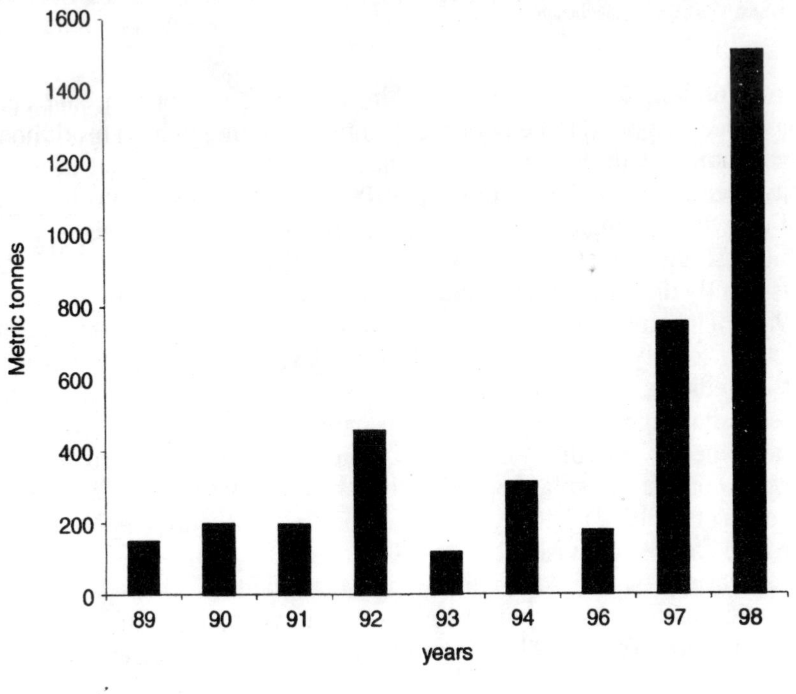

TABLE 4. Growth performance, survival and feed conversion rates recorded for *M. rosenbergii* under pond culture trials (adapted from Rangacharyulu 1999).

	Diet-I[1]	Diet-II[2]	Diet-III[2]
Pond area (m^2)	1,000	1,000	1,000
Stocking density/m^2	10	10	10
Mean initial body wt. (g)	0.070	0.076	0.073
Mean final body wt. (g)	24.26	32.36	31.29
Culture period (days)	186	186	186
Mean survival rate (%)	34.65	48.6	50.3
Gross total production (kg)	84.06	157.26	157.38
Feed conversion ratio	2.96	1.97	2.06
Specific growth rate (% per day)	1.44	3.25	3.25
Total net production (kg)	83.82	156.9	157.24

[1] Diet-I–Conventional mix. (Ground nut oil cake: Rice bran, 1:1).
[2] Diet-II and III (Differ in the proportion of ingredients)–Formulated diet (Ground nut oil cake, soybean meal, fishmeal, rice bran, vitamin-mineral mix.).

rified protein sources such as casein, gelatin, egg albumin, and zein have been reported to be poorly consumed resulting in high mortalities and poor growth performance (Kanazawa et al. 1982). Plant proteins have been found to be relatively poorly-utilized in crustaceans in terms of growth in comparison to protein of animal origin. However, digestibility studies in freshwater prawn have indicated that the species can efficiently digest both plant and animal protein sources (Ashmore et al. 1985). The omnivory of freshwater prawn permits the use of a wide variety of locally available feedstuffs including commercial by-products as ingredients in formulated diets. Protein sources such as mussel meat meal, squid meal, shrimp meal, fish meal, and earth-worm meal supported better growth, moulting frequency, and survival in freshwater prawn compared to plant protein sources such as various oil seed cakes (Leena et al. 1997). Silkworm pupae were also used as supplementary protein sources (Ravishankar and Keshavanath 1988).

A variety of non-conventional diet ingredients such as single-cell proteins and leaf meals have been used in the preparation of freshwater prawn diets (Tacon 1990). Poultry by-product meal, meat and bone meal, and blood meal can also be used as high-quality protein sources. By far, the commonest source of single cell protein currently used in prawn diet is *Spirulina* sp. This protein source cannot be used as the sole protein but can be effectively used as a supplementary protein in the diet of the post-larvae of freshwater prawn (James et al. 1992). Dry sugarcane yeast, *Saccharum officinarum*, a by-product of alcohol production from sugarcane, could be used at inclusion levels of up to 20% in a 30% protein grow-out diet with acceptable results (Zimmermann 1991).

The broodstock nutrition of freshwater prawn with regard to the protein content of the diet has been examined (Ganeshwaran 1989). Freshwater prawn fed a 40% protein diet with an energy level of 400 kcal/100 g attained higher fecundity over freshwater prawn fed a diet with 30% protein and 442 kcal/100 g diet (Das et al. 1996).

Dietary protein should provide sufficient quantity of essential amino acids which cannot be synthesized *de novo*, but are vital for growth and health of the species. The qualitative essential amino acid requirement for freshwater prawn is similar to that of most other animals (Watanabe 1975). It is reported that tyrosine is essential for freshwater prawn, whereas in teleost fish, it is non-essential and can be synthesized from phenylalanine. It was also observed that prepared diets with lower methionine and arginine resulted in poor growth in some prawn species (Deshimaru et al. 1985) indicating that arginine and methionine may have greater influences than other essential amino acids. Generally the essential amino acid profile of a particular species is used in the formulation of diets for the grow-out phase of that particular species (Poh 1985) so a required dietary amino acid profile may best be derived from the body amino acid profile of freshwater prawn. The requirement of amino acid for prawn is presented in Table 5.

Requirement for Lipid and Fatty Acids

Dietary lipids, besides serving as energy sources, also provide the essential fatty acids that are important dietary constituents for freshwater prawn and serve as a source of sterols and phospholipids necessary for growth, maintenance, functional integrity and proper functioning of many physiological processes (Kanazawa et al. 1977; Corbin et al. 1983). Quantitative lipid requirement ranges between 6-8% for scampi (Rangacharyulu et al. 1999). However, it is generally accepted that freshwater prawn cannot tolerate high levels of dietary lipid and more than 10% is not recommended (New 1980). The *de novo* synthesis of polyunsaturated fatty acids from either linoleic (n-6) and linolenic (n-3)

TABLE 5. Quantitative essential amino acid (% of protein) requirement for *M. rosenbergii* and essential amino acid (EAA) composition of prawn muscle tissue and prawn eggs.

EAA	Requirement (Ganeswaran 1989)	Muscle (Das et al. 1996)	Egg (Das et al. 1996)
Arginine	8.1	2.55±0.02	1.94±0.05
Histidine	2.3	2.89±0.06	2.34±0.02
Isoleucine	5.0	3.45±0.02	2.65±0.03
Leucine	7.5	3.46±0.04	2.40±0.03
Lysine	5.3	3.38±0.05	2.28±0.05
Methionine	3.0	1.97±0.03	1.57±0.04
Phenylalanine	3.9	2.31±0.06	1.76±0.04
Tyrosine	3.0	--	--
Threonine	3.4	2.33±0.00	1.81±0.03
Tryptophan	0.9	2.80±0.06	1.93±0.01
Valine	4.4	3.03±0.05	2.21±0.04

series is non-existent in freshwater prawn. The utilization of monounsaturated fatty acids such as oleic acid (18:1n-9) as an energy source and the effects of oleic acid levels and dietary soybean lecithin (SBL) on oleic acid utilization, growth and survival, lipid class, and fatty acid compositions of freshwater prawn juveniles have been determined (Querijero et al. 1997a).

The effect of dietary stearic acid (SA) and protein levels on the utilization of stearic acid by freshwater prawn juveniles has been evaluated which revealed that 2% SA + 35% crude protein level in diets showed good growth performances (Querijero et al. 1997b). The utilization of stearic acid (18:0) and carbohydrates by freshwater prawn juveniles has been estimated. The study showed that regardless of stearic acid level, different carbohydrate sources (starch and dextrin) were utilized effectively for energy sources by freshwater prawn and regardless of carbohydrate sources increase in dietary SA resulted in significantly higher content of SA and 18:1(n-9) in the body of freshwater prawn both in neutral and polar lipid fractions. The result also indicated the contribution of stearic acid for growth purposes rather than energy sources (Querijero et al. 1997c).

Monounsaturated fatty acids constitute the major moiety of fatty acids of stage 1 whereas polyunsaturated fatty acids are the dominant fatty acid group in later developmental stages. Moreover, polyunsaturated fatty acids are predominantly of the n-6 series in stage 1 and of the n-3 series for the rest of the larval stages. The major fatty acids of the stage-1 larvae are palmitic (16:0), oleic/vaccenic (18:1), and linoleic (18:2n-6) acids, whereas for the rest of the larval stages, palmitic (16:0), stearic (18:0), oleic/vaccenic (18:1), linolenic (18:3n-3), and eicosapentaenoic (20:5n-3) acids are the prominent fatty acids.

Polyunsaturated fatty acids tended to increase during larval development. Prawn larvae appear to have sufficient capability to convert palmitic (16:0) acid to stearic (18:0) acid. Dietary linoleic (18:2n-6) and linolenic (18:3n-3) acids appear to be converted to arachidonic (20:4n-6) and eicosapentaenoic acids to meet larval demand for such highly unsaturated fatty acids; whether docosahexaenoic acid (22:6n-3) is essential for larval freshwater prawn is yet to be ascertained. A high ratio of n-3 to n-6 fatty acids found after stage 1 may suggest the nutritional superiority of n-3 over n-6 for freshwater prawn larvae (Roustian et al. 1999). The effect of linoleic acid and n-3 highly unsaturated fatty acids (HUFA) on the reproductive performance and offspring quality of the freshwater prawn has been evaluated by Cavalli et al. (1999).

Fecundity was improved by the addition of higher amounts of 18:2 n-6 (from 3 to 13 mg/g dry weight). Eggs of females fed high levels of 18:2 n-6 and n-3 HUFA such as 20:5n-3 and 22:6n-3 showed increased hatchability. Larvae from females fed high levels of 18:2n-6 and n-3 HUFA as above tended to have better ammonia stress tolerance (Cavalli et al. 1999). D'Abramo and Sheen (1993) postulated that juvenile freshwater prawn is known to have an extremely limited ability to convert C18 fatty acids to fatty acids with 20 carbon or higher in the n-3 (linolenic) and n-6 (linoleic) families. But freshwater prawn have the ability to synthesize 20:5n-3 from 22:6n-3, although they lack the ability to bio-convert eicosapentaenoic acid to docosahexaenoic acid. There was no evidence of *de novo* synthesis of 18:3n-3 or 18:2n-6. Freshwater prawn require more n-6 PUFA; however, the optimum ratio of n-6:n-3 in prawn diets has not been investigated.

Cholesterol is thought to be a precursor of important steroids, brain and moulting hormone, and vitamin D in freshwater prawn. A cholesterol level in excess of 0.12% is not essential for juvenile freshwater prawn (Briggs et al. 1988). Similar results were obtained by Sherief et al. (1992) who were unable to demonstrate any requirement for dietary cholesterol in prawn post-larvae above the endogenous 0.12% level present in the control diet contributed by cod liver oil. However, D'Abramo and Daniels (1994) found that depriving freshwater prawn of all sterols caused death within 48 days. The cholesterol requirement was estimated as between 0.3 and 0.6% dry weight of the diet. Levels above 0.6% either did not increase weight gain or decreased it. Substitution with 0.6% ergosterol or stigmesterol gave lower survival and weight gain than 0.6% cholesterol. However, a mixture of phytosterols (sitosterol, campesterol, and dihydrobrassi-casterol) were found to be as effective as cholesterol.

Dietary phospholipids have beneficial effects on growth and survival of prawn. An inclusion level of 8% lecithin resulted in a significant improvement in growth in comparison to 0, 2, 4, and 6% levels (Reed 1987). On the contrary, results from other investigations suggest the non-essentiality of dietary phospholipids for prawn (Hilton et al. 1984; Briggs et al. 1988; Devresse et al. 1990; Kanazawa 1993; Querijero et al. 1997a).

A dietary source of soybean phosphatidylcholine has been found to be optimum for freshwater prawn larvae. Preliminary growth studies with soy lecithin as a 20 g/kg diet indicated better growth, survival, and carcass composition in prawn post-larvae than post-larvae fed a control diet (Srivastava and Parihar 1997). Tiwari and Sahu (1999) found maximum weight gain (164%) by supplementing 5% soy lecithin along with 1% cod liver oil and groundnut oil cake. The effect of increasing levels of dietary phospholipids on the reproductive performance, egg and larval quality and lipid composition of female prawn were studied by Cavalli (2000). A basal level of 0.8% dietary phospholipids was sufficient to meet the dietary demands of the prawn broodstock.

Requirement of Carbohydrates

Carbohydrates together with proteins and lipids form dietary sources of energy. They are also important in storage of dietary energy in the synthesis of chitin, steroids and fatty acids. Dietary carbohydrates such as cellulose facilitate the rate of passage of food through the gut while starch acts as a binder. A mixture of substrates have been found to be used for energy during starvation of various crustaceans. Fasting energy metabolism of prawn is dominated by carbohydrates, followed by lipids and proteins (Clifford and Bricks 1983). Prawn have much higher α-amylase and cellulase activities in the hepatopancreas, stomach, or intestine than do marine shrimp (Chauang et al. 1985). Complex carbohydrates are utilized more effectively than simpler di- or monosaccharides (Briggs 1991). Cellulose (often used as a non-nutrient bulking ingredient in diets) may contribute significantly to the nutrition of prawns (Briggs 1991). Fair et al. (1980) speculated that the presence of chitin is required for the exoskeleton which is shed and replaced during moulting (New 1980). Studies using glucosamine (an intermediary between glucose and chitin) in diets have resulted in enhanced growth rate in prawn. Chitinase activity has been reported in prawn. Clifford and Bricks (1979) found that for prawn a dietary ratio of lipid to carbohydrate of 1:3-1:4, with a 25% protein diet resulted in more efficient utilisation of dietary protein than did a ratio of 1:1 and 1:2.

Requirement for Minerals

Complete dietary mineral requirement of this species is not yet elucidated and knowledge of qualitative and quantitative requirements are scanty (Kanazawa et al. 1984). Various types and levels of mineral premixes have been used in studies of prawn with various combinations and levels of minerals (Table 6). Some authors suggested that prawn may obtain minerals from ambient water; however, the importance of dietary minerals has been supported by the demonstrated ability of prawn to absorb inorganic minerals from diets. It is found that levels of calcium and magnesium in the carapace peaked at inter-moult and declined during pre-moult. The mineral content in the carapace can be affected by the levels of minerals in the diet (Das 1999). There is a relationship between dietary calcium level and water hardness (Zimmermann et al. 1994). Highest survivability, average weight gain and biomass were obtained with 3% dietary calcium with 5 mg/L $CaCO_3$ in water whereas a 1.8% calcium diet performed better at the highest alkalinity level (74 mg/L $CaCO_3$). It is concluded that high dietary calcium combined with high alkalinity is detrimental (Zimmermann et al. 1994). Trace mineral zinc has a significant effect on growth promotion and survivability in freshwater prawn and 50-90 mg Zn gave better growth, survival, feed conversion efficiency, and specific growth rate (Rath and Kiron 1994).

Requirement of Vitamins

Like minerals the levels and types of vitamin premixes used in prawn diets vary widely (Table 6). The application of vitamins to diets without

TABLE 6. Composition of the vitamin-mineral premix used in the formulated diet of *M. rosenbergii* (adapted from Rangacharyulu 1999).

Thiamine hydrochloride	5 mg
Riboflavin	20 mg
Pyridoxine hydrochloride	5 mg
Choline chloride	500 mg
Nicotinic acid	75 mg
Calcium pantothenate	50 mg
Inositol	200 mg
Biotin	5 mg
Folic acid	5 mg
Cyanocobalamin	5 mg
Menadione	5 mg
Ascorbic acid	150 mg
Vitamin D_3	2000 IU
Tocopheryl acetate	40 mg
Vitamin A	5000 IU
Calcium lactate	2.5 mg
Di-calcium phosphate	70 g
Di-potassium hydrogen phosphate	8 g
Potassium chloride	5.5 g
Sodium chloride	6.5 g
Disodium hydrogen phosphate	2 g
Magnesium sulphate	2.5 g
Ferric citrate	1.0 g
Zinc sulphate	50 mg
Potassium iodate	100 mg
Cobalt chloride	50 mg
Cupric carbonate	10 mg
Selenium dioxide	5 mg
Manganese sulphate	100 mg
Aluminum chloride	50 mg

prior knowledge of requirements may be wasteful. If added in excess, vitamins can be toxic to or antagonistic in prawn. A quantitative dietary requirement of vitamin C has been determined for the juvenile freshwater prawn using two water soluble sources of ascorbic acid: ascorbyl-2-monophosphate calcium salt (AMP) and ascorbyl-6-palmitate (AP). The estimated dietary quantitative requirement for prawn is within the range of 60-150 mg vitamin C/kg diet, reported for several species of fish, and found in tissue of clam and adult brine shrimp as calculated by Conklin (1997). A quantitative requirement equivalent to 104 mg of vitamin C/kg diet was estimated by a quadratic surface analysis, using survival as the measured response to the different dietary levels (D'Abramo et al. 1994). A significantly positive effect of ascorbic acid had been demonstrated by means of a salinity stress test when prawn larvae were administered ascorbic acid-enriched *Artemia* nauplii (Merchie et al. 1995). The effect of vitamin C and vitamin E (tocopherol) on the maternal performance and offspring quality of prawn was investigated. The broodstock diets containing 60 μg ascorbic acid/g dry weight of diet and 300 μg α-tocopherol/g dry weight of diet are sufficient to ensure proper reproduction and offspring viability. However, feeding prawn females with higher dietary levels of both ascorbic acid and α-tocopherol (each around 900 μg/g dry weight of diet) might increase larval quality and higher tolerance to ammonia stress (Cavalli 2000). The effect of supplementary vitamin E (200, 400 and 600 mg/kg diet) on lipid peroxidation (LPX) and the antioxidant defence system in gills and hepatopancreas of the freshwater prawn has been studied by Dandapat et al. (2000). According to them, dietary vitamin E modulates the antioxidant defence system by decreasing LPX in the hepatopancreas in a comparatively lower dose than in the gill. Dietary vitamin E at 200 mg/kg diet modulated some of the antioxidants; increased amounts of vitamin E in the diet (400 mg/kg diet) rendered adequate the antioxidant function in both gills and hepatopancreas and reduced both functions *in vivo* and *in vitro* LPX. However, the highest dose of vitamin E (600 mg/kg) did not provide much additional protection against oxidative stress.

Factors Related to Diet Formulation for M. rosenbergii and Feeding Practices

The major factors that are of foremost consideration in the efficacy of a formulated diet are nutritional composition of raw material, nutrient digestibility, physical characteristics, handling and storage, feeding method, water quality, and critical standing crop. Diets prepared from poor quality material can have adverse value on prawn productivity. Quality control is thus of crucial importance. The basic information required for diet formulation includes nutrient requirements of the cultured species, availability of various ingredients to be used, their nutrient composition, cost, ability of the species to utilise the nutrients, and type of diet processing. Knowledge of nutrient digestibility and activity of important digestive enzymes should be available before proceeding to the formulation of diet. Fish meal, crustacean meal, squid meal, and mussel meat meal are some of the excellent ingredients used in prawn diet preparation. Ground nut oil cake, soybean oil cake, sunflower oil cake, and rice bran are also used as ingredients in diets.

The formulated diets for prawn must be pelletized so as to be fairly water stable. The bulk of the diet is required in the form of pellets which are manufactured by mixing together with all the finely ground ingredients with water to form a dough that can be passed through a die under great pressure. The spaghetti like strands formed are cut into small pellets. The pressure and consequent temperatures involved cause a change in certain ingredients that bind the dietary ingredients together and render the pellets water stable. Depending on the prawn size, pellet sizes range from 2 to 3 mm. The dust and undersized pellets can be reground and added to the next batch. Sometimes use of chemo-attractants in diets enhances growth in juvenile freshwater prawn (Harpaz 1997). Two biogenic amines (putrescine and cadaverine) and pheromones (crab urine and freshwater prawn green gland extracts) can be incorporated as feed attractants in basal diet. For monosexual culture, crab urine and prawn green gland extracts can be used (Mendoza et al. 1997). A variety of ingredients (Table 7) of vegetable and animal origin such as wheat flour, soy flour, corn flour, fish flesh, mussel, squid, hen's eggs, skimmed milk powder, and vitamin and mineral premixes are utilized to prepare larval diet. All the ingredients are mixed together with water in an electrical blender or mixer until the mixture is dough-like which is then steamed for 30 minutes. After cooling it is stored in a refrigerator and it is advisable to use this for only four days and feeding the larvae (Rao and Ayyappan 2000). The prepared diet is then sieved to obtain suitable particle size and is fed during daytime at intervals of 2 to 3 hours. Comparison of whole-body amino acid and fatty acid profile of prawn and natural food organisms indicate that zooplankton and oligochaetes may have the most appropriate biochemical composition as prawn food sources (Tidwell et al. 1997).

Some Recent Studies on Diet Development of M. rosenbergii Larvae and Post-Larvae

Mukhopadhyay et al. (1994) tested three formulated diets using different percentages of prawn meal, mussel meat meal, silkworm pupae, ground nut oil cake, soybean meal, and yeast with 35-36% crude protein, 10-11% lipid and 17 kJ/g gross energy. Diet I (prawn meal based), diet II (mussel meat meal based), and diet III (combination of prawn meal and mussel meat meal in a 50:50 ratio) were evaluated on growth and feed conversion ratio of prawn (FCR). Of these three diets, diet I elicited the best growth performance accompanied by least FCR and

TABLE 7. Ingredients used in prawns feed formulation and their proximate composition. NFE = Nitrogen free extract, 1US$ = 45.00Rs.

Ingredients	Protein (%)	Lipid (%)	Fiber (%)	NFE	Ash	Cost (Rs/kg)
Plant-sources						
Coconut cake	23-25	12-13	10-13	40-42	8-10	8
Groundnut cake (expeller)	40-43	4-8	6-7	30-33	7-8	10
Gingily oil cake	30-35	6-8	6-7	32-40	12-15	3
Cotton seed cake	27-30	6-9	15-18	36-40	6-7	8
Mahua cake	15-18	15-17	4-6	45-50	8-10	5
Rapeseed cake	30-35	10-12	6-9	35-38	8-10	7
Soybean cake	36-40	8-12	4-6	30-35	6-8	11
Soybean meal (solvent extracted)	50-55	1-2	4-5	30-32	5-6	13
Sunflower cake	28-32	4-6	16-22	40-45	6-8	7
Mustard oil cake	30-35	7-10	6-9	35-40	5-7	5
Animal sources						
Clam meat	12	1	0.5	-	82.7	-
Squid meal	59-64	6-9	0.1-0.5	-	14-15	-
Fish meal	40-45	8-12	2-3	5-8	20-40	12-20
Mussel meat meal *	45.0	9.0	5.0	17.0	26.0	-
Meat and bone meal	50-55	4-8	0	5-8	28-32	15
Prawn head meal	28-30	8-10	5-1	3-4	45-55	10
Silk worm pupae meal	60-65	18-20	3-5	3-5	4-8	16
Energy sources and filler						
Rice bran	8-10	12-16	10-15	40-45	5-8	6
Deoiled rice bran	12-16	1-2	15-20	40-45	8-12	4
Maize	9-11	4-6	2-3	70-80	2-3	12
Barley	8-10	2-3	4-6	70-80	3-6	12
Millet	13-15	2-3	3-6	70-80	3-6	11

*Mukhopadhyay and Das 1994; Rangacharyulu 2000.

highest protein efficiency ratio (PER) than either diet II or diet III. Shivananda Murthy and Naik (1997) evaluated the efficacy of soy flour as a source of dietary protein replacing fish meal in the diet for freshwater prawn. An indigenous feed has been evolved by Rangacharyulu et al. (1997) using locally available ingredients such as ground nut cake, fish meal, soybean meal, rice bran, and vitamin and mineral premix for commercial grow-out operations of freshwater prawn under pond conditions.

Two micro-encapsulated diets have been developed by Dutta et al. (1997) using gelatin and chicken eggs for larvae and post-larvae of prawn. Higher survival was obtained from the gelatin encapsulated diet but with less growth rate than the egg encapsulated diet which showed comparatively high growth rate with slightly less survival. The efficacy of five dietary protein sources such as fish meal, clam meal, beef waste meal, chicken entrails, and soybean meal have been evaluated for prawn. It was reported that beef waste meal can be used as a replacement of fish meal for the preparation of efficient practical diets of prawn (Hari and Kurup 2000). The effect of 33% animal protein and 67% plant protein-based diet and vice-versa was tested with different feeding frequencies (2 to 5). Diet with low levels of animal protein performed better when the feeding rate was 2-3 times per day, while higher levels of animal protein-based diets showed better performance with higher feeding frequencies of 5 times a day (Vasudevappa et al. 2000). Squilla meal can be used in partial replacement of fish meal also in the diet of prawn without affecting growth and production (Suman and Manissery 2000).

Environmental Issues Related to Nutrition and Feeding

Prawn farming is an aquaculture activity of significance in most parts of India. Considering the pace of its development in recent years, it is likely that use of feeds and various feed additives will continue to increase in the foreseeable future for this totally feed based production process. The organic wastes in the form of unutilized and undigested feed material and metabolic end products, which contribute to the pollution load can offset the natural recovery properties of the ecosystem. This can have serious repercussions unless strong emphasis is given on environmental quality monitoring, since production itself is dependent upon good water quality. It is therefore, of crucial importance to reduce the impact of potential dietary polluting elements particularly N and P excretion to achieve dependable freshwater prawn production. One possible way to improve the prevailing trend of rapid intensification of culture system for increasing productivity can be through optimization of dietary nutrient/energy balance in favor of increased voluntary feed intake since sustainability of any aquaculture development is directly related to the provision of nutritionally sound diets along with application of proper feeding strategies which improve growth, health and reproduction of the cultured species, and also have little impact on aquatic environment. This necessitates proper planning measures and resource management strategies in regard to the environmental carrying capacity of the aquatic habitat for the purpose of both higher sustainable economic production as well as social responsiveness. This bears particular relevance after the sudden collapse of shrimp farming industry in India due to uncontrolled and unregulated expansion for increasing productivity in an inconsiderate manner.

REFERENCES

Ahmed, I., and T.J. Varghese. 1992. Compatibility of *M. rosenbergii* (De man) for polyculture with major carps. Pages 197-199 *in* E.G. Silas, ed. Proceedings of the National Symposium on Freshwater Prawns (*Macrobrachium* sp.) Kerala Agricultural University, Trissur, India.

Ashmore, S.B., R.W. Standby, L.B. Moore, and S.R. Malecha. 1985. Effect on growth and apparent digestibility of diets varying in gram source and protein level in *M. rosenbergii*. Journal of the World Mariculture Society 16:205-216.

Behrends, L.L., J.B. Kingsley, and A.H. Price,1986. Polyculture of fresh prawns, tilapia, channel catfish and Chinease carps. Journal of the World Mariculture Society 16:437-450.

Biswas, S.N., U.K. Laha, and R.N. Das. 1992. Growth, survival and production of Proceedings of the National Symposium on Freshwater Prawns (*Macrobrachium* sp.) Kerala Agricultural University, Trissur, India.

Briggs, M.R. 1991. The performance of juvenile prawns, *M. rosenbergii*, fed a range of carbohydrate sources in semi purified diets. Journal of the World Aquaculture Society 22:16A.

Briggs, M.R.P., K. Jauncey, and J.H. Brown. 1988. The cholesterol and lecithin requirements of juvenile prawn (*Macrobrachium rosenbergii*) fed semi-purified diets. Aquaculture 70:121-129.

Cavalli, R.O. 2000. Broodstock Nutrition and Offspring Quality of the Freshwater Prawn *Macrobrachium rosenbergii* (De man). Doctoral thesis. Gent University, Gent, Belgium.

Cavalli, R.O., P. Lavens, and P. Sorgeloos. 1999. Performance of *Macrobrachium rosenbergii* broodstock fed diets with different fatty acid composition. Aquaculture 179:387-402.

Chakraborty, C. 1998. Seed production technology of giant freshwater prawn, *M. rosenbergii* (De man). Pages 27-34 *in* Short Term Training Programme on Fisheries and Aquaculture Engineering (organized by The Agricultural and Food Engineering Department), Indian Institute of Technology, Kharagpur-721302, India.

Chaung J.L., M.F. Lee, and J.S. Jenn. 1985. Comparison of digestive enzyme activities of five species of shrimp cultured in Taiwan. Journal of Fisheries Society of Taiwan 12:43-53.

Clifford, H.C. III, and R.W. Bricks. 1979. A physiological approach to the study of growth and bioenergetics in the freshwater shrimp *M. rosenbergii*. Proceedings of the World Mariculture Society 10:701-719.

Clifford, H.C. III, and R.W. Bricks. 1983. Nutritional physiology of the freshwater shrimp *Macrobrachium rosenbergii* (De man). I. Substrate metabolism in fasting juvenile shrimp. Comparative Biochemistry and Physiology 74A:561-568.

Conklin, D.E. 1997. Vitamins. Pages 123-149 *in* L.R. D'Abramo, D.E. Conklin, and D.M. Akiyama, eds. Crustacean Nutrition. World Aquaculture Society, Baton Rouge, Louisiana.

Corbin, J.S., M.M. Fujimoto, and T.Y. Iwai Jr. 1983. Feeding practices and nutritional considerations for *Macrobrachium rosenbergii* culture in Hawaii. Pages 391-412 *in* J.P. McVey, ed. CRC Handbook of Mariculture. Vol.1. Crustacean Aquaculture CRC Press, Boca Raton, Florida.

D'Abramo, L.R., C.A. Moncreiff, F.P. Holcomb, J.L. Montanez, and R.K. Buddington. 1994. Vitamin C requirement of the juvenile freshwater prawn *Macrobrachium rosenbergii*. Aquaculture 128:269-275.

D'Abramo, L.R., and S.S. Sheen. 1993. Polyunsaturated fatty acid nutrition in juvenile freshwater prawn *Macrobrachium rosenbergii*. Aquaculture 115:63-86.

D'Abramo, L.R., and W.H. Daniels. 1994. Sterol requirement of juvenile freshwater prawn *Macrobrachium rosenbergii*. Page 200. World Aquaculture, World Aquaculture Society. Baton Rouge. Louisiana (Abstract).

Dandapat, J., B.N. Chainy, and K.J. Rao. 2000. Dietary vitamin E modulates antioxidant defence system in giant freshwater, *M. rosenbergii*. Comparative Biochemistry and Physiology–Part C 127:101-115.

Dandapat. J., M. Sinha, T. Dutta, and K.J. Rao. 2000. Scientific Farming of *M. rosenbergii*. The Central Institute of Freshwater Aquaculture, Kausalyaganga, Bhubaneswar, India.

Das, D.N., B. Roy, and P.K. Mukhopadhyay. 1991. Rice-fish integrated farming in an open deepwater rice field. Pages 22-23 *in* Proceedings of the National Symposium on New Horizons in Freshwater Aquaculture, Central Institute of Freshwater Aquaculture, Kausalyaganga, Bhubaneswar, India.

Das, N.N., C.R. Saad, K.J. Ang, A.T. Law, and S.A. Harmin. 1996. Diet formulation for *M. rosenbergii* (De man) broodstock based on essential amino acid profile of its eggs. Aquaculture Research 27:543-555.

Deshimaru, O., K. Kuroki, M.A. Mazid, and S. Kitamura. 1985. Nutritional quality of compounded diet for prawn *Penaeus monodon*. Bulletin of the Japanese Society of Scientific Fisheries 54:1937-1944.

Devresse, B., M. Ramdhane, M. Buzzi, J. Rasowo, P. Leger, J. Brown, and P. Sorgeloos. 1990. Improved larviculture outputs in the giant freshwater prawn *M. rosenbergii* fed a diet of artemia enriched with n-3 HUFA and phospholipids. World Aquaculture 21(2): 123-125.

Durairaj, S., G. Chandrasekaran, and R. Umamaheswary. 1992. Experimental culture of giant freshwater prawn in Tamil Nadu. Pages-12-14 *in* E.G. Silas, ed. Proceedings of the National Symposium on Freshwater Prawns (*Macrobrachium* sp.). Kerala Agricultural University, Trissur, India.

Dutta, T., M. Sinha, and P.K. Mukhopadhyay. 1997. Preliminary studies on the development of micro-particulated diets for rearing larvae and post larvae of *Macrobrachium rosenbergii* (De man). Page 35, National Workshop on Fish and Prawn

Feeds. Central Institute of Freshwater Aquaculture, Kausalyaganga, Bhubaneswar, India (Abstract).

Fair, P.H., A.R. Frontner, M.R. Millikin, and L.V. Sick. 1980. Effect of dietary fibre on growth, assimilation and cellulase activity of the pawn *M. rosenbergii*. Proceedings of the World Mariculture Society 11:369-370.

FAO Year Book 2000, Fisheries Statistics, Aquaculture Production Vol. 86/2, FAO, Rome, Italy.

Ganeswaran, K.N. 1989. Reproductive Performance of Giant Freshwater Prawn *Macrobrachium rosenbergii* with Special Reference to Broodstock Age, Size, Nutrition, Egg Production, and Larval Quality. Doctoral dissertation, Stirling University, Stirling, United Kingdom.

Hari, B., and M. Kurup. 2000. Comparative evaluation of five dietary protein sources in *Macrobrachium rosenbergii* (de man). Page 53. Central Institute of Freshwater Aquaculture, Kausalyaganga, Bhubaneswar, India (Abstracts).

Harpaz, S. 1997. Enhancement of growth in juvenile prawn *M. rosenbergii* through use of chemoattractant. Aquaculture 156:221-227.

Hilton, J.W., K.E. Harrison, and S.J. Slinger. 1984. A semi-purified test diet for *M. rosenbergii* and the lack of need for supplemental lecithin. Aquaculture 37: 209-215.

James, T., P.M. Sherief, C.M. Nair, and P.M. Thampy. 1992. Evaluation of *Spirulina fusiformis* as a protein source in the diet of the post larva of *Macrobrachium rosenbergii*. Pages 234-237 *in* E.G. Silas, ed. Proceedings of the National Symposium on Freshwater Prawns (*Macrobrachium* sp.). Kerala Agriculture University. Trissur, India.

Kanazawa, A. 1993. Essential phospholipid of fish and crustanceans. Pages 519-530 *in* S.J. Kaushik and P. Luquet, eds. Fish Nutrition in Practice, INRA, St-Pee-Sur-Nivelle, Paris, France.

Kanazawa, A., S. Teshima, H. Sasada, and S. Abdel Rahman. 1982. Culture of prawn larva with micro-particulate diets. Bulletin of the Japanese Society of Scientific Fisheries 48:195-199.

Kanazawa, A., S. Tohiwa, M. Kayama, and M. Hirate. 1977. Essential fatty acids in the diet of prawn-1. Effect of linoleic and linolenic acid on growth. Bulletin of the Japanese Society of Scientific Fisheries. 43:1-14.

Kanazawa. A., S. Teshima, and M. Sakari. 1984. Requirement of juvenile prawn for calcium, phosphorous, magnesium, potassium, copper, manganese and iron. Memoirs of Faculty of Fisheries, Kagoshima University 33:63-71.

Lee, P.G., N.J. Blake, and G.E. Rodrick. 1980. A quantitative analysis of digestive enzymes for the freshwater prawn *Macrobrachium rosenbergii*. Proceedings of the World Mariculture Society 11: 392-402.

Leena, K.A., K. Reedy, and K. Dube. 1997. Effect of earthworm diet on the growth and maturity of freshwater prawn and fish. Page 41. National Workshop on Fish and Prawn Feeds, The Central Institute of Freshwater Aquaculture, Kausalyaganga, Bhubaneswar, India (Abstracts).

Ling, S.W., and J.J. Costello. 1976. Status and problems of *Macrobrachium* farming in Asia. Pages 66-71 *in* Marine Technology Society, Food Drugs from the Sea, Conference Proceedings. Mayaguez, Puerto Rico.

Mendozoa. R., J. Montomayor, and J. Verde. 1997. Biogenic amines and pheromones as feed attractants of the freshwater prawn *Macrobrachium rosenbergii*. Aquaculture Nutrition 3:167-173.

Merchie, G., P. Lavens, J. Radull, H. Nelis, A. De Lunheer, and P. Sorgeloos. 1995. Evaluation of Vitamin C enriched artemia nauplii for larvae of the giant freshwater prawn. Aquaculture International 3:355-363.

Mukhopadhyay, P.K., and K.M. Das. 1994. Effects of three compounded diets on growth and feed efficiency of the freshwater prawn *M. rosenbergii*. Pages 129-134 *in* S. Paul Raj, ed. Proceedings of the National Symposium on Aquaculture for 2000 A.D. Palani Paramount Publications, Tamil Nadu, India.

Murthy, R.C. 1977. Study of proteases and esterases in the digestive system of *Macrobrachium lamarri* Edwards (Crustacea:Decapoda). Journal of Animal Morphology and Physiology 24:211-216.

New, M.B. 1980. A bibliography of shrimp and prawn nutrition. Aquaculture 21:101-128.

Ninawe, A.S. 1994. Freshwater prawn farming: Developments in India. Indian Farming 44:13-16.

Padmakumar, K.G., J.R. Nair, and A. Krishnan. 1992. Farming of giant freshwater prawn *M. rosenbergii* (de man) in channels of coconut gardens in lower Kuttanad, Kerala. Pages 191-196 *in* E.G. Silas, ed. Proceedings of the National Symposium on the Freshwater Prawns (*Macrobrachium* sp.), Kerala Agriculture University, Trissur, India.

Poh, Y.T. 1985. Least Cost Formulation of Feed for Prawns with Particular Reference to *M. rosenbergii* (de man). Master's thesis. Universiti d' Pertanian, Selangor, Malaysia.

Querijero B.V.L., S. Teshima, S. Koshio, and M. Ishikawa. 1997a. Utilization of monounsaturated fatty acids (18:1n-9 oleic acid) by freshwater prawn *Macrobrachium rosenbergii* (De man) juveniles. Aquaculture Nutrition 3:127-139.

Querijero B.V.L., S. Teshima, S. Koshio, and M. Ishikawa. 1997b. Utilization of dietary stearic acid (18:0) and carbohydrate by freshwater prawn *M. rosenbergii* juveniles. Fisheries Science 63:971-976.

Querijero B.V.L., S. Teshima, S. Koshio, and M. Ishikawa. 1997c. Effect of dietary stearic acid and protein levels on the utilization of stearic acid and protein levels by freshwater prawn *M. rosenbergii* juveniles. Fisheries Science 63:1035-1041.

Rama Rao, P.V.A.N., K.V. Prasada Rao, R. Ramakrishna, and P. Haribabu. 1992. Studies on the growth of *Macrobrachium* sp. in monoculture. Pages 183-186 *in* E.G. Silas, ed. Proceedings of the National Symposium on the Freshwater Prawns (*Macrobrachium* sp.), Kerala Agriculture University, Trissur, India.

Rangacharyulu, P.V. 1999. Studies on the Nutrition and Diet Development of the Giant Freshwater Prawn. *Macrobrachium rosenbergii*. Doctoral thesis, University of Kalyani, West Bengal, India.

Rangacharyulu, P.V., M. Sinha, and K.J. Rao. 1997. Development of indigenous feed for grow out phase of *Macrobrachium rosenbergii*. Page 36. National Workshop on Fish and Prawn Feeds. The Central Institute of Freshwater Aquaculture, Kausalyaganga, Bhubaneswar, India (Abstract).

Rangacharyulu, P.V. 2000. Freshwater prawn nutrition. Pages 12-17 *in* Training Programme Manual on Fish and Shellfish Nutrition. The Central Institute of Freshwater Aquaculture. Kausalyaganga, Bhubaneswar, India.

Rao, K.J., and S. Ayyappan. 2000. Hatchery technology of giant freshwater prawn, *M. rosenbergii*. Fishing Chimes 19:112-117.

Rao, K.J. 2000. Technology Project in Mission Mode on Semi-Intensive Prawn Aquaculture, at CIFA: A Retrospection. Department of Biotechnology Government of India and Central Institute of Freshwater Aquaculture, Kausalyaganga, Bhubaneswar, India.

Rath, G.S., and D. Kiran. 1994. Role of zinc in promoting growth and survival of *M. rosenbergii*. Journal of Aquaculture in the Tropics 9:209-222.

Ravishankar, A.N., and P. Keshavanath. 1988. Utilization of artificial feeds by *Macrobrachium rosenbergii*. Indian Journal of Animal Sciences 58: 876-881.

Reed, L. 1987. Nutrition of Juvenile *Macrobrachium rosenbergii*: Evaluation of an Experimental System and the Effects of Dietary Lecithin and Protein Quantity. Master's thesis, Mississippi State University, Mississippi.

Roustaian, P., M.S. Kamarudin, H. Omar, C.R. Saad, and M.H. Ahmad. 1999. Changes in fatty acid profile during larval development of freshwater prawn *Macrobrachium rosenbergii* (De man). Aquaculture Research 30:815-824.

Santhanakrishnan, G., and M. Viswa Kumar. 2000. Present status and future prospects of freshwater prawn culture in India. Pages 13-16 *in* National Workshop on Aquaculture of Freshwater Prawn. Nellore, Andhra Pradesh, India.

Sherief, P.M., C.M. Nair, and V. Malika. 1992. The cholesterol requirement of larval and post larval prawn (*Macrobrachium rosenbergii*). Pages 213-217 *in* E.G. Silas, ed. Proceedings of the National Symposium on the Freshwater Prawns (*Macrobrachium* sp.), Kerala Agriculture University, Trissur, India.

Shivananda Murthy H., and R. Naik. 1997. Utilisation of soyaflour in the diet of giant freshwater prawn, *Macrobrachium rosenbergii*. Page 34. National Workshop on Fish and Prawn Feeds. Central Institute of Freshwater Aquaculture, Kausalyaganga, Bhubaneswar, India (Abstract).

Sierra, E., and F. Diaz. 1999. Dynamic bioenergetics of post larvae and juveniles of *M. rosenbergii* caused by unilateral eyestalk ablation. Journal of Aquaculture in the Tropics 14:113-119.

Srivastava, P., and A.S. Parihar. 1997. Preliminary growth studies on the use of soy lecithin in the diet of freshwater prawn, *Macrobrachium rosenbergii* post larvae. National Academy of Science Letters 20:22-26.

Suman, A.S., and J.K. Manissery. 2000. Studies on the growth of *Macrobrachium rosenbergii* fed squilla meal incorporated diets. Page 54. Central Institute of Freshwater Aquaculture, Kausalyaganga, Bhubaneswar, India (Abstract).

Tacon, A.G.J. 1990. Standard Methods for the Nutrition and Feeding of Farmed Fish and Shrimp. Vol. 3. Argent Laboratories Press, Redmond, Washington.

Tidwell, J.H., S.D. Coyle, C.D. Webster, J.D. Sedlack, P.A. Weston, W.L. Knight, S.J. Hill, L.R. D'Abramo, W.H. Daniels, and M. J. Fuller. 1997. Relative prawn production and benthic macro invertebrate densities in unfed, organically fertilized, and fed pond systems. Aquaculture 149:227-242.

Tiwari, J.B., and N.P. Sahu. 1999. Possible use of soyalecithin as a source of lipid in the post-larval diet of *M. rosenbergii*. Journal of Aquaculture in the Tropics 14:37-46.

Tripathi, S.D. 1992. Status of freshwater prawn fishery and farming in India. Pages 41-49 *in*. E.G. Silas, ed. Proceedings of the National Symposium on the Freshwater Prawns (*Macrobrachium* sp.), Kerala Agriculture University, Trissur, India.

Tyagi, A.P., and A. Prakash. 1967. A study on the physiology of digestion in freshwater prawn *Macrobrachium dayenum*. Journal of the Zoological Society of India 19: 77-83.

Vasudevappa, C., N. Bimalakumari, K.V. Mohire, and G.Y. Keshavappa. 2000. Effect of feeds and feeding frequencies on growth and production of giant freshwater prawn *Macrobrachium rosenbergii*. Page 54. The Fifth Indian Fisheries Forum. Central Institute of Freshwater Aquaculture, Kausalyaganga, Bhubaneswar, India (Abstract).

Watanabe, W.Q. 1975. Identification of the Essential Aminoacids of the Freshwater Prawn *M. rosenbergii*. Master's thesis. University of Hawai, Mayoa, Honolulu, Hawaii.

Zimmermann, S. 1991. Testing five levels of dry sugarcane yeast in feeds for freshwater prawns *Macrobrachium rosenbergii*, reared in nursery. Journal of the World Aquaculture Society 22:67-72.

Zimmermann. S., E.M. Leboute, and S.M. De Souza. 1994. Effects of two calcium levels in diets and three calcium levels in the culture water on the growth of the freshwater prawn *Macrobrachium rosenbergii*. Page 196. World Auaculture Society, Baton Rouge, Louisiania (Abstract).

Freshwater Pearl Culture Technology Development in India

K. Janakiram

SUMMARY. Development of science and technology of cultured pearl production in a freshwater environment in India is described. Distribution of Indian pearl mussels, pond mussel *Lamellidens marginalis*, paddy field mussel *L. corrianus*, and riverine mussel *Parreysia corrugata* in relation to environmental variables such as type of soil, nature of sediment substratum, presence or absence of macrophytes (*Eichhornia* sp., *Nechamandra* sp., and *Nymphaea* sp.) is described. Food and feeding of the mussels, together with density aspects in culture conditions are discussed. Mussel breeding including specificity in fish hosts, and mussel larval parasitic relationships are discussed. Basic steps involved in indigenous freshwater pearl culture technology are summarized. Pearl culture grafting procedures such as mantle cavity, mantle tissue and gonadal implantation along with different pearl products are also described. *[Article copies available for a fee from The Haworth Document Delivery Service: 1-800-HAWORTH. E-mail address: <getinfo@haworthpressinc.com> Website: <http://www.HaworthPress.com>*

KEYWORDS. Pearl culture, freshwater mussels, India

K. Janakiram, Central Institute of Freshwater Aquaculture, Bhubaneswar-751 002, Orissa, India.

[Haworth co-indexing entry note]: "Freshwater Pearl Culture Technology Development in India." Janakiram, K. Co-published simultaneously in *Journal of Applied Aquaculture* (Food Products Press, an imprint of The Haworth Press, Inc.) Vol. 13, No. 3/4, 2003, pp. 341-349; and: *Sustainable Aquaculture: Global Perspectives* (ed: B. B. Jana, and Carl D. Webster) Food Products Press, an imprint of The Haworth Press, Inc., 2003, pp. 341-349. Single or multiple copies of this article are available for a fee from The Haworth Document Delivery Service [1-800-HAWORTH, 9:00 a.m. - 5:00 p.m. (EST). E-mail address: getinfo@haworthpressinc.com].

INTRODUCTION

Pearl culture in a freshwater environment is developing in India. Pearl culture technology, which probably originated in China, is on the threshold of becoming a major aquaculture industry with an annual receipt of $2 (U.S.) billion in Japan and China. Realizing global trade potential of cultured freshwater pearls, other countries like Bangladesh, Korea, Philippines, Thailand, and Vietnam have initiated both research and industrial-scale projects in recent years (Fassler 1994). Chinese and Indian major carps are traditionally cultured in southeast Asian countries; however, a majority of fish farmers in these countries are presently looking at ways of supplementing traditional aquaculture of foodfish with value-added production systems for enhancing monetary returns. It is in this context that freshwater pearl culture assumes significance in the aquaculture sector.

The species of pearl mussels under the genera *Lamellidens* and *Parreysia* are widely distributed in southeast Asia. Anatomical aspects of some of freshwater mussel species were described by Patil (1976) and Thomas (1974); some of the earlier systematic works were by Janakiram and Radhakrishna (1984) and Rao (1989); food and feeding habits of the mussels were detailed by Patil (1974); aspects of reproduction and development were discussed in detail by Thomas (1974) and Kotpal (1995); technology of pearl culture has been described by Janakiram (1989), Anon. (1990), and Janakiram and Tripathi (1992); potential use of xenogenic grafts in pearl culture operations in terms of color of pearls has been described by Janakiram et al. (1994); specific relationships between mussel larvae and fish host species were indicated by Behera (1997); mussel feeding under captive culture conditions was dealt with by Gayatri et al. (1998); histology of mantle epithelium and pearl sac in mussels was investigated by Janakiram and Gayatri (1997); and bacterial biocoenoses associated with pearl mussels have been detailed by Kabi (1998) and Pattnaik (1998).

These contributions can be used as a starting point for development of the science and technology of pearl culture in India in terms of defining protocols in pearl mussel surgery for pearl production, captive culture, and mussel breeding. An attempt is made in this article to bring relevant observations together and to highlight contemporary investigations in India on the subject of pearl mussels and technology of pearl cultivation.

PEARL CULTURE

Distribution

Common freshwater mussels, pond mussel *L. marginalis,* paddy field mussel *L. corrianus*, and riverine mussel *P. corrugata* have been identified as important species for pearl culture operations in India (Janakiram 1989). They are widely distributed in the northeast, western, central, and southern states of India (Thomas 1974). Species of *Lamellidens* are described as inhabitants in stagnant to slow flowing habitats such as ponds and reservoirs up to a depth of 0.5-1.0 m, while *P. corrugata* is recorded in lotic habitats (Janakiram and Radhakrishna 1984).

Studies on distribution of pearl mussels *L. marginalis* and *L. corrianus* in the state of Orissa have indicated that mussels prefer alluvial soil areas and particularly ponds having soft sediment. Mussels also were recorded in greater abundance in water with green algal bloom (*Chlorococcum* sp. and *Scenedesmus* sp.). Environmental variables such as red loam soil areas, hard substratum of the bodies of water, presence of macrophyte *Eichhornia* sp. and algal blooms like *Euglena* sp. and *Microcystis* sp. appear to restrict the distribution of pearl mussels.

Food, Feeding, and Captive Culture

Gut content analysis of *L. marginalis* has shown the presence of algal spores, filaments of *Spirogyra* sp., colonial forms of *Oocystis* sp., and remains of unicellular algae *Chlorococcum* sp., *Scenedesmus* sp., and detritus. The mussels were also maintained on a variety of prepared and natural diets such as rice bran and oil cake (1:1), cultured green algae (*Chlorococcum* sp., *Scenedesmus* sp., *Kirchneriella* sp.), and blue green alga (*Spirulina* sp.) under 30 days of experimental captive culture conditions (Table 1). The mussels registered a shell length increment of 0.07 cm and wet weight increment of 0.586 g with cultured green algae (on a weekly ration of 10 L at a total cell density of 41,270 cells/mL) in the 3:2:1 ratio of *Scenedesmus* sp., *Kirchneriella* sp., and *Chlorococcum* sp. as food. Mussels have shown an average length increment of 0.11 cm and average wet weight increment of 1.02 g, when blue-green algae (on alternate day ration of 50 mL of *Spirulina* sp. at 0.176 million cells/mL density) was provided. Fish diets such as rice bran and peanut oil cake (at weekly ration of 10 g of 1:1 ratio of feed mixture) in finely powdered form (particle size below 0.5 mm) have given comparatively

TABLE 1. Growth of freshwater pearl mussel *L. marginalis* fed on selected diets for 30 days, 10 animals were used for each experiment.

Initial		Diet	Growth	
Length (cm)	Wet weight (g)		Length (cm)	Wet weight (g)
7.4	43.0	Green algae	+0.07	+0.59
7.4	42.9	Blue-green algae	+0.11	+1.02
7.5	41.1	Fish feed	+0.13	+4.0
7.4	39.9	Control (no feed)	+0.02	−0.274

better results in shell length increment (0.13 cm) and wet weight (4.0 g) of the mussels. Freshwater mussels are mucoid filter feeders and are known to subsist on natural phytoplankton, small zooplankters, and suspended particulate organic matter (Isom 1986). As the mussels are known to subsist on particulate organic matter, the finely powdered rice bran and peanut oil cake mixture has probably contributed as direct feed for mussels.

Mussels under pond culture conditions could be maintained with fertilization alone even at stocking densities of 0.2 million mussels/ha (Table 2). However, comparatively higher survival and growth indices were observed at a stocking density of 25,000 mussels/ha.

Mussel Larval Parasitism on Fish Host

The hatched glochidium larvae coming out of brood pouches of female mussels have been described as ectoparasitic in nature because they attached to gills and skin of certain freshwater fish hosts (Kotpal 1995). However, occurrence of glochidial larvae was recorded only on fins of host fish in recent captive mussel spawning trials. It was also observed that glochidial larvae metamorphose into juveniles in about 30-40 days, under controlled conditions. The mussel larvae appear to have some selectivity of host fish species when a choice of seven fish species is offered (Table 3). Fish hosts tried included as mirikali carp, *Cirrhina mrigala*; roho carp, *Labeo rohita*; sand gobi, *Glossogobius giurius*; catla, *Catla catla*; silver carp, *Hypophthalmitcthis molitrix*; common carp, *Cyprinus carpio*; and grass carp, *Ctenopharyngodon idella*, of which sand gobi, mirikali carp, and roho carp were the preferred species. These observations have probably application potential in commercial pearl mussel seed production strategies in future.

TABLE 2. Stocking density, average size, shell length, wet weight increment and survival of freshwater mussels in 0.02-ha ponds in a 12-month yield trial.

Stocking density	Average size of mussels (cm)	Shell length increment (cm)	Wet weight increment (g)	Survival (%)
4,000	4.7	0.7	6.2	84.9
2,000	4.7	1.0	8.8	75.3
1,000	4.9	1.1	8.5	66.2
500	4.9	1.7	16.2	95.1

TABLE 3. Fish host-mussel larval parasitic relationship data.

No.	Fish host	Glochidium larvae released per tank	Total no. of juveniles recovered per tank	Avg. no. of juveniles recovered per fish
1	Mirikali carp	3,250	270	12
2	Roho carp	3,200	230	10
3	Sand gobi	3,150	540	23
4	Catla	3,250	220	10
5	Silver carp	3,050	110	5
5	Common carp	3,200	80	4
7	Grass carp	3,100	90	4

Pearl Mussel Surgery

Pearl mussels *L. marginalis* and *L. corrianus* $\geq$ 8.0 cm in shell length and $\geq$ 35 g wet weight were hand-picked from natural habitats and brought to laboratory within 2 to 3 hours. For short-distance transport, mussels were brought in dry. However, for longer transportation, individual mussels were wrapped with wet cotton during transit in order to avoid dehydration and damage. Crowding of mussels (at the rate of one mussel/L) for 24 to 48 hours in ferro-cement or plastic containers (200 L capacity) ensured proper relaxation of the adductor muscles, facilitating easy opening of the shell valves for surgical implantation. Studies made at Central Institute of Freshwater Aquaculture (CIFA) have indicated that narcotizing procedures followed for marine oysters with menthol crystals (James 1991) were not applicable for freshwater pearl mussels. It was observed that freshwater mussels remained closed in the containers when menthol crystals were sprinkled over the water surface.

The surgical procedures followed were that of Janakiram and Tripathi (1992) and Janakiram (1997b). In the mantle cavity insertion method, the shell bead (6 ≥ mm diameter) was placed into the umbonal cavity in between the outer mantle layer and the inner surface of shell. In the mantle tissue implantation procedure, 2 to 4 pockets were made in the posterior aspect of the mantle tissue on both left and right mantle lobes of recipient mussels, and donor mantle grafts were placed in each pocket, ensuring secured positioning. Depending on thickness of the mantle tissue, a nucleus (2 mm diameter) was also implanted along with a graft. In the gonad implantation method, a donor mantle graft and nucleus (6 mm diameter) were placed in the middle of the gonad of recipient mussels. Mussel implantations were generally carried out throughout the year, except during hot (above 35°C) summer months (May to June), in order to minimize post operative mussel mortality and rejection of implanted grafts and nuclei. During spawning season (August to September) mussels were conditioned for 2 to 3 days to ensure evacuation of gametes prior to gonadal implantations.

The gonad-implanted mussels were maintained in post-operative care units (ferro-cement or plastic tanks of 200 L capacity) at the rate of 50 mussels/tank, with added antibiotic (chloramphenicol at 1 to 2 mg L/tank) support for 7 to 10 days to minimize rejection of implanted grafts and nuclei before transfer to pond culture units. Implanted mussels were placed in nylon bag nets or plastic crates suspended at 1.5 to 2.0 m depth in culture ponds. It has been observed that at 25,000 mussels/ha stocking density, survival rate was above 80%, compared to less than 60% at higher stocking densities of implanted mussels. Pond culture phase of implanted mussels is normally 12 months. It has been observed that rectangular fish culture ponds devoid of aquatic macrophytes and noxious algal blooms of the size > 0.4 ha are ideal for pearl culture operations (Janakiram 1997b). The quality parameters in the culture pond environment include water with pH 7.5 to 8.5, total alkalinity 75 to 150 ppm, total hardness 40 to 75 ppm, and dissolved calcium salts at 25 to 50 ppm, soil with 6.5 to 7.5 pH, organic carbon 1.0 to 2.5%, and available nitrogen 25 to 75 mg/100 g of soil. Presence of hydrogen sulphide in the sediment was observed to have adverse effect on mussel growth, survival and pearl formation.

Shell beads imported from Japan constituted main nuclear input in pearl culture operations in the country. Research on use of locally available, alternate nucleus material has indicated application potential (Velu et al. 1973; Janakiram 1993). Efforts to develop shell bead nuclei employing thicker shells of native molluscan species such as *Parreysia* spp. and *Velorita* spp. have yielded some success. Shell beads made from sea conch are being used as nuclei in pearl culture operations in the state of West Bengal (Anon. 1990).

The nature of secretory cells and events involved in pearl formation have been detailed previously (Janakiram and Gayatri 1997). It was observed that the outer epithelium of mantle graft is the important tissue in pearl formation. This outer epithelium of mantle graft leads to enveloping the closely placed nucleus in a pearl sac within fifteen days after implantation. It has been demonstrated that microvilli of pearl sac epithelium constituted the cellular basis in crystallization of calcium carbonate, the first step in pearl formation. The pearl-bearing standing crop of mussels are harvested at the end of 12 months of culture, and pearls are removed carefully from the pearl sacs without sacrificing the mussels. However shell-attached half-round and design pearls are cut away from the shell valves after sacrificing the mussels.

The success rate of pearl formation has been recorded at 60-70% in mantle cavity and in mantle tissue implantations and at 25-30% in gonadal implantation of the mussels. The color of the half-round and design pearls produced through mantle cavity insertion generally followed the color of the shell interior of the mussels employed. In the case of mantle tissue and gonadal implantations, color of pearls varied among silvery white, golden yellow to pink. It has been demonstrated that homogenic mantle grafts collected from smaller donor mussels (shell length $\leq$ 6.0 cm; wet weight $\leq$ 20 g) and xenogenic mantle grafts collected from a different genus (*Parreysia*) when implanted in recipient mussels *L. marginalis* have yielded poor quality pearls and deep pink pearls, respectively (Janakiram et al. 1994).

The trade of imported raw freshwater pearls in India is estimated to be $ 22 (U.S.) million per year. As the technology of freshwater pearl culture is fairly new in the country, yields from some of the recently started commercial units are not yet standardized. The economics of newly evolved technology of freshwater pearl culture have been recently discussed (Janakiram 1997a). It has been projected that for a small-scale commercial venture (culture area: 0.4 ha; period of culture: 2 years) the input costs would be about $ 9,000 (U.S.); the expected returns on sale of pearl being $ 14,000 (U.S.), with locally appraised value of pearls at $ 5 (U.S.)/g of non-nucleated pearls and $ 2 (U.S.)/design pearl.

Freshwater pearl culture is a developing technology in India. Considerable progress has been made in areas such as identification of suitable local pearl mussel species, evolution of appropriate surgical implantation procedures, and captive culture of pearl mussels. Currently, attention is focused on post-harvest value addition to culture pearls, pearl mussel seed production, and, importantly, on training and demonstration to develop technical expertise in the country.

ACKNOWLEDGMENTS

The author thanks S. Ayyappan, Director CIFA, Bhubaneswar and National Agricultural Technology Project (NATP), New Delhi for the facilities and financial assistance; K. K. Bhanot, Senior Scientist, CIFA for providing information on mussel culture; G. Misra for the help in preparation of the manuscript; Chandrabhanu Maharathy, Bijaya Kumar Behera, Sophia Pattnaik, and Rashmi Kabi, CIFA, for providing information on mussel distribution, breeding, bacterial associates, and antibiogram tests.

REFERENCES

Anon. 1990. Freshwater Pearl Culture in West Bengal. Government of West Bengal, Calcutta, India.

Behera, B. K. 1997. Breeding of Selected Species of Indian Freshwater Pearl Mussel *Lamellidens marginalis* (L.). Master's thesis, Orissa University of Agriculture and Technology, Bhubaneswar, India.

Fassler, C. R. 1994. Pearls '94. International Pearl Conference, Honolulu, Hawaii, 14-19 May1994. Journal of Shellfish Research 13:325-354.

Gayatri, M., K. Kumar, and K. Janakiram. 1998. Role of selected feeds in captive culture of Indian pearl mussel *Lamellidens marginalis* (Lamarck). Pages 241-243, P. C. Thomas, ed. Current and Emerging Trends in Aquaculture. Daya Publishing House, New Delhi, India.

Isom, G. B. 1986. Systems culture of freshwater shellfish (Bivalves). Pages 1-28, Proceedings EIFAC/FAO, Symposium on Selection, Hybridization and Genetic Engineering in Aquaculture of Fish and Shellfish for Consumption and Stocking. Bordeaux (France), EIFAC/86/Symposium. E53, Decatur, Alabama, USA.

James, P. S. B. R. 1991. Training Manual on Pearl Oyster Farming and Pearl Culture in India. Training Manual No. 8. Central Marine Research Institute, Tuticorin, India.

Janakiram, K. 1989. Studies on cultured pearl production from freshwater mussels. Current Science 58:474-476.

Janakiram, K. 1993. Alternate nuclear material for pearl culture. *In* M. Randhir, ed. Proceedings of the National Meet on Aqua-Farming Systems, Practices and Potentials. CIFA/OUAT, Association of Aquaculturists, Bhubaneswar, India.

Janakiram, K. 1997a. Project Completion Report. National Centre for Freshwater Pearl Culture. Department of Biotechnology, New Delhi, India.

Janakiram, K. 1997b. Freshwater pearl culture in India. NAGA, The International Center for Living Aquatic Resources Management Quarterly 20(3&4):12-17.

Janakiram, K., and M. Gayatri. 1997. Preliminary studies on histology of pallial mantle and pearl sac of Indian freshwater pearl mussel *Lamellidens marginalis* (L.). Journal of Aquaculture 5:95-98.

Janakiram, K., and S. D. Tripathi. 1992. A Manual on Freshwater Pearl Culture. Manual Series 1, Central Institute of Freshwater Aquaculture, Bhubaneswar, India.

Janakiram, K., and Y. Radhakrishna. 1984. The distribution of freshwater mollusca in Guntur District (India) with a description of *Scaphula nagarjunai* sp.n. (Arcidae). Hydrobiologia 119:49-55.

Janakiram, K., K. Kumar, and M. Gayatri. 1994. Possible use of different graft donors in freshwater pearl mussel surgery. Indian Journal of Experimental Biology 32: 366-368.

Kotpal, R. L. 1995. Mollusca. Rastogi Publication, Meerut, India.

Patil, V. Y. 1974. Food and feeding in freshwater mussel. Journal Shivaji University, India 7:33-38.

Patil, V. Y. 1976. Functional morphology of the mantle of freshwater mussel. Journal Shivaji University, India 16:79-80.

Rashmi, K. 1998. Identification and Control of the Associated Pathogenic Bacterial Flora of Pearl Mussels Before and After Implantation. Master's thesis. Orissa University of Agriculture and Technology, Bhubaneswar, Orissa, India.

Sophia, P. 1998. Pearl Mussels and Associated Bacterial Flora Before and After Implantation and Characterization. Master's thesis. Orissa University of Agriculture & Technology, Bhubaneswar, Orissa, India.

Subba Rao, N. V. 1989. Handbook of Freshwater Mollusc of India. Zoological Survey of India, Calcutta, India.

Thomas, E. I. 1974. The bionomics, anatomy and development of the freshwater mussel, *Lamellidens marginalis* (Lamarck). Annals of Zoology Quarterly (Agra, India) 10:71-169.

Velu, M. K. Alagarswami, and S. Z. Qasim. 1973. Technique of producing spherical shell beads as nuclei for cultured pearls. Indian Journal of Fisheries 20:672-676.

Index